Tree Crops

A PERMANENT AGRICULTURE

CONSERVATION CLASSICS

Nancy P. Pittman, Series Editor

With the Conservation Classics, ISLAND PRESS inaugurates a new series to again make available books that helped launch the conservation movement in America. When first published, these books offered provocative alternatives which challenged established methods and patterns of development.

Today, they offer practical solutions to contemporary challenges in such areas as multiple-use forestry, desertification and soil erosion, and sustainable agriculture. These new editions include valuable introductions from the leaders of today's conservation movement.

The inaugural titles in the series are:

BREAKING NEW GROUND
by Gifford Pinchot

Introduction by George T. Frampton, Jr.

PLOWMAN'S FOLLY
and
A SECOND LOOK
by Edward H. Faulkner

Introduction by Paul B. Sears

TREE CROPS
A Permanent Agriculture
by J. Russell Smith

Introduction by Wendell Berry

ISLAND PRESS
WASHINGTON, D.C.   ∾   COVELO, CALIFORNIA

TREE CROPS

A PERMANENT AGRICULTURE

J. RUSSELL SMITH, Sc.D.

ISLAND PRESS

Washington, D.C.   □   Covelo, California

Library of Congress Cataloging-in-Publication Data

Smith, J. Russell (Joseph Russell), 1874-1966.
   Tree crops.

   (Conservation classics)
   Reprint. Originally published: New York : Harcourt, Brace. c1929.
   Includes index.
   1. Tree crops. 2. Tree crops--United States.
I. Title. II. Series.
SB170.S65 1987   634.9   87-82037
ISBN 0-933280-44-0 (pbk.)

Manufactured in the United States of America
10 9 8 7 6 5 4 3 2 1

This book is dedicated to

CHARLES C. HARRISON
EDGAR F. SMITH
JOSIAH H. PENNIMAN

THREE PROVOSTS OF THE UNIVERSITY OF PENNSYLVANIA
WHOSE SYMPATHETIC UNDERSTANDING AND PRACTICAL
AID WERE OF GREAT ASSISTANCE IN THE PROSECUTION
OF THE RESEARCHES WHICH RESULTED IN THIS BOOK

Acknowledgments

As I send this book to the press, I feel a great sense of gratitude for the large amount of aid that I have received.

I wish to express my feelings of deep appreciation to a group of Philadelphians, who, under the leadership of Mr. J. Levering Jones, formed the University of Pennsylvania Tree Crops Exploration Committee. This committee assisted in making possible some of the journeys incident to gathering the facts that have helped to make this book: Mr. and Mrs. J. Levering Jones, Morris L. Clothier, Eckley B. Coxe, Jr., Cyrus H. K. Curtis, George W. Elkins, Charles C. Harrison, Charles E. Ingersoll, J. Bertram Lippincott, J. Franklin McFadden, Mr. and Mrs. Louis C. Madeira, Joseph G. Rosengarten, William Jay Turner, John Weaver, J. William White, G. Searing Wilson, Stuart Wood, Dr. George Woodward, Charlton Yarnall.

Professor Howard H. Martin of the University of Washington spent many weeks searching for material.

Valuable letters have been sent to me by many private individuals; also by agricultural scientists in the employ of the United States Department of Agriculture, the state universities, the state experiment stations, and foreign experiment stations. From among these many I must make especial mention of David Fairchild, H. P. Gould, C. A. Reed, Walter T. Swingle, William A. Taylor, and Raphael Zon, all of the United States Department of Agriculture. Dr. J. Eliot Coit has enlarged my knowledge of the carob, and I am indebted to Dr. H. H. Bennet and many other members of the U. S. Soil Conservation Service for valuable data.

For the troublesome job of reading chapters of another man's

vi

manuscript, I am indebted to Dr. W. C. Deming, of Hartford, Connecticut; J. Ford Wilkinson, of Rockport, Indiana; J. M. Westgate, Director, Agricultural Experiment Station, Honolulu; Robert Forbes, Ex-Director, Agricultural Experiment Station, Tucson, Arizona; H. J. Webber, Director, Citrus Substation, Riverside, California; and Dr. George A. Clement, Los Angeles Chamber of Commerce.

As a member of the Northern Nut Growers' Association, I have for many years been absorbing information from T. P. Littlepage, Dr. W. C. Deming, Dr. Robert T. Morris, Willard Bixby, E. A. Riehl, J. F. Jones, and many others.

More than a hundred members of this Association have given me detailed reports of their experiments. Their reports are discussed in Chapter 25 of this book and appear in full in the Annual report of the Northern Nut Growers' Association for 1949 (J. C. McDaniel, Secretary, Tennessee Department of Agriculture, Nashville 3).

Mrs. Bessie W. Gahn, late of the U. S. Department of Agriculture, has assisted in botanical research and also prepared the manuscript for the printer.

I am indebted to Mr. John W. Hershey in a very especial manner. He had a lot of material gathered for writing a book on nut growing but the pressure of business and rather poor health prevented his writing the book. Instead he gave me free access to his material—a rare courtesy.

Miss Myra Light, my secretary, assisted in many skillful ways.

Professor S. S. Visher of Indiana University has read the galley proofs, making many valuable suggestions.

I remember with pleasure the warm and helpful interest in this, my avocation, that has been shown by my departmental colleagues, John E. Orchard, Louis A. Wolfanger, and George T. Renner.

J. RUSSELL SMITH

Swarthmore, Pa.
June 1, 1950

Foreword

HOW TO READ THIS BOOK

This book can be read in three ways—depending on your hurry or your interest.

First, look at the pictures and the legends and you have the essence of it.

Second, read the first three short chapters and you have the idea. This might be followed by chapters 24 and 26 to get a similar general statement of the applications.

Third, if you are still interested, the table of contents or the index will guide you into the main body of supporting data.

Contents

A Practical Visionary

I first read J. Russell Smith's *Tree Crops* perhaps fifteen years ago, and since then I have returned to it many times. Having just paid it another visit, I am pleased to say that I still like it as much as I ever did, and am just as much convinced by it. This book contributed to a fundamental rethinking of agriculture in our century. It is a predecessor of Sir Albert Howard's *The Soil and Health*, Wes Jackson's *New Roots for Agriculture*, and Masanobu Fukuoka's *The One-Straw Revolution* in its perception that in order for humans to know how to use the land of a particular locality, they must look to see how nature uses it. Agriculture, he wrote, must be "adapted . . . to physical conditions"; he insisted *"that farming should fit the land."*

Smith begins his thinking with a fact that agricultural industrialists have made a convention of ignoring: that most land is vulnerable to erosion, and much land, especially hilly land, is extremely so. The normal condition of farming is not that of the deepest, flattest soils of the American grain belt. Much of agriculture, the world over, must take place on land that is not safely adaptable, or is not adaptable at all, to industrial assumptions and procedures. It is remarkable that in most times, but never so much as in modern times, agricultural methods have presupposed optimum conditions. The result, as J. Russell Smith saw, was a catastrophic impropriety: "Man has carried to the hills the agriculture of the flat plain." The catastrophe was, and is, soil erosion. The cycle of hill agriculture has thus too often been a one-time cycle: "Forest-field-plow-desert."

What, then, are we to do with the enormous acreage of potentially usable and perhaps much-needed land that is too steep for the plow? The practices and plans of the agricultural industrialists apparently offer us only two choices: to abandon it entirely, or to continue to treat it as a minable resource until it has been completely ruined. There is, of course, a good argument for abandonment—not all land should be used—but it is a limited argument because, for good economic and political reasons, we cannot afford to abandon all but the very best agricultural land. The argument for ruination, though it is never made, is implied in attitudes and methods, economies and technologies. The counterargument is a simple one: we have no right to ruin land, and we cannot afford to ruin it. We are stuck, therefore, with two daunting agricultural obligations: to use vulnerable land, and to preserve it.

The best solution to this dilemma that I know about is in Smith's fundamental perception that "Trees are the natural crop plants for all such places." Smith understood that the annuals, as agricultural plants, have a great fault in that they leave the soil exposed to erosion for much of the year. The steeper the land, the worse it is served by annuals. Trees, on the other hand, are perennials; they "are a permanent institution." For that reason, Smith said, "the crop-yielding tree offers the best medium for extending agriculture to the hills, to steep places, to rocky places, and to the lands where rainfall is deficient."

He espoused, furthermore, the attractive idea of a "two-story agriculture" for the flatter lands, an agriculture consisting of "trees above and annual crops below." This idea is attractive because it would diversify both the agricultural and the natural life of the arable farmlands; it would make them more productive, more healthy, and more beautiful. It proposes a countryside as far as possible unlike the monocultural deserts that we have made in the prime lands of the present day.

But Smith's mind was mainly on the hill lands, and the agriculture he proposed for them was "two-story" as well: trees above and pasture below, both stories here being perennial. He had a vision of what our hill lands should be. I would like to quote him at length on this, for it is a vision at once practical and lovely:

I see a million hills green with crop-yielding trees and a million neat farm homes snuggled in the hills. These beautiful tree farms hold the hills from Boston to Austin, from Atlanta to Des Moines. The hills of my vision have farming that fits them and replaces the poor pasture, the gullies, and the abandoned lands that characterize today so large a part of these hills.

These ideal farms have their level and gently sloping land protected by mangum terraces and are intensively cultivated—rich in yields of alfalfa, corn, clover, legumes, wheat, and garden produce. This plow land is the valley bottoms, level hill tops, the gentle slopes, and flattened terraces on the hillsides. The unplowed lands are partly shaded by cropping trees—mulberries, persimmons, honey locust, grafted black walnut, grafted heart nut, grafted hickory, grafted oak, and other harvest-yielding trees. There is better grass beneath these trees than covers the hills today.

The trees would produce food both for people and for foraging animals, protecting the slopes while increasing their yield. Toward the realization of this vision, Smith contributed the incontrovertible argument "that farming should fit the land," an agenda of work to be done, and a host of examples of successful tree-cropping systems from all over the world. (One of the pleasures of this book is in its abundance of such information. It is a book to rummage in.)

Smith's vision is inherently and necessarily a democratic one. What he is proposing is not simple; he would protect the hills and make them more productive, not by the mechanical and chemical simplifications that are associated with industrial agriculture, but by complicating the biological pattern of human use. For this, great care, knowledge, and skill would be needed. The good agricultural solution thus presupposes the need for democratic land-ownership, not the plutocracy always implied in the economic and technological determinism of the industrialists. As Smith saw it, "the presence of the landowner is also needed. This is not a job for tenants. Let the tenant go down to the level land which carelessness cannot ruin so quickly."

To describe this book, even so minimally as I have done here, is to reveal the reasons why its influence, so far, has been small. The minds that have dominated agriculture since 1929 when *Tree Crops* was first published, have been little interested in conserving either the land or the people on the land. They have, Heaven knows, seen no visions of "a million neat farm homes snuggled in the hills." A farming system in which millions of small landowners would manage devotedly and skillfully a diversified, locally-adapted system of tree crops, pastures, animals, and row crops has been simply unthinkable to them.

Tree Crops nonetheless sets forth a thinkable possibility. It is a book full of common sense, and of great love for a kind of knowledge and a kind of life, addressed forthrightly "to persons of imagination who love trees and love their country." In its last pages, it pleads for "a new patriotism," expressing amazement at "patriots" who are willing to fight and perhaps die to defend from foreign threat a country that they themselves are destroying. The general run of patriots have always understood patriotism as love of a government. *Tree Crops* is one of the best American attempts to define it—and enact it—as love of a country.

Wendell Berry
Port Royal, Kentucky
July 1987

PART ONE

The Philosophy

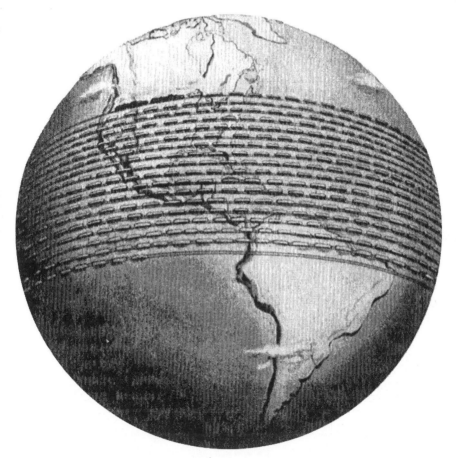

Fig. 1. The U. S. Soil Conservation Service reports that the soil washed out and blown out of the fields of the United States each year would load a modern freight train long enough to reach around the world eighteen times. If it ran twenty miles an hour continuously, it would take it nearly three years to pass your station. We began with the richest of continents, but . . .

CHAPTER I

How Long Can We Last?

I stood on the Great Wall of China high on a hill near the borders of Mongolia. Below me in the valley, standing up square and high, was a wall that had once surrounded a city. Of the city, only a few mud houses remained, scarcely enough to lead one's mind back to the time when people and household industry teemed within the protecting wall.

The slope below the Great Wall was cut with gullies, some of which were fifty feet deep. As far as the eye could see were gullies, gullies, gullies—a gashed and gutted countryside. The little stream that once ran past the city was now a wide waste of coarse sand and gravel which the hillside gullies were bringing down faster than the little stream had been able to carry them away. Hence, the whole valley, once good farm land, had become a desert of sand and gravel, alternately wet and dry, always fruitless. It was even more worthless than the hills. Its sole harvest now is dust, picked up by the bitter winds of winter that rip across its dry surface in this land of rainy summers and dry winters.

Beside me was a tree, one lone tree. That tree was locally famous because it was the only tree anywhere in that vicinity; yet its presence proved that once there had been a forest over most of that land—now treeless and waste.

The farmers of a past generation had cleared the forest. They had plowed the sloping land and dotted it with hamlets. Many workers had been busy with flocks and teams, going to and fro among the shocks of grain. Each village was marked by columns of smoke rising from the fires that cooked the simple fare of these sons of Genghis Khan. Year by year the rain has washed away the loosened soil. Now the plow comes not—only the shepherd is here, with his sheep and goats, nibblers of last vestiges. These four-footed vultures pick the bones of dead cultures in all continents. Will they do it to ours? The hamlets in my valley below the Great Wall are shriveled or gone. Only gullies remain —a wide and sickening expanse of gullies, more sickening to look upon than the ruins of fire. You can rebuild after a fire.

Forest—field—plow—desert—that is the cycle of the hills under most plow agricultures—a cycle not limited to China. China has a deadly expanse of it, but so have Syria, Greece, Italy, Guatemala, and the United States. Indeed we Americans, though new upon our land, are destroying soil by field wash faster than any people that ever lived—ancient or modern, savage, civilized, or barbarian. We have the machines to help us to destroy as well as to create. The merciless and unthinking way in which we tear up the earth suggests that our chief objective may be to make an end of it.

We also have other factors of destruction, new to the white race and very potent. We have *tilled* crops—corn, cotton, and tobacco. Europe did not have these crops. The European grains, wheat, barley, rye, and oats, cover all of the ground and hold the soil with their roots. When a man plows corn, cotton, or tobacco, he is loosening the earth and destroying such hold as the plant roots may have won in it. Plowing corn is the most efficient known way for destroying the farm that is not made of level land. Corn, the killer of continents, is one of the worst enemies of the human future.

We in America have another factor of destruction that is almost new to the white race—the thunderstorm. South Europe has a rainless summer. North Europe has a light rainfall that comes in gentle showers. The United States has the rippling torrent that follows the downpour of the thunderstorm. When the American heavens open and pour two inches of rain in an hour into a hilly cornfield, there may result many times as much erosion as results from two hundred inches of gentle British or German rain falling on the wheat and grass.

I asked county agents in a number of counties in the hill country of North Carolina the following question: "What is your estimate of number of cultivated crops secured on steep land after clearing and before abandonment of cultivation?" The answers from ten counties were as follows: 5; 20; 12; 10; 5 to 10; 10 or 12; 10 or more; 12; 5, extremely variable and 10. (See Figs. 2, 3, 4, 6, 9.) Ten tilled crops, and ruin has arrived! How long can we last?

Even Oklahoma, newest of the new, so recently wrested from the Indian, who did not destroy it, has its million miles of gullies and a kingdom of good land ruined and abandoned.

Five years ago there was not a gully on the place . . . now it is badly cut by gullies . . . all the top soil washed away, leaving nothing but the clay. . . . If not terraced . . . the gullies [will] cut deeper until the rocks are touched or until all the clay soil is gone. . . . Five years ago it could have been saved by spending less than three dollars an acre to have it terraced. Today it will cost five times as much, in addition to getting nothing from it for at least two years." (*Oklahoma Extension News*, January, 1928.)

To the surprise of a past generation, Oklahoma proved to be good land; so we pushed the Indian out and sent him on to the deserts farther west. The whites entered in two great rushes, 1890 and 1893. One of my students made a study of a typical county. (See Russell W. Lynch, "Czech Farmers in Oklahoma," published June 1942, as No. 13 of Vol. 39 of Bulletin, Oklahoma Agricultural and Mechanical College.)

The white "civilization" (?) grew cotton in summer and oats in winter on rolling land, and by 1910 land abandonment had begun. Without any doubt, the American is the most destructive animal that ever trod the earth.

For decades, reports of ruin have come out of the hill section of the American Cotton Belt—thousands of square miles of ruin.

Some counties were reported one-third worn out before 1850. Worst of all is the plight of the loess lands east of the Mississippi. This layer of rich, wind-blown soil, half as wide as the State of Mississippi, reaching nearly all the way from the Ohio to the Gulf, is a kind of thin veneer lying on top of coastal plain sands. It is extremely rich and erodes very easily.

E. W. Hilgard, the great pioneer writer on soils (*Soils,* p. 218), says:

The washing away of the surface soil . . . diminished the production of the higher lands, which were then (at the time of the Civil War) commonly "turned out" and left without cultivation or care of any kind. The crusted surface shed the rain water into the old furrows, and the latter were quickly deepened and widened into gullies—"red washes". . . .

As the evil progressed, large areas of uplands were denuded completely of their loam or culture stratum, leaving nothing but bare, arid sand, wholly useless for cultivation, while the valleys were little better, the native vegetation having been destroyed and only hardy weeds finding nourishment on the sandy surface.

In this manner, whole sections and in some portions of the State [Mississippi] whole townships of the best class of uplands have been transformed into sandy wastes, hardly reclaimable by ordinary means, and wholly changing the industrial conditions of entire counties, whose county seats even in some instances had to be changed, the old town and site having, by the same destructive agencies, literally "gone downhill."

Specific names have been given to the erosional features of this district; a "break" is the head of a small retrogressive ravine; a "gulf" is a large break with precipitous walls of great depth and breadth, commonly being one hundred or one hundred and fifty deep; a "gut" is merely a road-cut deepened by storm wash and the effects of passing travel.

In this way we have already destroyed the homelands fit for the sustenance of millions. We need an enlarged definition for treason. Some people should not be allowed to sing "My Country." They are destroying it too rapidly.

Field wash, in the United States, Latin America, Africa, and many other parts of the world, is the greatest and most menacing of all resource wastes. For details see the U. S. Soil Conservation Service. It removes the basis of civilization and of life itself. It is far worse than burning a city. A burned city can be rebuilt. A field that is washed away is gone for ages. Hence the Old World saying, "After the man the desert."

H. H. Bennett, Chief of Soil Conservation Service, United States Department of Agriculture, says:

We have in this country about 460 million acres of available and potentially available good productive cropland. This includes from 80 to 100 million acres which need clearing, drainage, irrigation, or other improvements to condition or fit it for cultivation. All but about 100 million acres of this remaining good productive cropland is subject to erosion.

If present rates of erosion were allowed to continue—although I cannot conceive that we will permit any such thing—we would lose or severely damage a tremendous acreage of good productive cropland within the next 20 years. Nearly 90 percent of our farmland which is still subject to erosion is yet without the needed protection of effective, acre-by-acre soil conservation treatment.

Nearly 25 percent of our cropland is being damaged at a rapid rate of erosion. This is an area of something over 100 million acres of cropland. The productive capacity of much of this highly vulnerable land will be permanently damaged and around 500,000 acres a year ruined for further cultivation unless and until it has the benefit of sound conservation farming within the next 10 to 15 years, preferably by 1960. On another large area (around 115 to 120 million acres) of cropland, erosion is taking place somewhat less rapidly but still at a serious rate. (Letter, October 29, 1947.)

Can anything be done about it? Yes, something can be done. Therefore, this book is written to persons of imagination who love trees and love their country, and to those who are interested in the problem of saving natural resources—an absolute necessity if we are to continue as a great power.

CHAPTER II

Tree Crops—The Way Out

Again I stood on a crest beneath a spreading chestnut tree, and scanned a hilly landscape. This time I was in Corsica. Across the valley I saw a mountainside clothed in chestnut trees. The trees reached up the mountain to the place where coolness stopped their growth; they extended down the mountain to the place where it was too dry for trees. In the Mediterranean lands, as in most other parts of the world, there is more rainfall upon the mountains than at sea level. This chestnut orchard (or forest, as one may call it) spread along the mountainside as far as the eye could see. The expanse of broad-topped, fruitful trees was interspersed with a string of villages of stone houses. The villages were connected by a good road that wound horizontally in and out along the projections and coves of the mountainside. These grafted chestnut orchards produced an annual crop of food for men, horses, cows, pigs, sheep, and goats, and a by-crop of wood. Thus, for centuries, trees upon this steep slope had supported the families that lived in the Corsican villages. The mountainside was uneroded, intact, and capable of continuing indefinitely its support for the generations of men.

Why are the hills of West China ruined, while the hills of Corsica are, by comparison, an enduring Eden? The answer is plain. Northern China knows only the soil-destroying agriculture of the plowed hillside. Corsica, on the contrary, has adapted agriculture to physical conditions; she practices the soil-saving tree-crops type of agriculture.

8

Man lives by plants. Plants live in the soil. The soil is a kind of factory in which the life force of plants, using plant food from earth, air, and water, and assisted by bacteria and the elements of the weather, changes these natural elements into forms that we can eat and wear, manufacture and burn, or use for building material. This precious soil from which we have our physical being is only a very thin skin upon the earth. Upon the hills and mountains it is appallingly thin. In some places there is no soil at all, and rocks protrude. Sometimes the earth mantle may be only a few inches in depth; rarely does the soil on hill or mountain attain a depth of many feet. Often soil is so shallow that one great rainstorm can gash and gully a slope down to bare rock. Where man has removed nature's protecting cover of plants and plant roots, the destroying power of rain is increased a hundredfold, a thousandfold, even at times a millionfold, or perhaps even more than that.

The creation of soil by the weathering of rock is a very, very slow process. In some places, centuries and millennia must have passed in making soil that, if unprotected, may be washed away in an hour. Therefore, today an observer in the Old World might see myriad landscapes once rich with farms where now only poverty-stricken men creep about over the ruined land, while their sheep and goats, scavengers and destroyers, pick the scanty browse that struggles for life in the waste. A handful of men are now living uncomfortably where once there were prosperous villages. Similar examples, even of large areas, can be found in almost any hill country with a long history of occupation by agricultural man.

Syria is an even more deplorable example than China. Back of Antioch, in a land that was once as populous as rural Illinois, there are now only ruin and desolation. The once-prosperous Roman farms now consist of wide stretches of bare rock, whence every vestige of soil has been removed by rain.

Geikie, in modern times writing of a section of Palestine, gives a similar example, which shows the ruination that man can create:

The ride from Eriha to the Jordan is about five miles over a stony plain, on which there is no vegetation. Year by year the winter rains sweep down the slope and wash away a layer of the wide surface, carrying it to the Jordan, there being little to check them but copses of the Zukkum tree and Apina Christi. Yet seven monasteries once stood on this now desolate tract, three of them still to be identified by their ruins. Until we reach the edge of the Jordan, only the stunted bushes I have mentioned, unworthy of the name of trees, and a few shrubs with dwarfed leaves are to be seen after leaving the moisture of Sultan's Spring. Not a blade of grass softens the dull yellow prospect around. (Quoted from *Gila River Flood Control*, p. 18, Secretary of the Interior, 1919.)

Greece, once so great, is shockingly ruined by soil wash. Its people undernourished and living (1949) on the American dole (Marshall Plan). Wolf von Schierbrand, in his book *Austria Hungary* (Chap. 14), says that in parts of Europe, people even pound stone to get a little bit of loose material in which plant roots can work.

In our own South, millions of acres are already ruined. "Land too poor for crops or grazing, such as old abandoned fields, of which Brazos County [Texas] alone has thousands of acres." (H. Ness, Botanist, Texas Experiment Station, *Journal of Heredity*, 1927.) The same destructive agency has caused ruin and abandonment of land in Ohio, Illinois, Indiana—indeed, in every one of our States. "In many sections of Iowa, Missouri, Nebraska, and other Corn Belt states, water erosion has a tendency to form deep, steep-sided ravines which will sometimes make farming almost impossible in a field as large as twenty or forty acres." (Letter, Ivan D. Wood, State Extension Agent, Agricultural Engineering, University of Nebraska, July 19, 1923.)

This is double danger. First, water erosion destroys soil. Secondly, it cuts up the remnant so that it cannot be used by the new machinery which cries aloud for *room* and good surface. And yet, as human history goes, we came to America only yesterday.

If we think of ourselves as a race, a nation, a people that is to occupy its country generation after generation, we must change some of our habits or we shall inevitably experience the steadily diminishing possibility of support for man.

FLAT-LAND AGRICULTURE GOES TO THE HILLS

How does it happen that the hill lands have been so frightfully destroyed by agriculture? The answer is simple. Man has carried to the hills the agriculture of the flat plain. In hilly places man has planted crops that need the plow; and when a plow does its work on lands at an angle instead of on flat lands, we may look for trouble when rain falls.

Whence came this flat-land agriculture of grass and grains? The origin of wheat, barley, and most of our important food plants is shrouded in mystery; but we know that our present agriculture is based primarily on cereals that came to us from the unknown past and are a legacy from our ancient ancestress —primitive woman, the world's first agriculturist. Searching for something to fill little stomachs and to hush the hunger cries of her children, primitive woman gleaned the glades about the mouth of her cave. Here she gathered acorns, nuts, beans, berries, roots, and seeds.

Then came the brilliant and revolutionary idea of saving seed and planting it that she might get a better and more dependable food supply. Primitive woman needed a crop in a hurry, and naturally enough she planted the seeds of annuals, like the ones that sprouted around the site of last year's campfire. Therefore, we of today, tied to this ancient apron string, eat bread from the cereals, all of which are annuals and members of the grass family.

As plants, the cereals are weaklings. They must be coddled and weeded. For their reception the ground must be plowed and harrowed, and sometimes it must be cultivated after the crop is planted. This must be done for every harvest. When we produce these crops upon hilly land, the necessary breaking up of the soil prepares the land for ruin—first the plow, then rain, then erosion. Finally the desert.

CAN WE GET ECONOMICS INTO BOTANY?

Must we continue to depend primarily upon the type of agriculture handed to us by primitive woman? It is true that we have improved the old type. Many of the present-day grains, grasses, and cereals would scarcely be recognized as belonging to the families that produced them. Indeed the origin of corn, and of some others, is still in doubt. Present-day methods of cultivation but dimly recall the sharpened stick in the hand of primitive woman. But we still depend chiefly on her crops, and sad to relate, our methods of which we are so proud are infinitely more destructive of soil than were those of the planting stick in the hands of Great-Grandmother ninety-nine generations back.

We are now entering an age of science. At least we are scientific in a few respects. It is time that we made a scientific survey of the plant kingdom—still the source, as always, of a very large proportion of that which is necessary to the existence and comfort of man and without which we would all be dead in a year. We should carefully scrutinize types of agriculture in relation to environment. Agricultural America should scientifically test the plant kingdom in relation to potential human use and do it as carefully and patiently as industrial America has tested cement. We test cement in every possible way, make it of all possible materials, mix all possible combinations, test it by twisting it, pressing it, pulling it; test it thousands of times, hundreds of thousands of times, perhaps even millions of times, and in a few years our whole physical equipment is made over by reenforced cement made possible by these millions of tests.

THE TREE AN ENGINE OF NATURE—PUT IT TO WORK

Testing applied to the plant kingdom would show that the natural engines of food production for hill lands are not corn and other grasses, but trees. A single oak tree yields acorns (good carbohydrate food) often by the hundredweight, some-times by the ton. Some hickory and pecan trees give us nuts by the barrel; the walnut tree yields by the ten bushels. There are bean trees producing good food for cattle, which food would apparently make more meat or milk per acre than our forage crops now make. It is even now probable that the king of all forage crops is a Hawaiian bean tree, the keawe. (See Chap. V.)

These wonders of automatic production are the chance wild trees of nature. They are to be likened to the first wild animal that man domesticated and to the first wild grass whose seed was planted. What might not happen if every wild crop-bearing tree was improved to its maximum efficiency? Burbank and others have given us an inkling of what may result from well-planned selection, crossing, or hybridizing.

The possibilities, at present quite incalculable, that lie in such work are hinted at in one almost unbelievable statement of the great authority, Sargent, who says of the English walnut, which we all know is so good and meaty:

The nut of the wild tree is small, with a thick hard shell and small kernel, and is scarcely edible; but centuries of cultivation and careful selection have produced a number of forms with variously shaped thin shells, which are propagated by grafting and budding. (*Silva*, Vol. VII, p. 115.)

The "careful selection" to which Sargent refers must be the habit of the (perhaps) illiterate farmer in picking out good nuts to plant. The crossing of such plants standing side by side is real plant breeding, and the result is our splendid English walnut. I have grown Chinese chestnut trees that bore nuts, and good ones too, in their second growing season. The origin of the seed was similar to that suggested for the above men-tioned walnuts.

We now know how to breed plants. In the short space of a few years we can surpass the results of centuries of chance breeding. The plant kingdom has become almost as clay in the hands of the potter. Where we now have one good crop plant, we may some day have five or ten. We need to start in earnest to apply some of our science to producing genius trees—trees that are to other trees as human geniuses are to other men.

Genius trees produced either by chance or design can be propagated a million or ten million times, as was done with the one chance navel orange tree.

If you want to get a look-in on plant breeding, see the Yearbook, United States Department of Agriculture, for 1936. You will find hundreds of pages of it.

So much for better trees, but I wish to suggest a little-explored line of experimentation that may improve the productivity of the trees we now have—girdling, ringing, or otherwise injuring the tree in such a way that it will recover the injury but will, because of it, yield a large quantity of fruit. This is a regular practice of the Greek growers of a grape that enters the world market under the name of currant. It appears that the quantity of fruit a tree bears is in part a matter of habit. One tree bears every year. Another bears every other year. One tree, say the York Imperial apple, will average twenty bushels per year, while a Spitzenberg, just as large and standing beside it, may possibly average five bushels. Girdling makes many trees produce. I have no idea that most trees bear all the fruit they are physiologically capable of producing. Careful experimentation along this line might be very productive.

THE TREE A BETTER CROP PLANT

We need a new profession, that of the botanical engineer, which will utilize the vital forces of plants to create new mechanisms (crop-yielding trees) as electrical and mechanical engineers use the forces of electricity and the elements of mechanics to create new mechanisms for the service of mankind.

This creation of new types by plant breeding depends upon three facts—first, the variation of different offspring from the same parents; second, the varying combinations in offspring of the qualities of the parents; and third, the appearance in offspring, especially hybrid offspring, of qualities possessed by neither parent.

First, variation of offspring. Look at the children of almost any family you know. *This tendency to variation runs deep into both animal and plant life.* For example, Texas Agricultural Experiment Station, Bulletin 349, "Variation in certain lint characters in a cotton plant and its progeny," shows that the average length of lint in the individual plants of the progeny of a certain boll (seed pod) varied from 19 millimeters to 28.5 millimeters, a variation of 50 percent. This is very suggestive of the way by which, through a selection of parents, we have changed the cow so marvelously for milk production. The object of selection here is to find desirable strains that produce uniform progenies. It is fortunate that tree breeding has a more easily attainable objective—namely, *one* good specimen.

Second, varying combinations in offspring of qualities of parents. A hybridization of hazels and filberts (Fig. 19) produced plants ranging from twelve inches to twelve feet in height—suggestive of variations in great degree for each quality a plant can have.

Third, the appearance in hybrid offspring of qualities possessed by neither parent. Some of the above-mentioned hazel x filbert hybrids bore larger fruit than either parent. It is common for occasional plant hybrids to exceed either parent in speed of growth, size, earliness of fruiting.

Further experiments with cotton breeding show the dynamic and creative tendency of hybrids.

Not only was there all manner of recombination of the characters of the parent types, but many of these characters were expressed in an exaggerated form. Moreover, numerous characters not observed in either parent appeared in the second generation, some of these having been decidedly abnormal. Many individuals were so strikingly different from either upland or Egyptian cotton that a botanist unaware of their hybrid origin would take them to represent new species. . . .

A remarkable character was the bluish white color of the practically glabrous foliage. There was no suggestion of this color in either parent and it does not occur in any cultivated cotton known to the writer. (Extracts from "A Hybrid Between Different Species of Cotton," by Thomas H. Kearney, U. S. Department of Agriculture, published in the *Journal of Heredity*, July 1924.)

For experiments in breeding, the tree has one *great advantage* over most of the annuals. We propagate trees by twig or bud, by grafting or budding. Therefore, any wild, unstable (though useful) freak, any helpless malformation like the navel orange which cannot reproduce itself, can be made into a million trees by the nurseryman. The parent tree of the Red Delicious variety of apple grew, by chance, in an Iowa fence row. A representative of the Stark Nursery Company saw the apple at a fair and raced with all speed to the tree, bought it, and reproduced it by the million, an easy process if you really need a million trees. With corn, oats, or alfalfa, the breeder must produce a type true to seed before the farmer can use it.

Not only is the tree the great engine of production, but its present triumphant agricultural rivals, the grains, are really weaklings.

All plants require heat, light, moisture, and fertility. Give these things and the tree raises its head triumphantly and grows. But in addition to these requirements the weakling grains must have the plow. A given area may have rich soil and good climatic conditions, but be unsuitable for grain if the land happens to be rocky. Nor are steep lands good farm lands for grains. Trees are the natural crop plants for all such places.

Moreover the grains are *annual* plants. They must build themselves anew for each harvest. They may, therefore, become victims of the climatic peculiarities of a certain short season. It is rain in July that is so vital to the American corn crop. The rains of June cannot bring a good crop through. Also, if most of the rain due to fall in July happens to come in August, it comes too late. The corn has shot its bolt; it cannot be revived. Trees are much better able than the cereals to use rain when it comes.

They can store moisture much better than the annuals can store it, because they thrust their roots deep into the earth, seeking moisture far below the surface. They are able to survive drought better than the annual crops that grow beside them. For example, a drought that blasts corn or hay or potatoes may have little influence on the adjacent apple orchard. Trees living from year to year are a permanent institution, a going concern, ready to produce when their producing time comes.

Therefore, the crop-yielding tree offers the best medium for extending agriculture to hills, to steep places, to rocky places, and to the lands where rainfall is deficient. New trees yielding annual crops need to be created for use on these four types of land.

TWO-STORY AGRICULTURE FOR LEVEL LAND

The level plains where rainfall cannot carry soil away may continue to be the empires of the plow, although the development of two-story agriculture (trees above and annual crops below) offers interesting possibilities of a greater yield than can be had from a one-story agriculture.

This type of agriculture is actually in practice in many Mediterranean lands. In the Spanish island of Majorca I estimated as a result of several journeys across the island that nine-tenths of the cultivated land carries an annual crop growing beneath the tree crops. I recall a typical farm planted to figs in rows about forty feet apart. Beneath the fig trees was a regular rotation of wheat, clover, and chick-peas, one of the standard articles of Mediterranean nutrition. The clover stood two years and was pastured by sheep the second year.

Other two-story Majorca farms had, for a top crop, almonds, one of the staple exports of the island. Other lands were in olives, and a few were in the sweet acorn-bearing oak. The people said that the farmer did not get the greatest possible crop of wheat or the greatest possible crop of olives or figs, but that he got about a 75-percent crop of each, making a total of 150 percent. It is like the ship which fills three-fourths of her

tonnage capacity with pig iron and five-sixths of her cubic
capacity with light wood manufactures.

The two-story type of agriculture has another advantage. It
divides the seasonal risks which everywhere beset the farmer. If
frost kills the almond, it probably will not injure the wheat. If
drought injures the wheat, the almond may come through
with a bumper crop.

In some cases the Majorcan landlord rents the ground crops
out to a tenant for a share and keeps all of the tree crops for
himself, the tenant having contributed no labor in their production.

A VISION FOR OUR AMERICAN HILLS

We have large areas of hilly land where the climate is good.
We have such an area of great beauty, with excellent climate
and good soils, reaching from Maine to Alabama, from Alabama
through Kentucky and Tennessee to central Ohio, from central Ohio through southern Indiana and Illinois into Missouri
and Arkansas. Again, such an area appears on the foothills of the
Rockies and the mountains of the Pacific Coast. Then too,
there are hilly bits of land in nearly all sections of our country.
When we develop an agriculture that fits this land, it will become an almost endless vista of green, crop-yielding trees. We
will have plowed fields on the level hilltops and strip crops on
the gentle slopes. The level valleys also will be plowed, but the
steeper slopes will be productive through crop trees and will
be protected by them—a permanent form of agriculture. When
we have done all this, posterity will have a chance. Under the
trends of 1930 it had but little opportunity, and it is only a
little better now. Good luck to the U. S. Soil Conservation
Service!

SOME CROPS FOR THE HILLS

Chestnuts and acorns can, like corn, furnish carbohydrates
for men or animals. To many it may seem ridiculous to suggest
that we moderns should eat acorns, and I hasten to state that

the chief objective of this book is to urge new foods for animals rather than for men. Food for animals is the chief objective of the American farmer. Our millions of four-footed brethren who bellow and neigh, bray and squeal, bleat and butt, eat much more than the two-footed population consumes. Their paunches receive the crop from about four-fifths of our farm acres.

When tree agriculture is established, chestnut and acorn orchards may produce great forage crops, and other orchards may be yielding persimmons or mulberries, crops which pigs, chickens, and turkeys will harvest by picking up their own food from the ground. Still other trees will be dropping their tons of beans to be made into bran substitute. Walnut, filbert, pecan, and other hickory trees will be giving us nuts for protein and fat food.

Even this partial list of native tree products, *now producing in convincing quantity,* shows nearly all of the elements necessary to man's nutrition save bulk, and for that we have leaf greens in plenty, and all that without introducing a single new species from foreign countries where dozens of new crop trees are waiting for the time to come when they can be made useful in American agriculture.

This permanent agriculture is much more productive than mere pasture, or mere forest, the only present safe uses for the hill fields. Therefore, tree crops should work their way into the rolling and sloping lands of all sections. New crop trees need to be created. Extensive scientific work at exploration and breeding in the plant kingdom should begin at once.

DOUBLING THE CROP AREA

As the deep-rooting, water-holding trees show their superior crop-producing power in dry lands, we may expect some of our now-arid lands to become planted with crop trees. Thus by using the dry land, the steep land, and the rocky land, we may be permitted to increase and possibly double our gross agricultural production and that, too, without resort to the Oriental miseries of intensive hand and hoe labor. Tree crops also have a special advantage in their adaptability to a field reservoir system of irrigation which is at the same time of great promise as a means of flood control. (See Chap. XXII.)

The great question is, how can we shift from the grain type of agriculture and ruin to the permanent tree agriculture in those localities where the change is necessary to save the land from destruction? In the next chapter the attempt is made to find an answer to this question.

FIG. 2. Scene in the Carolina mountains showing stream now 100 yards wide, which once ran close to the hill at the right but is now destroying good meadow because of increased floods due to bare hillsides on its drainage basin. Note the steepness of the slope of the fields, the patch of corn at the right, and bad gullies above it. Cane Creek near Bakersville, N. C. (U. S. Forest Service.)

FIG. 3. *Top.* The author, a six-foot man, stands in the corn beneath the arrow lower right. The worthless stalks by the hat measure the ruin of the hill—typical of forty-five American states. FIG. 4. *Center.* South Central Ohio. Nobody loves this land. Therefore it goes swiftly to economic Hades. It was a good field, and is now typical of hundreds of thousands of hillsides. (U. S. Forest Service.) FIG. 5. *Bottom.* No gully, but plucking raindrops have carried away many feet of top soil in this Algerian wheat land. (Photo J. Russell Smith.)

CHAPTER III

The Plan—An Institute
of Mountain Agriculture

THE PLAN—AN INSTITUTE OF MOUNTAIN AGRICULTURE

Those who have read this far in this book will at once appreciate that a vast work is proposed. It is nothing less than the deliberate creation of a whole new set of crop trees and then to make a new agriculture based upon the use of these new crop trees.

This is a work in which some results may be had in a few years. It is also a work which, once started, might enrich each of many succeeding generations.

It could employ the full time of a man, of ten men, of a hundred men, of five hundred men. After it really got under way, it might employ as many men as it takes to man and operate a battleship. Oh, for a battleship! That is to say the money required to build, maintain, and operate one. What millions of acres of land it might save! What billions of wealth it might make! What tens of thousands of homes it might teach us how to support in places where now the chief crop is gullies, waste, and struggling briars!

THE KINDS OF WORK TO BE DONE

Such an institute would have a variety of work to do:

1. Finding parent trees from which to start crops;

2. Hybridizing to produce better parent trees;

3. Maintaining testing grounds to try out the new trees as trees—testing them by ten thousands to find the good *trees*.

4. Operating experimental farms to try out the new *trees* as *crops*.

5. Making studies in *farm management*. Farm management is a great art. I have seen well-educated persons take Soil Conservation Service devices and use them in a wrong way, so that the good thing could not succeed. There is a great gulf between knowledge and judgment (gumption). Judgment telleth how to use knowledge. There are thousands of college and school courses in knowledge, but I do not recall hearing of one on judgment.

6. Carrying on publicity work to get the new idea into the conservative mind.

1. Finding Parent Trees

The one original parent Baldwin apple was born by chance in a fence corner. It was propagated by grafting until there were millions of Baldwin apple trees. If this same process were applied to the best wild crop trees now growing in the world, I am sure that the hill fields from Massachusetts to California could become more valuable than much of the best nearby farm land, and be covered with orchards of bearing trees like the best wild

(a) Shagbark hickory trees and their natural hybrids,
(b) Shellbark hickory trees and their natural hybrids,
(c) Pecan hickory trees and their natural hybrids,
(d) Butternut trees,
(e) Black walnut trees,
(f) Chinese walnut trees and natural hybrids,
(g) Chestnut trees and already created hybrids,
(h) American persimmon trees,
(i) Chinese persimmon trees.

These crops would be for humans. For animals the much wider expanses of hill land should be covered with

(a) Acorn-yielding oaks
(b) Honey locust trees
(c) Chestnut trees
(d) Persimmon trees
(e) Mulberry trees
(f) Many, many others.

Professor C. C. Colby, of the University of Chicago, says the Highland Rim County of East Tennessee is "alive with food in the fall—butternuts, walnuts, hickorynuts, chestnuts, persimmons, and pawpaws, all in great abundance and lying on the ground."

It is easy to speak of an orchard of best shagbarks or acorn-yielding oak, but where are the best wild parent trees for these orchards? No one knows where they are. The task of finding the parent trees may be long and difficult. We see how great it is when Sargent mentions fifty species of oak trees as being native to the United States. There are also natural hybrid oaks scattered about the United States, many of them, but no one has any idea how many or where most of them are. Sudworth, United States Forest Service, said there were one hundred and seventy species of oaks. There are many varieties of hickory. The honey locust tree and the persimmon, species of great promise for crop production, are each growing on a million square miles of land. How would you find the best tree? When it comes to persimmons we need a half dozen best trees; one ripening in August, one in September, one in October, one in November, one that drops its fruit in December, and yet another which drops it in January. There are such wild persimmon trees, fruitful trees too, already growing in the United States. This is a tree of astonishing fruitfulness, growing wild on much of the poorest land in the United States.

Yet more! Foreign countries need to be carefully searched in the quest for parent trees. Something has already been done by the U. S. Department of Agriculture in a more active period of the past, but the task is only begun. It may be that the best crop trees are growing in some little valley in Spain, Portugal,

Jugoslavia, Asia Minor, Persia, the Himalayas, or the remote interior lands of southwestern China, which seems to be such a wonderland of trees. Nor should we forget South America, Africa, and Australia.

2. *Hybridizing to Produce Better Trees*

There is little doubt that we could start a good tree agriculture by merely propagating the best wild trees, but we must remember that a new force came into the world with the turn of the century. That force, known as Mendel's Law, reduced plant breeding to a science—genetics. Plants became as clay in the hands of man. If agricultural science has worked out any result it is this: The purposeful hybridization by man of existing trees can produce trees that are much better for agricultural purposes than those nature has produced. This work should use both native and introduced trees. A dozen men could at once go to work on the various oak species, a half dozen on the chestnuts, and another half dozen on the hickories. The magnitude of such an undertaking may be grasped when we realize that one species of apple has given rise *in America* to over seven thousand named varieties, and new ones and better ones are being made every year.

The hybridization work of an institute of mountain agriculture should amount to hundreds and thousands of cross pollinations every year. Each hybrid seed would have to be planted and grown to fruiting age unless it demonstrated worthlessness at an earlier stage. A man in Louisiana hybridizes sweet potatoes. Then he sprays the hybrids with the germs of all diseases. If they die, he knows it early.

3. *Testing Grounds*

The best hybrids should be tested to determine their possibilities as crop producers. This would cover many acres of ground and need a considerable staff of men.

4. Experimental Farms

It is one thing to tell the farmer that here are good black walnuts or chestnuts or acorn-yielding oaks or honey locust trees, and it is quite another matter to organize these into an effective farm. That is a matter of agricultural economics and farm management; so the institute should have a number of farms in which tree crops were worked out into a system to make a well-balanced and profitable use of land, well-balanced use of a man's time, and good, safe living for the family who depended upon it. Before the farmer plants a new thing he wants to know what an acre of it will yield over a term of years.

5. Publicity

An institute of mountain agriculture, as outlined above, needs an expert in publicity on its staff. In the beginning years he could be chiefly employed in the very difficult task of getting the attention of almost every landowner in the United States, of every tree lover, and of every hunter, vacationist, botanist, and Boy Scout, so that we might find those rare parent trees that are standing in fence corners, back fields, distant pastures, and remote mountainsides.

Another task for the publicity expert would be to attract the attention of the American farming public to the fact that there could be such a thing as tree-crop agriculture, and that some crops are now ready to use. This is just as much a task for a publicity expert as is a press campaign for a candidate for the Senate or a corporation that wishes to raise fares or put the public to sleep. The sedate government bulletins already published are helpful, but experience shows that they do not fill a very big place. These bulletins are so safe, so sound, so dead, so unhuman, so many times edited by the man higher up, that they are usually only raw material for a person who can write. And if a bureaucrat could write and *did*, ———? The task of spreading a new idea in agriculture is most difficult.

The history of agriculture shows a conservatism probably unequaled in any other phase of human activity.

MANY INSTITUTES OF MOUNTAIN AGRICULTURE NEEDED

Suppose a man who started an institute of mountain agriculture lived in North Carolina and worked out a North Carolina agriculture; the interested farmer from Massachusetts discovers that few of the North Carolina varieties and practices fit his conditions. This shows at once that we need several institutes of agriculture in locations where their findings are adaptable for considerable areas. (See Chap. XXVI and World Map, Fig. 138.)

6. Erosion Survey

The objects of an institute of mountain agriculture might be classed as twofold—to create a new wealth through new crops and to save our soil resources from destruction by erosion. For the latter purpose we need to know what the danger really is. The thoughtful part of the American public might be shocked into doing something about it if they could be made to understand what soil erosion has done to China, Syria, Greece, South Carolina, Mississippi, and Ohio. A few well-planned and well-manned expeditions making an erosion survey of foreign lands and our own country would bring back material which might be one of the scientific sensations of the day. In the hands of good publicity experts it might make this reckless American people see that we are today destroying the most vital of our resources (soil) faster and in greater quantity than has ever been done by any group of people at any time in the history of the world. If our people could be made to *feel* this, they would try to stop it. Here are several highly suggestive facts: (1) The sight of the ruin of central and western China by erosion burnt the soul of Walter C. Lowdermilk; it set him on fire. He wound up in the U. S. Soil Conservation Service. (2) Lowdermilk spent several months studying the ruin of the Mediterranean world by erosion. (3) The war stopped him in mid-career. (4)

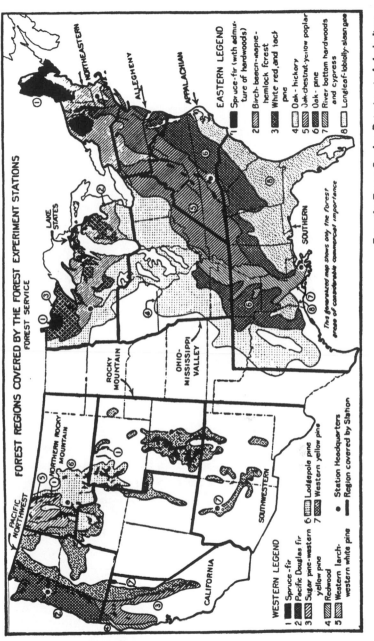

FOREST REGIONS COVERED BY THE FOREST EXPERIMENT STATIONS
FOREST SERVICE

WESTERN LEGEND
1 Spruce-fir
2 Pacific Douglas fir
3 Sugar pine-western yellow pine
4 Redwood
5 Western larch-western white pine
6 Lodgepole pine
7 Western yellow pine
• Station Headquarters
— Region covered by Station

EASTERN LEGEND
1 Spruce-fir (with admixture of hardwoods)
2 Birch-beech-maple-hemlock forest
3 White red, and jack pine
4 Oak - hickory
5 Oak-chestnut-yellow poplar
6 Oak - pine
7 River bottom hardwoods and cypress
8 Longleaf-loblolly-slash pine

This generalized map shows only the forest areas of considerable commercial importance

From the Forest Service, Department of Agriculture

This map shows fifteen different types of forests having areas of considerable commercial importance. Each type of forest represents a type of tree growth and might require a special experiment station to work out tree crop possibilities for that type of area. This same map shows that our Federal Government has 11 forest experiment stations covering the 15 types of forest. Be not confused. Tree crops experiment and forestry experiment have the same relation that white has to black.

The Federal economy wave clipped the Soil Conservation Service; its staff was nonpolitical. Lowdermilk, more valuable for the ultimate national defense than a regiment, went out—but look, look up the results of his foreign survey.

A PLEASANT RECREATION

As outlined above, this work can employ heavy endowments and large staffs. On the other hand, experiments with trees can be on almost any scale. Two trees, for example, might produce great (hybrid) results. There are thousands of individuals who can experiment and perhaps do something of great value to the human race and at the same time have recreation and pleasure.

I have altogether something like a hundred varieties of walnuts, hickories, pecans, persimmons, pawpaws, and honey locust on test on my rocky hillside, and I find that I am having an amount of fun out of it that is as perfectly unreasonable and genuine as is the joy that remains for a month or two after making a good drive at golf, catching a big fish, or shooting a deer—not that I do all of these things.

Experimentation with nut trees is especially to be recommended for people in middle age and upward. One of the pains of advancing years is the declining circle of one's friends. One by one they leave the earth, and the desolating loneliness of old age is felt by the survivors. But the man who loves trees finds that this group of friends (trees) stays with him, getting better, bigger, and more lovable as his years and their years increase. This perhaps explains the delightful enthusiasm of some of the septuagenarian and octogenarian tree lovers whom I know and have known, such as the late E. A. Riehl and Benjamin Buckman, both of Illinois, who were plunging ahead in their eighties as though they were in their forties.

Mr. Riehl began nut-tree pioneering on some Mississippi bluffs near Alton, Ill., at the age of sixty-three and actually made money out of nuts. He was really just getting started when he died at the age of eighty-seven. I knew him for eleven years.

It was a great pleasure to associate with such a youthful and enthusiastic spirit. He was one of the youngest old men I ever knew, still living in the future, not in retrospect as is so common with old age.

PART TWO

Some Facts About Some Crop Trees to Suggest Some of the Possibilities That Lie in This Proposed Work

Should I begin with the nut crops, the well-known pecans, walnuts, and hickories with which almost every American is somewhat acquainted?

This book is primarily an attack upon the gully. To succeed in this we must have millions of acres of tree crops replacing the destructive plow crops. Now, the nuts that people eat are fine and worthy of much improvement, but a few hundred thousand acres of them would glut the market. Not so with stock food. Once we get a cow-feed tree crop established we have a guaranteed outlet, and twenty or thirty million acres will not glut the market. We would simply convert thirty or forty million acres of our hundred million acres of corn to a more profitable and soil-saving crop.

Therefore, I start this part of the book with stock foods.

There is another reason also. Some of the stock-food crops seem to be in the class of sure things with which the farmer can safely begin without waiting for a lot of scientific work to be done.

Then, too, stock foods start on an honest-to-goodness basis. They don't begin five prices high, as human food novelties usually do, and then come down bumpety-bump as soon as a few carloads are produced.

CHAPTER IV

Some Stock-Food Trees—
The Producers of Bran Substitutes

Who has not pitied John the Baptist because he had to eat insects (locusts) in the wilderness, and the Prodigal Son because he was brought so low that he would gladly have fed upon the husks that the swine did eat?

But the locusts that John the Baptist ate may not have been insects; they may have been the pods of the carob tree, sometimes still spoken of in the Near East as "locust." The husks coveted by the Prodigal Son were not the dried husks of corn, or "maize," as the American farmer naturally believes. They, too, were likely the pods of the carob bean. Corn was discovered with America by Columbus.

Like our own maize, the carob bean is food for animals. It has been used for that purpose throughout the Mediterranean region for several thousands of years. Like maize, it is also used for human food. My son, when three years of age, ate, without invitation, the samples of carob that I had brought home. Carob beans are regularly sold for human food by little shopkeepers and pushcart vendors on the streets in some parts of New York and other American cities occupied by Mediterranean peoples. They have long been used in American factories to add both flavor and nutriment to certain patent calf foods, livestock conditioners, and dog biscuit. U. S. Department of Agriculture Bulletin No. 1194 states that four hundred tons per year are used in the United States to flavor chewing tobacco.

A carob gum stiffens the consistency of cream cheese.

The trees and the beans have been introduced into California, and it is an astonishing fact that promoters of Los Angeles claimed years ago that their products make good food for the American table. John the Baptist and the Prodigal Son might easily have fared worse.

The bean-producing carob tree is one member of a group of leguminous trees, some of which are equipped to gather nitrogen from the air and make sugars from earth materials and air and to make forage for beast (if not food for man) in nearly all the climates that circle the globe in the latitude of Washington, St. Louis, Santa Fe, Los Angeles, Shanghai, Kabul, Teheran, Jerusalem, Rome, and Gibraltar. In the southern hemisphere, a similar band of bean-tree climate covers large areas in Australia, South Africa, Argentina, Chile, and other areas.

Three entirely different types of climate are found in this bean zone. (See map in this book.)

FIRST CLIMATE TYPE

The Mediterranean climate, on the western coasts in middle latitudes. This climate has a mild, slightly frosty winter with some rain which is followed by a hot, dry summer. This is the climate of southern and central California, the fruit climate *par excellence*. Here the carob thrives.

SECOND CLIMATE TYPE

On the eastern coasts of the continents in middle latitudes, the American Cotton Belt, South China, also southeastern Australia, southeastern Africa, and southern Brazil, is a climate with a frosty winter and rainy, humid, hot summer. This is the climate of the honey locust tree, *Gleditsia*, bearing beans a foot or more in length and good for forage.

THIRD CLIMATE TYPE

Between these two type regions, California and the Cotton Belt, is an area of arid interior, typified by Arizona, New Mexico, and western Texas. In the Old World this between-region of drought includes large areas of Syria, Persia, and Afghanistan, with similar regions of large size in Argentina, Australia, and South Africa. This is the climate of the mesquite, native to America, and of certain allied species which thrive amazingly in the arid lands of both North and South America. (For map, see Fig. 138.)

All of these bean bearers have very ingeniously bedded their seeds in a sugary pod which is greedily eaten by many ruminants. The seed itself no beast can bite, bruise, or digest. It passes with the excreta and is dropped on every square rod of pasture land and bedded down in fertilizer to help it start its new life. Nature is indeed ingenious!

All of these beans and their pods are much alike in food service and in food analysis. In nutritive value, both protein and carbohydrate, they are much like wheat bran—that standard nutrient of the dairy cow. (See table, Appendix.) Therefore, it seems fair to call these bean trees "bran trees," because some are already used as bran substitutes and others may be made to afford a commercial substitute for bran. This gives the possibility of their being major crops of American agriculture.

CHAPTER V

A Stock-Food Tree, the Keawe, or Hawaiian Algaroba

I shall begin my discussion of the stock-food trees with the Hawaiian algaroba, commonly called keawe in those islands. I start with keawe because the facts about it have been worked out by American agricultural officials and American businessmen and accountants; because the evidence that these men have produced is official; and because of the astonishing, convincing, and yet almost unbelievable nature of that evidence. I shall present much of the material in the exact words of officials of the American Agricultural Experiment Station in Honolulu.

E. V. Wilcox, Special Agent in charge, Agricultural Experiment Station, Honolulu, said (Press Bulletin No. 26, Hawaii Agricultural Experiment Station, *The Algaroba in Hawaii*):

The algaroba, or keawe, *Prosopis chilensis,* is commonly recognized as the most valuable tree which has thus far been introduced into the Territory of Hawaii.

There are eighteen or more species of Prosopis, the natural habitat of which is in tropical and semitropical America. The algaroba occurs from Texas to Chile and in the West Indies. . . .

Algaroba or keawe thrives best at low altitudes but is everywhere in Hawaii gradually extending to higher levels, and it is found in some localities at altitudes as high as two thousand feet.

Dr. J. M. Westgate has not seen it above 1,000 feet and reports 600 feet to be its usual upper limit on the island of Oahu. Apparently it is gradually becoming acclimated to higher altitudes, but it bears most abundantly at lower levels. On the whole, its distribution has been largely accomplished by stock. The flowers of algaroba furnish the most important source of pure honey known in the Territory. The bee raisers of the Territory have shown an active interest in securing the rights of placing apiaries so as to utilize to the fullest extent the algaroba forests. The yield of honey is recognized as large and important and occurs at two seasons, there being two crops of flowers and pods annually.

As a forage crop, algaroba is of far greater financial value. The pods are everywhere recognized as one of the most important grain feeds of the islands and are greatly relished by all kinds of live stock, including chickens. The quantities of pods produced by the algaroba forests cannot be estimated even approximately, for a large proportion of the pods are allowed to fall on the ground and are eaten by cattle, hogs, and horses without being previously picked up. Wherever the belts of algaroba timber are large, it has been found possible to maintain stock for a month or two of each season without any other forage than algaroba beans.

Practically all of the islands have enormous belts of algaroba forest extending from the seashore on the leeward side up to an altitude of eight hundred or one thousand feet.

There are few trees which are distinctly useful for more purposes than is the algaroba.

The botanists have long had a hot dispute as to whether this Hawaiian tree was American or Chilean in its origin. Certainly it was introduced about a century ago, and there are at least three good rumors as to its source.

I have become very much interested in the identity of this species and have succeeded in obtaining related species from many parts of the world and have these growing in seedling form on our station here in Honolulu. I feel pretty sure that ours is a South American species, possibly *Prosopis chilensis*. I am quite certain that it is neither the true mesquite nor the true carob. (Letter signed by J. M. Westgate, Agronomist in Charge, Agricultural Experiment Station, Honolulu, Hawaii, July 5, 1916.)

See Figs. 26 and 38 for pictures of beans of keawe and carob. It is one of the jokes on horticulture that they have been called the same—even in Bailey's good *Encyclopaedia of Horticulture*.

Some of the rough areas unsuited to cultivation on the windward side of the island are being planted to algaroba since the demand for the meal is gradually increasing. (Letter, March 12, 1913, E. V. Wilcox, Special Agent in Charge.)

The algaroba in Hawaii occupies at least 50,000 acres of land, growing satisfactorily with a rainfall of from 12 to 15 inches per year. (Letter from E. V. Wilcox, Special Agent in Charge, Agricultural Experiment Station, Honolulu, March 12, 1913.)

This tree grows under rainfall conditions varying from ten inches per annum to as high as fifty inches. In altitude it grows from the sea level to about eight hundred feet. . . .
A great many of the algaroba trees grow where the soil is two feet deep, being underlaid with solid lava rock. (Letter from J. M. Westgate, Agronomist in Charge, July 5, 1916.)

Algaroba wood also constitutes one of the best and chief sources of fuel in the Territory. Its growth is comparatively rapid and the larger trees can be removed for fuel, thus making room for the growth of another generation of trees. In addition to these uses of the algaroba, it might also be stated that the bark contains tannin, and the gum is suitable for use in varnish. Being a legume and of remarkable penetrating power in the soils, it is also a soil maker of some importance. As a shade and ornamental tree it is highly appreciated. The form of the tree is graceful and spreading. The small branches furnish excellent material for making charcoal. Piles made from algaroba are relatively free from the attack of the teredo.

Moreover, since the pods contain a high percentage of sugar, they may be used in the manufacture of denatured alcohol and vinegar. (Hawaiian Agricultural Experiment Station, Press Bulletin 26, p. 4, *The Algaroba in Hawaii.*)

The algaroba bean industry is getting to be a very important one here in the Islands, the ground pods being regarded as equal to barley or oats for feeding purposes, pound for pound. It is rather difficult to get any definite figures on the production as so many of the companies utilize their own grinding plants and feed the ground beans to their own stock. Thousands of head of stock in certain seasons of the year are allowed to eat the pods as they fall from the trees in the native pastures. (Letter from J. M. Westgate, Agronomist in Charge, Agricultural Experiment Station, Honolulu, July 5, 1916.)

"The business which has developed this year from the sale of algaroba bean meal amounts to about $350,000." (Letter signed E. V. Wilcox, Special Agent in Charge, March 12, 1913.) Mr. Wilcox further states, "It has been estimated that approximately 500,000 bags of the beans are annually picked up and stored, particularly for feeding horses and cattle. On two or three estates, at least 15,000 bags of beans are annually stored for this purpose."

As to yield, Mr. Wilcox said, "It has been found that the yield per acre varies from two to ten tons. This yield varies but little from year to year and occurs in two crops per year, the figures given covering the sum of both crops."

In another connection Mr. Wilcox said, "The yield of beans per acre of good algaroba forest is about four tons per acre." (Letter, June 18, 1912.)

In 1927, Dr. Westgate, then Director of the Hawaiian Agricultural Experiment Station, weighed the algaroba beans from a yard tree, 17 inches in diameter, 30 feet high, 60 feet spread—500 lbs. of beans. (Letter, February 1, 1928.) And they have two crops per year!

L. A. Henke, Animal Husbandman at the University of Ha-

waii, wrote (Oct. 22, 1947) that in 1930 they selected five
trees on the campus of the University and picked up the beans
between June 7 and December 2. Yields of the five trees in air-
dry pods were 255, 439, 154, 54, and 122 pounds. Average
yield (omitting an extra tree yielding 146 lbs.) was 205 pounds.
At 20 trees to the acre, the yield would be about 1½ tons. The
value? See the price of ground barley in your town. However,
the higher wages of 1947 have reduced the bean-meal business
in Hawaii.

J. E. Higgins, Horticulturist at the Honolulu Station, said,
"It has taken possession of large tracts of otherwise unoccupied
land, prospers where the soil is too dry for any other crop, produces a verdure and shade where otherwise there would be an
almost barren waste, and yields a pod of high feeding value."
(Letter, July 3, 1916.)

Mr. Wilcox, above mentioned, said, "Thus far there has
been no cultivation of algaroba. The large areas of trees which
we have stand, for the most part, in rock soil where cultivation
would be practically impossible.

"We therefore have no evidence as to whether the increased
yield obtained by cultivation in tillable soil would pay for the
added cost of cultivation. At present the cost of gathering and
grinding is about eighteen dollars per ton, allowing ten dollars
per ton for gathering, five dollars for grinding, and three dollars for transportation. The regular price paid for picking up
the pods from the ground under the trees and delivering them
in bags at the roadside is one-half cent a pound. At this rate,
men, women, and children make from one dollar and twenty-
five cents to one dollar and seventy-five cents per day."

Speaking of the value of the beans, Mr. Wilcox said, "The
feeding value of twenty-five dollars which I have placed upon
algaroba meal is to be compared with bran or rolled barley at
about forty dollars per ton at present prices in Honolulu. At
these prices it is worth more than twenty-five dollars per ton,
since its analysis shows it to be practically equal except for the
fact that the crude fiber is a little more."

The two sets of beans require two sets of blossoms, all of which are rich in honey, which Mr. Leslie Burr estimates at two and a half pounds per year for a tree with thirty-foot spread. (See *Gleanings in Bee Culture*, January 1917.) Honey by the ton is one of the products of a Hawaiian algaroba pasture.

THE ALGAROBA MEAL INDUSTRY

Owing to the fact that no animal can digest the bean, it is estimated that most of the protein value (see table of food analysis, Appendix) is lost. Some estimate that about forty per cent of food value is wasted when the animals eat the beans under the trees. This loss, combined with the desire to have food in the off season, led to attempts to grind the beans. Owing to the fact that the sugar of the pods has the consistency of molasses, it stuck to the parts of the grinding machine and looked like vulcanized rubber. For a long time it interfered with attempts at successful grinding. After years of work a technique for grinding was evolved.

The Hawaii Experiment Station recommends a fine spray of water on the rolls, which prevents the meal from sticking, yet does not wet the meal enough to cause it to spoil.

Mr. Ben Williams of the Hawaiian Commercial & Sugar Company on the Island of Maui worked out a different technique. He heats the beans to a temperature of 600° to 800° F. by superheated steam in a rotary kiln. This turns the molasses to a white powdered sugar. The beans are then ground in a swing-hammer sand machine, a machine made to crush soft sandstone.

In 1926 Mr. Williams, then ranch manager on the above-mentioned estate on the Island of Maui, told me that he was having one thousand to one thousand four hundred tons of beans per year picked up. Women picked up eight or ten forty-pound sacks of beans a day and received an average of one dollar for the day's work. Many women picked up twenty sacks daily, thus making two dollars to two dollars and twenty-five cents a day, which was more than they made in the sugar fields.

It cost Mr. Williams twelve dollars and fifty cents per ton to have the beans picked up and put in the storehouse, five dollars to take them from the storehouse and have them ground and bagged; the bags cost two dollars; total cost was nineteen dollars and fifty cents. Allowing ten percent for shrinkage, the meal cost a little over one cent per pound.

Six men ground ten tons per day with electrically driven machinery and one-quarter of a barrel of oil. The total cost was three dollars per ton, itemized as follows:

Costs $\left\{\begin{array}{l}\text{Labor} \dots\dots\dots\dots\dots\dots\dots\dots\dots\dots\dots\dots \$12.00 \\ \text{Power} \dots\dots\dots\dots\dots\dots\dots\dots\dots\dots\dots\dots\ 12.00\end{array}\right.$

Capital charge. 6.00

It could be done for less if the plant were worked more continuously.

Hawaii Agricultural Experiment Station reported:

The feeding test made by this Station showed that the seeds thus cracked are completely digested by horses, mules, and cattle.

The keeping quality of the meal is quite sufficient for the ordinary demands of the trade. When kept in sacks or open containers, it retains its original odor and flavor, without change, for six or eight months; and the meal is no more subject to the attacks of insects than is any other grain feed.

E. V. Wilcox said (letter, August 12, 1913):

The algaroba beans have been formerly shipped to Japan as a food for cavalry horses, but the product is now all used in Honolulu. It has been adopted as a part of the rations for army horses of this Territory.

The evidence in the passages quoted above is truly astonishing when compared with that of the standard crops of the American field. In considering these facts one should remember the necessary restraints and conservatism of statement which must and do mark the official representatives of the Department of Agriculture.

Even more astonishing evidence was given me by Mr. Ben Williams (mentioned above), who was a Welshman, fifty-five years of age, and ranch manager of the Hawaiian Commercial & Sugar Company's estate on the Island of Maui. This estate has fifteen thousand acres in cane land, twenty-one thousand three hundred acres classed as waste and pasture, of which none is real waste and none real pasture. The area involved, thirty-six thousand three hundred acres—nearly sixty square miles—is as large as some of the smaller counties in the United States. Its population is in thousands, and its organization is a vast group of industries, well subdivided, with a superintendent of cane fields, a factory superintendent, a chief engineer in the sugar mill, a foreman of machine shop, and the ranch manager, Mr. Williams, in charge of two hundred dairy cows, eight hundred cattle, seven hundred horses and mules, and two hundred and fifty pigs.

Of Mr. Williams's ranch domain, eight to nine thousand acres are in algaroba, planted by the cattle as they scattered the beans from one tree that stood by the windmill and its attendant drinking trough. By rule of thumb observation Mr. Williams said that one hundred pounds of keawe meal, when fed to pigs, horses, and cattle, were about equivalent to eighty pounds of good barley, which had long been the standard horse feed of the islands. Mr. Williams said that each acre of good keawe will fatten at least six head of cattle. If the animals weigh six hundred pounds when turned in, in ninety days they will weigh over eight hundred pounds. I expressed my doubts. "I have seen it done," declared Mr. Williams. "I know this, that you can take cattle, lean ones, that do weigh five hundred pounds and should weigh seven hundred pounds; and if you put six of them on an acre of good keawe, they will average better than two pounds a day on raw beans which they pick up for themselves. You can take the season from the middle of July,

and the six cattle will gain twelve hundred pounds and sometimes will go to sixteen hundred pounds of beef per acre on
land with rainfall of twenty inches a year."

Mr. Williams reiterated these figures and saw me write them
down. It should be mentioned that this man has a corps of accountants who keep books on every field for crops, for pur-
chases, for sales, and for every bit of labor employed, just as
you would expect to find in the Standard Oil Company, or in
other highly organized corporations.

When one considers that a good acre of Kentucky bluegrass
pasture or of the rich pasture of old England will produce one
hundred and fifty pounds of mutton per year, and an Illinois
farm in corn and alfalfa will make about four hundred and
fifty pounds of beef and pork per acre per year, the keawe
bean tree looms up as one of the king crops of the world.

In explaining the prodigious yields, Mr. Williams pointed
to a particular tree (see Fig. 24) which he said would yield
from two to three tons. I measured the tree. It had a reach of
eighty-four feet. The tree hung full of beans (see Fig. 25).
Many were dropping, and the tree was still blooming. It was
then the fourth of August, but Mr. Williams assured me the
tree would keep on blooming for some time and that it would
drop beans for five months, from July to December.

We walked through the copses of keawe where the sand was
merely held in place by protection of keawe root and keawe
tops, which kept the wind from getting the sand. Before the
reader jumps to the conclusion that this was a desert waste, however, he should consider the geologic origin of the sand. It
was volcanic sand blown up from the shore of the sea and mixed
with perhaps ten to fifteen percent of shells. There is no richer
soil combination known on the face of this earth than certain
volcanic sands and lavas mixed with limestone. This richness
of the fresh unleached lava soil of Hawaii should be kept in
mind when one thinks of applying Hawaiian facts to many other
areas of semiarid frostless lands in which the algaroba may

probably find a suitable climate. (See further discussion of this point in Chap. VIII on the Mesquites.)

In Cuba, where I lived a number of years, one of the locusts, called a "guasima," furnishes considerable food for cattle, and they are introducing from the Hawaiian Islands a locust with a sweet substance in the pod as a food for horses, cattle, and swine. I believe this tree is called an algaroba. (Letter, N. S. Mayo, Animal Husbandman, Virginia Agricultural Experiment Station, Blacksburg, Virginia, May 22, 1913.)

The Hawaiian keawe seems truly tropical, but the genus to which it belongs is by no means limited to lands without frost. (See the findings of Dr. Walter S. Tower in the next chapter.)

The surprising performances reported in this chapter may almost without exception be said to be products of wild trees, although the process of thinning out may sometimes leave the better ones. It should be noted that nine thousand acres of trees on the sugar plantation mentioned above were scattered by cattle from one chance tree at the windmill. What a shame that it was not an exceptionally good tree! Or was it? Said Professor E. V. Wilcox, mentioned above, "There is a great difference in the yielding capacity of different trees. They begin to bear profitably at four or five years of age."

There is every reason to think that the keawe produced by chance is capable of much improvement by selection and breeding. Then the propagation of orchards from the best trees should give a still better crop than is at present obtainable.

The chief value of the keawe in this book is that it is an example—a successful tree crop in a world that needs many other tree crops to fit particular places and keep its scanty soil upon its rocky ribs. The keawe and the mesquites are growing well in South Africa and Australia, where large areas of land with light rain await better use by some such tree. (See world map at end of this book.) Apparently many other semiarid areas might benefit.

CHAPTER VI

A Stock-Food and Man-Food
Tree—The Carob

AN INDUSTRY, AGE-OLD AND WIDESPREAD

The carob, the food of Mediterranean people, of the Mediterranean farm animals, and of the calves and dogs of America, also fed the cavalry of Wellington in his Peninsular campaign and that of Allenby in Palestine during World War I. (John S. Armstrong, *Orchard and Farm*, February 1919.)

University of California Publication, "Feeding Dairy Calves in California," Bulletin No. 271, September 1916, p. 32, by F. W. Woll and E. C. Voorhis, has this suggestive statement on the carob and its uses:

According to Pott (*Futtermittellehre*, Vol. II, pp. 453, 455), the crushed carob pods are frequently used in England for fattening sheep, and for ewes with lambs, also in connection with other concentrates for fattening steers. It is used in France as a feed for milch cows and young stock, and in southern Italy and other countries as a concentrate for horses and for growing pigs. British horses are at times fed as much as three kilos (6.6 pounds) per head of carobs daily, either cooked and mixed with cut straw or raw. Fattening steers are also fed preferably cooked carobs towards the end of the fattening period. For horses it is not even necessary to crush the pods. In southern Italy nobody would think of doing it, although

46

the strong ponylike horses do not receive any other concentrates and are fed only hay or green feed in addition.

Carob beans are sold in many American cities where Mediterranean peoples live. They are eaten from the hand as are apples, peanuts, and chestnuts. In Sicily they serve as candy. Almost any American child will eat them if he gets a chance. "Fruit of this tree is variously known as carob, carob bean, algaroba, algarroba, karoub, caroubier, locust, sweet bread, sugar pod, and St. John's bread." (*Scientific American,* January 11, 1913.)

The tree has been cultivated in the Mediterranean region from an unknown antiquity, and both the wild and cultivated trees still grow throughout that region. I have seen carobs in South Portugal overlooking the Atlantic; in Valencia, eastern Spain, overlooking the western Mediterranean; in Majorca, overlooking the northern Mediterranean; in Algeria, the south Mediterranean. On the stony slopes of Mount Carmel in Palestine I saw gnarly old specimens and young newly grafted trees overlooking the eastern Mediterranean. Because the carob is easily injured by frost, it hugs close to the seashore in Mediterranean lands and is especially important (90,000 tons per year) in Mediterranean islands—Sardinia, Cyprus, and Sicily. It even rises to the point of chief export of the Mediterranean island of Cyprus, where the Director of Agriculture says, "Wild carob trees abound all over the island," and where its per capita export value in 1924 ($4.00) was greater than that of grain and grain products and forest products from the United States. In 1938 its export value, £170,000, was exceeded only by the output of a copper mine booming to supply war industries.

The carob is an evergreen tree with rich, glossy, evergreen foliage. It blooms in the autumn and, like the orange, carries the young fruit to the end of the next summer.

This tree is in itself an example of two parts of the tree-crops thesis: the tree is sometimes the best means of getting harvests from *steep land* and also from *arid land.*

THE CAROB'S PLACE IN MEDITERRANEAN AGRICULTURE

Everywhere, the carob takes second-class land, either rocky or dry. On the plains of Valencia the irrigable land is in oranges and garden crops, but ten feet above the last irrigation ditch the carob and the olive begin making a crop on the rocky hillside of the semiarid land. This is the case in Majorca, in Cyprus, in Algeria, on Mt. Carmel, and in most Mediterranean lands. M. Trabut, the government botanist at Algiers, told me that he had seen carob-tree roots at a depth of sixty feet on the hills of northern Algeria. Sometimes carob trees cling to hillsides which seem to be almost pure rock. In Sicily it is an indispensable shade tree.

In Tunis I have seen carob trees in arid locations where the rainfall was about ten inches.

Unfortunately, the carob is often injured by winter temperatures of 20° F. or even a little above. The *Origin of Cultivated Plants*, by Alphonse de Candolle, says, "It does not pass the northern limit beyond which the orange cannot be grown without shelter. This fine evergreen tree does not thrive where there is much humidity."

The following statement seems to show that carobs vary in resistance to frost:

Eighteen degrees of frost do not injure the carob to any extent. Frost conditions that did marked damage to citrus trees made no impression on carobs growing within a few feet of them. (Monthly Bulletin, State Commission of Horticulture, Sacramento, California, Vol. V, No. 8, "The Carob," p. 292.)

It is generally considered that the carob and the orange occupy approximately those lands where the temperature is suitable for the orange. However, as the orange is a water lover, it requires good irrigable land, while the carob is a drought resister, and it therefore occupies the rocky land above the irrigated vales. The climatic relationship of the carob and the orange is well illustrated in the Valencia district of Spain, which contains four-fifths of the Spanish orange acres and two-fifths of Spanish carob acres.

An unpublished report of American Consul C. I. Dawson at Valencia, Spain, January 28, 1913, states:

The carob is a leguminous growth, indigenous to the shores of the Mediterranean Sea and particularly to the east coast of Spain, where for centuries it has been the principal forage crop of this intensely cultivated region. The tree is not frequently cultivated as a crop of primary importance. Except for a few well-kept plantations, it usually occupies the least valuable parcels of land in the irrigated plain.

The tree apparently flourishes equally well in any soil except stiff clays or other compact formations.

The flowering period begins at the tenth or twelfth year, but forty or more years pass before the tree is in full bearing. Then under normal conditions the hardiest varieties will yield crops with little variation from year to year, for generations and even hundreds of years.

The average annual yield of carobs per tree is placed at 110 pounds and in cultivated plantations 24 trees are set out to the hectare (2.47 acres). At the current market price of the fruit— 76 cents a bushel (60 lbs.)—the crop would return $33.44 gross per hectare. Cultivation is estimated to cost $14.00 per hectare, leaving a net income of $19.44, to which may be added $2.20 for the prunings (which are sold as fire wood). This gives a profit of 8.65 percent on an estimated investment of $250 per hectare, including the cost of the land, budded stock, cultivation, and compound interest on the actual outlay until the tree begins to bear profitable crops. In 1910, 271,000 acres in this consular district were reported to yield an average of 1,180 pounds of carobs per acre.

In the vast irrigated plain of Valencia, there exist a few important plantations which produce per tree far in excess of the 110-pound average above stated. Individual trees frequently yield 600 to 900 pounds of carobs every year, and instances are known where crops of two and three times these figures were gathered from single trees.

Cultivated plantations are quite profitable to the owners and amply demonstrate the possibilities of the carob tree under the most favorable conditions of care and cultivation.

The carob is commonly used in conjunction with fresh and dry alfalfa as fodder for draft animals in heavy agricultural and industrial work and less extensively as forage for sheep, goats, cows, and hogs. It undergoes no process of manufacture or treatment whatever, being fed to the stock as gathered from the trees. Sometimes the meat of the finer varieties is ground and used with wheat flour in the daily diet of the poorer classes.

Despite the economic value of the carob tree, its easy and inexpensive cultivation in soils often valueless for more remunerative crops, the regularity of yield, and the simplicity of the harvest, it is doubtless true that both acreage and production are declining. The reason is said to be the improved conditions in the orange and olive industries, the extensions of which are made at the expense of the carob.

Nearly everywhere, in the Mediterranean countries, the carob is a supply crop. It is like corn on the American farm, something to be fed to the farm animals. A few localities export it. A few thousand tons are exported from Algeria, but it reaches its greatest commercial importance in Cyprus, where the carob furnished twenty percent of the exports in 1924. The export next in importance was animals, which are in part the product of carob food. In 1938, carobs were twenty-three percent of the exports from the land, and animal products were fourteen percent from an island with an unusually diverse list of agricultural exports.

THE YIELD OF THE CAROB TREE

How much do carob trees yield? It is very difficult to get reliable figures of yield of crops. This is true even for apples in the United States. Since most Mediterranean carobs are grown helter-skelter in chance and irregular locations, most figures are estimates, and in making estimates it is easy to let the influence of the phenomenal tree run away with the lead pencil. However, an article in the *Scientific American*, January 13, 1913, says, "The yield of these pods per tree is often very great. Some trees frequently produce as much as 800 or 900 pounds."

A circular on cultural directions for the carob issued by the U. S. Department of Agriculture, January 21, 1908, speaks of trees that may reach a sixty-foot height, seventy-five-foot spread, and a yield of three thousand pounds of beans.

United States Consul Dawson, at Valencia, got a Spanish figure of one thousand one hundred and eighty pounds per acre as the average annual production.

As the result of a conference between a leading dealer and the local agronomist official in the town of Faro in South Portugal, I was told that the ordinary carob tree would yield one hundred to one hundred and thirty pounds, a very good tree three hundred to four hundred pounds, and an unusual tree sixteen hundred to eighteen hundred pounds. They further said that ordinary land with carob would bear on the average about forty-four hundred pounds per acre, and that a good stand of carob tree raised the value of rocky ground from three hundred to seven hundred dollars per hectare (2.47 acres).

In this locality the carobs were very common. They were almost always in sight and almost invariably standing at random in rocky (and I presume) unfertilized pastures where the tree had sprung up by chance and then had been grafted. Personally, I prefer to cut this figure of average yield in two.

Mr. Louisides of Larnaca, Cyprus, says, "We very often see large trees yielding nine to ten hundredweight," and he claims that there are farms in Cyprus that make more than one hundred hundredweight per acre per year on the average. This gentleman, who was a leading steamship agent of Larnaca and an esteemed correspondent of the American Consul at Beirut, reiterated this and other strong statements in writing to me after I had expressed some doubt about the statements being accurate. He wrote:

As the carob tree in Cyprus is self-planted, it is rather difficult to estimate the exact yield per acre.

Generally there are fifty to sixty trees in an acre of land at the age of fifty to five hundred years, and each tree brings in one to eight cwt. a year according to the age of the tree and fertility of the ground.

After a careful inquiry in the matter we find that these trees yield seventy to eighty cwt. in an acre each season, and this quantity is considered out here quite normal.

Trees which are grown in fields yield much more fruit than those on the mountains, and we very often see large trees yielding nine to ten cwt.

The quality of carobs produced in the fields is much inferior to those on the mountains.

It occurs that the crop is sometimes less than the usual outcome, but at any rate there is always a crop. We do not take into consideration the large trees which yield more than two cwts., and we can assure you that there are farmers who earn more than one hundred cwt. per acre every year.

P. S. One cwt. is equal to one hundred and twelve pounds.

Locust Tree, Supplementary Report submitted by U. S. Consul at Beirut, dated April 24, 1913, reports:

The annual rainfall of the region in Cyprus producing the locust beans varies from 21.88 to 27.25 inches. As we have already said, the locust tree can be planted in any land except in marshy places. It grows in rocky places and in limestone too, and withstands the driest weather.

The locust tree in Cyprus is found self-planted on unplowed rough land.

ACTUAL FIGURES AND A TWO-STORY AGRICULTURE

Of only one regularly planted carob orchard have I had an absolutely measured record of which I am reasonably sure. The owner was an educated Frenchman, M. Chouillou, living a few miles up the river from Bougie on the northern coast of Algeria. His trees, which were twenty years old, were planted on well-drained alluvium. They were interplanted with grapes. There was still almost a full stand of grapes. Often only one vine was missing where the carob tree stood. From the sixteenth to the twentieth year, in addition to a full crop of grapes this orchard produced on the average eight hundred and seventy-five pounds of carob beans per acre. The selling price was the same as that of corn, and taking fifty-six pounds to the bushel, it figures up to 15.6 bushels per acre. This compares favorably with certain American yields of corn, and don't forget the grapes.

5 Yr. Average	N. Carolina Bu. Per Acre	S. Carolina Bu. Per Acre	Georgia Bu. Per Acre
1921–5	19.7	14.3	12.3
1941–5	22.3	15.3	11.8

The above figures of yield, eight hundred and seventy-five pounds, should be compared with the eleven hundred and eighty pounds per year reported by Mr. Dawson for the Spanish crop. And do not for an instant forget that these Algerian trees stood in a productive vineyard and were the top story of a two-story agriculture.

I wish to emphasize the great value of this French testimony as to accuracy and also as to its significance as an example of the two-story type of agriculture. The plantation was run by an intelligent Frenchman assisted by his educated grown son. Their plantation of grapes, interspersed with carob trees at a distance of fifty feet, was as regular as a geometric diagram and as clean as a Chinese garden. The grapes were sprayed by American machinery. I asked how they could speak so definitely about the yields of the carob. The men took me into a neat stone building Near the door stood a good platform scale suitable for weighing sacks or packages. We went to the office in one corner, and there they showed me books in which were recorded tables of yields for a dozen years. A druggist could not have seemed more exact.

They emphasized the fact that barley grew well right under the carob trees, and they said the yield of grapes was absolutely the same as that of similar lands alongside which did not have carob trees. The appearance of the grapevines gave support to the claim. The effects of the open top of the carob trees and the blazing Algerian sun need to be considered. Fifteen hectares of newly planted carob demonstrated the satisfaction of these French farmers with the twenty years' experiment. Their method of planting was: Get seeds from a manure heap. Plant them in pots. Transplant to the field when one meter high. Bud the next year. Get the first fruit four or five years later. (It should be noted that their plantation (Fig. 31) was only twenty years old, which would be considered young for carob.)

They told me that work mules doing full labor did well on the straw of oats and all the carob beans they would eat.

As I reread this Algerian story in 1949, I keep thinking of my experimental Millwood (grafted) honey locust tree standing in a hillside pasture, its feathery tops black with one hundred pounds of long, fat pods carrying thirty percent sugar and eagerly awaited by the pasturing animals below, who now have only grass to eat, but who will gobble these pods at the first opportunity.

THE ABILITY OF THE CAROB TO STAND ABUSE

After having traveled through the carob districts of Spain, Portugal, Algeria, Tunisia, and to some extent in Palestine, I wonder that many of the trees can yield at all, so shocking is the treatment to which the soil has been subjected for centuries. The trees almost invariably were scattered at random on rough land that was pastured. Pasturage is a steady removal of fertility. In most cases the good soil has been removed by the erosion of tillage, by hoof beats, and by the pattering of rain drops. Often little but a rocky framework remains. The bean crop is also usually carried away from the trees—another removal of fertility.

Certainly the soil of an average Mediterranean carob plantation has been treated worse than the test plats of Rothamsted Experiment Station, England, which have been continuously in wheat for generations in order to test the fundamental and enduring fertility of the soil, which in that case proved to be about enough to yield eight bushels of wheat per acre, a quarter of the average English yield.

THE CAROB IN CALIFORNIA—WILL IT BE A HUMAN FOOD?

After these many centuries of Old World experience, the carob has joined the procession of Mediterranean crops emigrating to that section of the United States (California) having the Mediterranean climate. California suddenly discovers that it has carob trees, much carob land, and the possibility of an industry.

In hustling California this millennial crop tree is still in the introductory stage, but it has received the scrutiny of the inventive Yankee mind, and discoveries which may help to revolutionize the industry have already been made and put to work.

One of these discoveries applies to the growth of the trees, and others to the use of the beans. For years I have been stating that it was possible that tree crops such as chestnuts and acorns might be made into acceptable human foods by machine manufacture in factories. Nevertheless, it was with some surprise that I found some Californians in 1927 turning out acceptable factory-made food products from the carob beans imported from Europe. In 1927 one Los Angeles company claimed an output of many loaves of carob bread a day.

I call this rainbow bread. I could never quite find the end of it. As the story came to me first it was one thousand loaves per day by one company, and I saw an airplane picture of a vast factory labeled "The Home of Carob Bread." It looked almost as large as an automobile factory. I tried to verify. I could never find that factory, but I was credibly informed that the bread industry had been primarily brought about by persons trying to sell new-planted carob land—new bait in an old, old, yet ever new game. Latest reports (January 1928) indicated that the land-selling idea had collapsed and the bread industry was surviving, 1,200 one and one-half pound loaves per day—25 percent carob in the recipe. In 1947, the bread was not find-able—no land for sale.

Lest I should appear unsympathetic, I wish to state my be-lief that carob beans are good material for human food. This is true of at least two hundred other materials not now used to any large extent (or not used at all) for food in the United States. The question is, who can make us eat these new things? A pig or calf eats what is set before him. People in rich America eat what pleases them, and one of the last things any reformer can do is to change food habits. Apparently there is no reason other than inertia why carob bread should not come to great impor-tance. Food factories now open the way (remember the rise of the peanut), but don't forget inertia.

It is said that the carob makes excellent cereal, candy, and syrup—a pound of syrup from a pound of beans—a fact that is almost staggering. The candy, which seemed to have coconut in it as well as the easily recognized carob flavor, was an in-stant success in my family, and we liked the flavor of syrup made from carob.

The analyses of carob (Appendix), with its very high sugar content, show remarkable food values.

Human foods from carobs must stand at present on the list of perfectly good possibilities. Meanwhile there is an open door for their use as stock food. Carob stock food has outstanding and perfectly established qualities, and has long been in use in a small way in the United States as an appetizing element in proprietary stock foods.

THE EARLY CAROB TREES IN CALIFORNIA

The carob tree is demonstrating itself in California in much the way that so many other Mediterranean crop plants have come to the front. Early plantings in the first few decades of American occupation resulted in fruiting trees by 1885. As a result the California State Horticulturist reported in 1890, "No tree distributed by the stations is more likely to make a popular shade or ornamental tree for dry, rocky situations."

H. J. Webber, Professor of Subtropical Horticulture in the University of California and Director of the Citrus Experiment Station at Riverside, says in a letter of February 14, 1927:

The carob has been planted more or less all over southern California, largely as a street tree, but in some places commercial plantings have been made. The tree has proved hardy and very drought-resistant. After it is once started it thrives fairly well without irrigation, which indicates that it is quite drought-resistant, as it is very few plants, for instance the pepper tree and eucalyptus, that manage to survive at all here in southern California without irrigation.

In 1912 Dr. Aaron Aaronson, having visited California, reported that individual carob trees in Palestine produced three to five hundred pounds, and five tons to an acre might be produced.

This statement seems to have caused the Californian mind to begin examining the Californian carobs. The Experiment Station reported (Bulletin 271) that carob beans tested out a little better than barley when mixed with milo and fed to calves for thirteen weeks as a grain ration, supplementing the milk and alfalfa-hay ration fed in addition.

C. W. Beers, Horticultural Commissioner at Santa Barbara, went up and down the State studying carob trees. He wrote me in June 1917 that twenty-year-old seedlings, fourteen feet across the tops, had produced one hundred fifty pounds of beans for three consecutive years.

In his letter of June 14, 1917, Mr. Beers stated:

The carob trees that are bearing one hundred and fifty pounds each are about twenty years old. They have never had any attention since having been planted and have fought their way in land well grassed over, never having been irrigated. They are about twelve feet to fourteen feet across and about the same in height.

In a subsequent letter from Mr. Beers, dated June 30, 1917, he said:

The carob mentioned as bearing one hundred and fifty pounds a year has been very regular in this production for the past three years, which is as long a period as I have been observing them. I believe they can be considered as regularly bearing this quantity.

The ground upon which these trees are growing is overlaid with a very heavy deep clay hard pan, which precludes the probability of subirrigation. The rainfall is about fourteen inches a year. The land is sloping towards the ocean and is a heavy soil, but is probably not over two hundred feet elevation above sea level. It is about three miles to the ocean front.

The remark about being grassed over may need this explanation, namely, that the grass is green only through the winter and early spring, while at this time of the year it is brown and apparently dead.

CALIFORNIA CAROB POSSIBILITIES

The geographic possibilities for carob culture in California seem to me to be excellent. The carob grows in orange temperatures, and California has a large area with orange climate. As a thermal belt above the frosty valley floors, it stretches for a great distance along the eastern edge of the Great Valley. It also rings around much of the shore and lowland between San Francisco and San Diego. Since the orange trees are water hogs and California is a land of almost rainless summer, the orange can be grown only where irrigation is possible. Naturally this is possible on but a small fraction of the land with orange temperature. Therefore, the major part of California land having orange temperature cannot become orange land, but much of it may become carob land, since this tree can survive and can even bear a light crop in the rainless summer. Sample plantings years ago have proved that the carob will thrive over an area much larger than the possible orange area.

University of California publications, *The Carob in California*, by I. J. Condit, Bulletin No. 309, June 1919, says:

Experience has shown that the trees when young are no hardier than orange trees. When once established, however, the carob is more frost-resistant than the orange. . . . Even if the blossoms escape injury from cold and rain, the developing fruit is liable to be killed by frost later on. For this reason the successful production of carob pods in the interior valleys is practically limited to the citrus belts along the foothills. The carob tree thrives in regions of intense heat, such as the Imperial and Coachella valleys where the winters are mild.

Mr. G. P. Rixford, Physiologist at U. S. Department of Agriculture Field Station, San Francisco, California, said in a letter of February 6, 1917:

It frequently happens that the flowers which are produced in late fall or early winter are destroyed by frost, which does not affect the tree itself but prevents its fruiting.

There are trees near Centerville, Alameda County, planted by the late Professor E. W. Hilgard, University of California, that are now thirty years old and are annually producing regular crops of pods. The cold wave of January 1913 was the severest frost in thirty years. Trees growing as far north as Biggs, Butte County, were somewhat injured but have fully recovered. I think this may be considered, perhaps, the northern range of the tree in California. In most parts of California, the tree must be planted in the least frosty localities, which are usually not far from the sea, or in the citrus belt of the Sierra foothills. The tree will certainly endure as much frost as the orange, but for the reasons mentioned above may fail to produce its crop of pods. However, it is a beautiful tree and worth growing for ornament and can be successfully grown over a large part of California. Where the conditions are favorable it will be a very profitable producer of pods, which are equal in nutrients to barley for all kinds of stock and even for poultry when ground.

Mr. G. P. Rixford, in another letter, February 26, 1917, said:

I have no doubt there are large areas about and above the citrus belt where the tree could be planted with reasonable expectations of success.

Can California have a vast carob industry? Probably, but it will take years of experiment to prove it. The carob boom that seemed in the offing in the 1920s did not materialize. Carobs are too slow for the first-generation orange growers, and there is a lot to learn even about transplanting an old industry. Perhaps the carob must stay near the ocean where humidity is high and evaporation therefore low.

In favor of the carob is the fact that Californians know little of the tree-crop possibilities of their unirrigated land. They have not fully tested complete conservation and use of all rainfall. (See Chap. XXIV on farm practice, especially water-pocket irrigation.)

If carob trees grow and thrive and bear in California, the next question is—can we harvest the crop at a profit in our land of high wages? I think so, but it needs to be proved. We need a lot of small experimental plantings of varieties suited to particular localities.

THE SURE THING IN CALIFORNIA CAROBS

The carob is a beautiful evergreen tree with dark, glossy leaves. It is a good shade tree. It endures drought. It looks like a sure advantage to plant them widely on that rather large area of California land which cannot be irrigated and which is considered as possible home sites. Nothing will be lost by using carobs for shade trees. Much may be gained.

Walter M. Teagle, General Manager of the Limoneira Company, Santa Paula (near the sea) wrote, February 6, 1948:

These (carob) trees were planted (in 1917) on a rather steep hillside which is covered with sagebrush and is of a very poor type of soil. The trees were not planted with the idea of any kind of commercial proposition being involved, but were put there simply as an experiment and with the idea of making the hillside more attractive. The trees were watered for the first two or three years after planting, but have had no irrigation since. They have survived surprisingly well, and I believe that any trees that have died have done so because of damage from gophers.

We have been having an extremely dry season this year, and during the past thirteen months have had only a little over three inches of rainfall. The trees at the present time show some effect of the lack of moisture, and the color of the leaves is not too good, but from past experience, they probably will survive. The test to which they have been put in this particular case is certainly a most severe one, and the survival of most of the trees is a testimonial to the hardiness of the carob and its ability to survive in a climate which has a rather low rainfall.

It was reported in 1947 that people pick up the pods of chance seedling street carob trees in California and carry them away for food or feed.

SOME ECONOMICS OF A CAROB INDUSTRY FOR CALIFORNIA

In the words of J. Eliot Coit, Ph.D. (formerly Professor of Citriculture and Head of the Department of Citriculture, University of California, previously Superintendent of the Citrus Experiment Station at Riverside, and now a horticulturist in private practice):

I have long been anxious to see a carob industry developed here in California. We have many thousand acres of good plow land or barley land which is well up out of the frost where there never can be any irrigating water. Barley farming is very cheap and careless and results in very serious erosion of the hillsides. Carob culture would not only add another large source of income, but would largely prevent the serious losses from erosion.

The way California is sending its future into the sea by way of soil-robbing, gully-washing culture of hill lands in barley and beans!—"After me the desert."

The conservationist should note that since the carob tree lends itself admirably to the small reservoir system of water conservation described in Chapter XXIV, we may expect two checks on erosion—one with its roots and one with its field reservoirs. This latter might be one of its greatest advantages to the State through increase of water supply (page 324).

A NEW METHOD OF PLANTING TREES

A very important Yankee discovery in connection with the carob is a greatly improved system of transplanting the trees. The young carob, like most trees of arid lands, has a root several times as long as the top. Thus it survives drought. But transplanting becomes a problem in a climate of 100° F. in summer, where there is no rain from April until October or November, and then only twelve to twenty inches in a season of winter rain and occasionally much less. Planting the little seeds in place is slow and difficult, and the rabbits eat the leaves of seedlings if they can reach them. Growing in a pot is exasperating and at times injurious to a tree that yearns to send roots straight into the ground and needs such a root when it gets to the field. Some California genius invented the so-called splint system. The tree is grown in a little tube of earth an inch or more square and two or three feet long, walled in by four plastering laths. These lath tubes are arranged in banks inclined at an angle of sixty degrees. One lath is soaked in nitrate of soda solution, and to this the roots of the little tree cling as ivy to a pole. Thus the little four-inch tree with two feet of roots sticking fast to the lath may have its whole long root system inserted into a crow-bar hole deep in the ground.

This discovery alone may make success where before a very high percentage of loss might have meant delay and greatly increased cost.

There is little reason to believe that we are growing in California carobs of the best varieties obtainable. The slowness with which many of the best varieties come into bearing would indicate that proper breeding might produce much more precocious strains. Like many other trees, they vary in resistance to cold—offering the possibility of more frost-resistant varieties.

The table of analyses (Appendix) shows remarkable variation in content. See especially, in that table, the figures in No. 2201 and 2371, and the protein variation between the maximum and minimum of the whole bean (pods and seeds); these offer

interesting possibilities of breeding carobs of such special quali-
fications as "high in sugar for sugar manufacture" or "high in
protein for milk-making and growth-making foods."

Carob improvement offers two lines of work:

(a) Crossing carobs.

(b) Perhaps hybridizing carobs with some of the numerous
allied species, possibly some of the American mesquites and the
South American algaroba (page 86), which have very much
greater resistance to frost. A strain of the Hawaiian algaroba
might add both precocity and productivity. This is work for
individual enthusiasm, for private endowments, and for State
experiment stations supported by legislatures with vision and
manned by staffs with vision. Where are these stations?

Here is an unexplained mystery. Why did not the California
Agricultural Experiment Station start, ten, twenty, or thirty
years ago, a series of small plantings of carobs of assorted varieties in locations with a variety of climates?

In January 1949 Walter Rittenhouse, address 4575 58th
Street, San Diego 5, California, published and began to distrib-
ute a booklet by J. Eliot Coit, "Carob Culture in the Semi-Arid
Southwest." He had been bitten by the sight of California's soil
loss and the successful carob tree here and there.

I quote from Coit's booklet:

In the interest of scientific agriculture, Dr. Walter Rittenhouse
of San Diego has provided funds for a small demonstration carob
orchard to be planted and cared for as a public exhibit of what the
carob can do, when given a fair chance, in the way of soil conservation and the production of cattle food. This orchard is being planted
on dry land which has suffered from some erosion. The location is
on Buena Vista Road, one-fourth mile east of Pechstein Lake, between Vista and Twin Oaks Valley, in Northern San Diego County.
The elevation is approximately eight hundred feet. The average
rainfall is about 17 inches. The frost hazard is considered quite low
for carob, as there are avocados and bearing citrus growing in
the immediate vicinity. The general public is invited to observe
the development of this orchard.

This orchard will be managed in cooperation with the University of California and the Soil Conservation Service. It is planned for a period of thirty years.

This may be of service to much of northwestern Mexico.

One Californian is said to have devoted a private fortune to breeding avocados. I am sure he had a lot of fun and rendered a great service.

CHAPTER VII

A Stock-Food Tree—
The Honey Locust

A TREE OF WIDE RANGE AND GREAT POSSIBILITIES

The algaroba (keawe) of Hawaii is limited to the arid and semiarid lands of the frostless tropics, and apparently it must be situated close to sea level. The carob apparently is limited to the orange-growing sections of regions with the Mediterranean type of climate, or to warm subtropics such as South Africa. But the Creator did not neglect the humid East of U. S. A. The honey locust, *Gleditsia triacanthos,* a cousin to the carob and keawe, offers a great crop possibility to a million square miles of the eastern United States in the climates of corn and cotton, and presumably also to the similar climatic regions in four other continents, where it is also surviving the climate. (See Climate Regions Map, Fig. 138 in this book.) This promising tree is native from New York to Nebraska, from Louisiana to Minnesota, and it has proved its adaptability beyond this area. It does well in California, for example, and therefore in the other regions having that climate. Some strain of this tree will certainly thrive in ninety percent of the area of the United States, except the really arid lands.

Of all the species tested in many parts of western Kansas, the honey locust is the most conspicuous success.

65

Its rate of growth is only moderate, but the rate is maintained for many years. A large proportion of the trees planted have good form, and they are strong in stem and branch, not often injured by wind or ice storms.

In a demonstration block planted eighteen years ago and neglected for so long a time that the buffalo sod had gained a secure foothold, the honey locust has made a very creditable growth. The best trees have reached a height of twenty-three feet and a diameter of six inches. At Dodge City the honey locust trees have done very well indeed. (Kansas State Agricultural College Experiment Station, Bulletin 165, pp. 316, 317.)

> Ft. Collins, Colorado,
> September 11, 1916.

It will stand as much exposure as any other forest tree that we have. We have some groves in this section doing well without irrigation with a normal rainfall around fourteen inches.

> (Signed) E. P. Sandsten.

This species is particularly adapted to the West, as it stands considerable drought and low temperature. It grows very well on the high table land in western Nebraska, where the annual rainfall is only about sixteen inches. More than almost any other tree that we have, it will also grow in a soil that is alkali. While we do not have a great deal of alkali soil in the West, there are some of the valleys where the honey locust will do very well and where other trees will fail. (Letter, Chet G. Marshall, Marshall's Nurseries, Arlington, Nebraska, March 24, 1928.)

More recently an Extension Farm Forester, of Kansas, reported:

Observations of honey-locust planting after our severe droughts of 1935 and 1936 verify the fact that honey locust is a very drought-hardy species and is very good for planting in western Kansas.

THE GREAT PLAINS SHELTER BELT PROJECT

In 1932 Franklin D. Roosevelt was wreck-bound before an unsheltered, denuded slope near Butte, Montana. As he sat there, he thought out the Shelter Belt Plan, which he was able to start in 1934. It was kept up until 1943 and reported in *The Journal of Forestry*, April 1946, pages 237 to 257:

About 19,000 miles of belts were planted on approximately 33,000 farms. A survey of the plantings made in 1944 covered 1079 belts on a random sample of 3.6% of all belts planted. The greatest growth per tree was made by the plains cottonwood, 19.1 feet; with the black locust second, 16.3 feet; Siberian elm, 14.7 feet; honey locust 11.5 feet. The highest percentage of survival was with the honey locust, 79.0% against 78.3% for black locust and 59.4% for the cottonwood.

This suggests very considerable possibilities of a bean crop, especially as the shelter belts themselves break the wind and pile the snow so that a place of rainfall of 15 inches might have an effective soaking from drifts equivalent to two or even three times as much rainfall on this spot. A kind of automatic irrigation—snowdrift basis.

A VALUABLE STOCK FOOD

Like the algaroba and the carob, the pods of the honey locust are greedily devoured by farm animals and sometimes are eaten by children.

Speaking of customs on the Georgia plantation of her youth, a Southern woman wrote:

One of those customs was in regard to the honey locust. Not only the pigs and cows ate the pods that dropped from the honey locust, but the Negroes and the white children ate them as well. I can't say if all the grown-up white people ate them, but I know that my mother approved of our eating them, for she liked to eat them herself. (Letter, Mrs. J. W. Carlin, Alva, Oklahoma, August 3, 1914.)

In addressing a garden club in northern Virginia with twenty-three ladies present, I heard eleven report that, as children, they had eaten the sweet part of honey-locust pods. It is likely that this could be duplicated in a thousand rural locations in the United States of America.

Much interesting correspondence about honey locust came into the office of the *Journal of Heredity* in response to offers for prizes for the best honey locust tree. There seems to be a widespread conviction that the beans are prized by nearly all kinds of American livestock. Several excerpts from these letters are as follows:

"The cows eat the beans as fast as they fall."

"And all bear an awful big crop of beans which the stock like so well that they will break down the fence to get them."

"The cattle ate the pods I gave them with great relish." (*Journal of Heredity*, Vol. XIX, p. 223.) On page 460 of the *Journal of Heredity* for October 1927, there is an interesting discussion of the wide geographic range of the tree and the amazing variation in the content and quality of the pods.

Compare the analysis of *selected* honey locust bean and pod with that of wheat bran and corn in the Appendix.

Speaking of his locust beans, a farmer said affectionately, "They hang on the tree close as your fingers. They hang on there till a freeze comes, and then they turn black and fall off, and the cows get in then and eat 'em. Maybe they don't like 'em!"

He pointed to a honey locust tree in his field that had filled a two-horse wagon full of beans, which was about ten bushels. The tree had a girth of sixty-four inches and a spread of forty feet. An acre would hold twenty such trees.

John W. Hershey, Downingtown, Pennsylvania, reports:

A letter from a dairyman in North Carolina states, "I mix an ideal feed of equal parts of honey locust, bran, and shorts, run through a hammermill."

A dairyman in northern Georgia who has been using the honey locust for cattle forage for years states:

I let the cattle pick them up where they can; and where they can-not graze, the beans are gathered and fed to them. My herd of heifers get a great part of their winter pasture from the honey-locust pods.

I found one man who had planted honey locusts that his cattle might eat the beans in the field, and also that he might harvest the beans for winter forage. That man was the late Lamartine Hardman, Governor of Georgia.

I also found a group of farmers at Sanford, North Carolina, who have unusually good wild honey locusts; they report a regu-lar practice of grinding the beans and mixing them with grain or alfalfa meal as part of the dairy-cow ration. The samples looked good and smelled good.

HONEY LOCUST FOR HUMAN FOOD

Mr. John W. Hershey reports that he ground a meal combina-tion of two-thirds yellow corn and one-third honey locust of the Milwood variety and distributed it to about two dozen people, suggesting that they use it for muffins.

Fifteen replies were received. Thirteen were enthusiastic about it, some said it was too sweet. One used it for johnnie-cakes and wanted a regular supply. One sent it to the State Secretary of Agriculture and caused him to recommend it to a baker who planted five acres of trees and made tests with it for flavoring.

THE FINE QUALITIES OF THE HONEY LOCUST TREE

This tree has a remarkable list of qualities.

1. It is beautiful and is also a good timber tree, with a strong, durable, and beautiful wood.

This wood is heavy, hard, and strong, generally ranking with or above white oak in these properties but somewhat below those for black locust. Honey locust is not important from the lumberman's standpoint, although a few logs are occasionally sawed into lum-ber. The principal use of the wood is for posts and railroad ties,

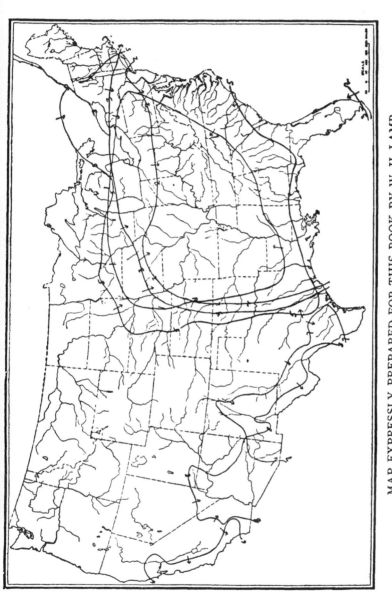

MAP EXPRESSLY PREPARED FOR THIS BOOK BY W. H. LAMB
Forest Service, U. S. Department of Agriculture

Lines enclose areas showing natural ranges for the following species:

1. Honey Locust, *Gleditsia triacanthus*
2. Persimmon, *Diospyros virginiana*
3. Mulberry, *Morus rubra*
5. Eastern Black Walnut, *Juglans nigra*

6. California Black Walnut, *Juglans californica*
7. Mesquite, *Prosopis chilensis* (syn. *juliflora*)
8. Screwbean, *Strombocarpa odorata*

as it is quite durable in the ground. The lumber is generally used for furniture and inside finish of houses and is said to be often mixed with sycamore for these uses. (Letter, **H. S. Betts**, Engineer in Forest Products, U. S. Department of Agriculture, Forest Service, Washington, January 28, 1927.)

2. It is a rapid grower.

An annual increase of two feet in height and one-half inch in diameter is not uncommon in favorable locations for a score or more years, and in less favorable locations it will generally add a foot or more in height, and in diameter fully one-third of an inch. (Elliott, *The Important Timber Trees of the United States*, pp. 324, 325.)

3. Like the carob and the algaroba, it is a legume, but the question of its nitrogen-gathering ability is unsettled. It has not the root nodules common to many legumes.

4. It is an open-top tree, through which much light can pass to crops below, thereby favoring a two-story agriculture, like the carobs of Algeria (page 52). This is especially valuable for pastures. It is probable that in many situations a pasture might be better with honey locusts than without them.

An editorial in the *Breeder's Gazette*, 1926, emphasizing this point, led to the following detailed report about a hillside planted to the *black* locust, *Robinia pseudoacacia*, a tree with agricultural characteristics greatly like the honey locust:

The ground on which the locusts were planted consisted of a hillside, sloping abruptly to the south. It has two rough gullies in it. The land is unsuitable for cultivation both because of the roughness of the two gullies and the steepness of the slope. The virgin timber had been removed before the farm was acquired by my father. The soil was covered by a good bluegrass sod; but the exposure to the sun was such that the grass dried up early in the summer. It was with a view to making this land valuable for pasture that the locusts were planted. Young locusts were planted early in the spring and the ground covered with straw to hold moisture until the trees became properly rooted. The locust trees soon shaded the ground, and the pasture on that hillside has been excellent ever since. Aside from the pasture this tract has yielded a large number of fence posts. (Letter, Llewellyn Bonham, the Bonham Engineering Company, Oxford, Ohio, February 23, 1929.)

This black locust is a noted pasture improver, because its surface roots have nitrogen-bearing nodules.

5. It is a productive tree. One correspondent tells me of a tree producing six consecutive crops estimated at twenty bushels each.

This tree has borne uniformly large crops for six years consecutively to my knowledge. I have never measured or weighed the pods but I have thought that twenty bushels would be a minimum estimate.

It is a big nuisance as far as the lawn is concerned.

As the pods are falling through a long season, many lodge on the roofs and are blown to quite a distance sometimes. (Letter, S. P. Thomas, Ashton, Maryland, August 4, 1913.)

J. M. Preston of the Branch Experiment Station at Hays, Kansas, gathering seeds for planting, reports a tree with seven-inch or eight-inch trunk, producing eighty pounds of beans, and another tree in Manhattan, Kansas, from which he gathered "about four hundred pounds of pods." He reported that this tree had a trunk of two and one-half feet in diameter, broad top, and bore in 1911 and 1912, but failed in 1913.

6. Frequent reports of consecutive crops seem to indicate that some honey locusts are regular bearers as trees go, and apparently they can be expected to produce crops with greater regularity than do most fruit or nut trees.

In regard to the honey locust, will say that the fruit of this tree is rarely injured by spring frosts, and the result is we have a heavy crop every year. (Letter, C. C. Newman, The Clemson Agricultural College of South Carolina, Clemson College, South Carolina, July 17, 1913.)

It has so many beans that the branches bend down like a fruit tree. (L. F. Quisenberry, R. No. 2, Moberly, Missouri.)

The tree grows in a Bermuda pasture and is always loaded in the fall with luscious fruit. The cows and hogs stand under it, always ready to devour every pod that falls. The tree is very large and very, very beautiful. The cows improve in milk and the hogs in weight when the locusts ripen, for there are always bushels and bushels on the tree. (Ellen Williams, Goldworth Farm, Villa Rica, Georgia, R. No. 4.)

7. The honey-locust bean is of large size, and this should make the crop easy to harvest. The beans are often one foot or more in length, and seventeen prize beans in the *Journal of Heredity* contest, 1927-1928, weighed a pound when bone-dry after weeks in the house.

The beans or pods are flat and black and measure as long as sixteen inches and about one and one-half inches wide; the honey lying in the thicker portion of the pod and the beans in the thin. I have seen them [trees] over eighty feet in height. (Letter from F. F. Bessoe, Natchez, Mississippi, *Scientific American*, February 22, 1913.)

The large size of the pods, in addition to their tendency to curl up, should make them easy to harvest, possibly even with a hand rake or a horse rake. I have not tried it, but I think that large quantities could be raked up in a short time by hand, even on ground that was not entirely smooth. Perhaps a special rake would need to be devised. Apparently, the beans could then be picked up with a pitchfork (perhaps a special fork) and handled somewhat like hay.

The North Carolina pioneers (p. 68) who have learned how to grind these beans and pods, despite the sticky molasses quality, by mixing them with meal in a hammermill, have already shown us how to make them into meal. Thus another concentrated food can be added to the dietary of the American farm animals.

8. Once a good honey locust tree has been found, it is easy to propagate either by grafting or by the still more simple system of root suckers, which bear the characteristics of the mother tree.

The inclination of this tree to sprout whenever the roots are broken or cut by the plow or cultivator would indicate that it could be readily propagated from root cuttings, and this has now been done.

So far as is known to us, all trees propagate as true to parent type by root propagation as through budding or grafting. (Letter, William A. Taylor, then Assistant Chief of Bureau of Plant Industry, U. S. Department of Agriculture, Washington, D. C., February 21, 1913.)

9. S. B. Detwiler, Soil Conservation Service, U. S. Department of Agriculture, compiled 180 pages (mimeographed) of valuable honey-locust data, from which I select the following eight paragraphs.

Seedlings of honey locust less than 3 months old and only 8 inches high were found by Clements, et al., to possess a taproot system 40 inches deep and widely spreading horizontal branches 6 to 18 inches long. (Biswell, *Botanical Gazette*, June 1935; Detwiler, p. 8.)

This species possesses a strong tap root with many large laterals and penetrates the soil deeply. . . . The root system of honey locust was the most profusely branched of all, and this readily explains the ability of this tree successfully to endure drought. (*Plant Competition*, by Clements, Weaver, and Hanson; Carnegie Institute, Washington, 1929; Detwiler, p. 11.)

It [the honey locust] is very intolerant of shade, but has a tendency when not crowded to produce long branches near the ground, and hence it is an excellent tree to plant for hedges and windbreaks. In the Middle West the honey locust is equaled in drought-resisting power only by the Russian mulberry and the Osage orange, and it endures severe winters to which these species would ordinarily succumb. The roots do not sprout unless they are injured by cultivation. (U. S. Department of Agriculture *Forest Service Circular 74*, 1907; Detwiler, p. 12.)

The honey locust is one of the hardiest trees for upland planting in the semiarid regions of the Middle West. (U. S. Department of Agriculture *Forest Service Circular 74*, 1907; Detwiler, p. 13.)

It is the writer's opinion that one of the most overlooked trees for Colorado plantings is the thornless honey locust. Few insects attack this tree, about the only one worth mentioning being the Red Spider. Other than the above, it is practically insect free. It thrives and grows under the most trying conditions, on either extreme of wet or drought, and pretty generally takes care of itself if given any chance at all. I consider it one of the most useful, as well as graceful, of all the trees common to this locality. There should be more of these trees planted in the Colorado area. (S. Wilmore, "Four Good Trees for Colorado," *Green Thumb*, November-December, 1946. Detwiler ***.)

New England: Hardy throughout New England, grows in any well-drained soil, but prefers a deep, rich loam; transplants readily; grows rapidly; is long-lived; free from disease. (Dame and Brooks, *Trees of New England*, 1902. Detwiler, p. 119.)

Arizona: With proper handling on the mesas, honey locust requires but a small amount of irrigation, while in cultivated, alluvial soils it does fairly well without watering. (*Arizona Station, Hints for Farmers No. 83*, 1910. Detwiler, p. 152.)

Tobacco plants growing under a honey locust tree that stood in a large tobacco field near Brown Summit, N. C., were darker green, taller, and of better grade as compared to the rest of the field. (Observation by Taylor Alexander, August 28, 1939, reported to S. B. Detwiler by letter, September 26, 1939. Detwiler, p. 42.)

THE URGENT NEED OF SELECTION, TEST, AND BREEDING

In view of the above facts about the honey locust and the established industrial facts about the carob and keawe, I suggest that the honey locust tree is worthy, more than worthy, of extensive experiments and further investigation. If the hill farmer has a chance to rake up a one-ton or two-ton crop of bran material per acre from his bluegrass pasture at the same time that he gets good pasture and an annual increment of wood, such possibilities should be tested in a thousand communities quickly. Fortunately, there has been *one* test, and that has amazing significance.

My prize honey-locust pods of 1926 had 29 percent sugar. A decade later, the enterprising Mr. John W. Hershey, who worked for a time with the TVA when it was young and imaginative, found the following honey locusts: Milwood, 36 percent sugar; Calhoun, 38.9 percent sugar.

Detwiler, of Soil Conservation Service, took these up, and the Service persuaded *one* (only one) experiment station (at Auburn, Alabama) to give a complete honey-locust trial. The results are more astounding than anything I ever dared mention or even to think. Here are some excerpts.

The 53rd Annual Report (1942) of the Alabama Agricultural Experiment Station contains the following account by O. A. Atkins (p. 54):

The 5-year-old trees of the Milwood variety in 1942 produced an average of 58.30 pounds of pods per tree (dry-weight basis), which at 48 trees per acre would be the equivalent of 2,798 pounds of pods. The four highest-yielding trees averaged 12.5 feet in height and 3.6 inches in trunk diameter and produced an average of 65.62 pounds of pods or 3,150 pounds per acre. . . .

The estimated yield of 3,150 pounds per acre which came from trees only five years of age had a feed value equivalent to 105 bushels of oats or 56 bushels of corn.

From J. C. Moore, Alabama Agricultural Experiment Station, in project work in 1945 (p. 55):

In experimental studies initiated by the Hillculture Section of the Soil Conservation Service, Alabama Agricultural Experiment Station, a combination of honey locust and *Lespedeza sericea* (*cuniata*) has been very effective in controlling erosion on rough land and at the same time has furnished large amounts of feed for livestock. . . .

For two years preliminary feeding tests with dairy cows at Auburn, Alabama, have shown that ground honey-locust pods are equivalent in feed value to oats, pound for pound. . . .

The average yield for the Milwood selection for the four years 1942-1945 inclusive, was 2,923 pounds of pods, dry weight, equivalent to 97 bushels of oats per acre. As the trees increase in size these yields should greatly increase.

In addition to yields of pods from the trees the *Lespedeza sericea* (*cuniata*) has averaged 2½ tons of hay per acre per year, besides giving practically 100 percent protection to the soil. . . .

Pods stored three years in the dry are in fair shape. Some weevil injury is noted, but the sugar content is still high. . . .

Two mules, one cow, and one hog have been fed unground honey-locust pods for 30 days on the Hillculture Farm at Auburn, Alabama, with excellent results. These animals have never failed to consume their daily allotment, and no injurious effects have been noted.

An increase in milk flow and butterfat has been observed as a result of honey-locust pods in the cow's ration. . . .

Having a combination of honey locust and *Lespedeza sericea* (*cuniata*), the following benefits are derived over a period of years:

1. Soil is completely protected.
2. A concentrate and hay can be produced on the same area.
3. A good grazing and feeding-out program can be maintained.
4. Low seed and management costs over a period of years.
5. Weed control.
6. Low labor requirement.
7. Maximum production from the soil.

From Mr. Moore (quoted above), 1946 Progress Report on Work Item R-1-1-3(h), page 57:

This year some of the 8-year-old trees produced over 250 pounds of pods per tree. With 35 trees per acre, this would be 8,750 pounds of concentrates or the equivalent of 275 bushels of oats per acre.

AN INTELLECTUAL MIRACLE

These almost unbelievable results at the Alabama Station suggest an agricultural revolution. With a little publicity, the place should have become a center of state, national, and international pilgrimage. But—believe it or not, but apparently you must believe it—the management cut down every last one of those most promising trees. That's the miracle—wiped out what was probably the most promising experiment the Auburn Station had ever made—possibly the most significant single experiment any southern, or any U. S., Agricultural Experiment Station ever made.

And they planted peaches in their place!!?? I have no comment. You say it.

All these above-mentioned astounding harvests on the honey locusts in Alabama were produced on grafts from some wild trees that we happened to find. The ones now in hand originated in Georgia, Tennessee, and the mountains of Alabama. They have proved themselves good croppers as far north as southern Pennsylvania and southeastern Iowa (Soil Conservation Service). They are hardy much farther north. We need at once to search for the best trees in higher latitudes, and at the same time we need to breed better varieties, because this species, like many other species, has much variation among its millions of specimens.

There are in certain localities in Kansas, and probably in other States, strains of the honey locust which are almost entirely thornless, and some of these are quite productive of fruit. (Letter, S. C. Mason, Aboriculturist, United States Department of Agriculture.)

In youth some of these trees are astonishingly thorny, having thorns five or six inches long. As the tree gets large and does not need such protection, the thorns are less conspicuous. The Tennessee Valley Authority reports that a thornless cion from the top of an erstwhile thorny tree will produce a thornless tree. This is important. *But* as luck would have it, we already have thornless trees producing heavy crops of good, sweet beans.

Breeding from among selected specimens should produce much better strains. Then there is the indefinite but suggestive field of hybridization with other species, possibly carob, keawe, or the many species of mesquite and screw bean. (See Chap. VIII.)

James Neilson, Professor of Horticulture, Fort Hope, Ontario, wished to hybridize the honey locust with the Siberian pea shrub (*caragana arborescens*), because this hedge plant is very hardy at Winnipeg and bears small pods of beans.

How fine it would be if a million hills now gullying with corn or cotton or tobacco could be held in place by the roots of honey locust trees and an attendant crop of grass, while the landscape was made green and beautiful by the feathery tops of these trees, which at the same time yielded their ton or two of bran substitute per acre per year and also added to the accumulating sawlogs and firewood.

And yet Secretary of Agriculture Jardine told me personally that his Department could not get funds for experimental work necessary to test such a possibility. The Soil Conservation Service, under Franklin D. Roosevelt, Henry Wallace, Hugh Bennett, and Rexford Tugwell, made a splendid start at honey locusts and many other things, but an economy streak (some say a personal grudge) sent all this work into coma in 1947, with the records in the attic and the trees left to the rabbits and brush fires. The Congressional pork barrel had better luck. The good honey locusts, also the unborn or undiscovered honey locust trees, so valuable to an intelligent posterity, if we should have one, have no vote this year.

The diligent research of Mr. Detwiler, of the Soil Conservation Service, previously mentioned, yielded the following interesting facts about honey locust:

1. Ripening fruit in London and survived smoky atmosphere.
2. Doing well on the steppes of South Russia.
3. In Tunisia, pods 40-50 cm. long, with pulp richer in sugar, and especially in protein, than the carob.
4. In Australia, a more or less useful fodder for cattle and pigs.

There is no foreign report of selection of parent trees, and if we wonderful Americans, with all the agricultural paraphernalia of the United States, have neglected it thus long, how long will these others let it lie virtually dormant?

CHAPTER VIII

A Group of Stock-Food Trees— The Mesquites

Choice spots in southern California may have the carob for their bran tree, the Corn and Cotton Belts have the honey locust for their bran tree, and the area between has the honey locust and the frost-resisting mesquites with their crops of beans. When one considers the ancient use of the mesquite, its present use, and its remarkably useful and promising qualities, it becomes difficult to understand why it also has been so greatly neglected by the scientific world.

VALUE TO INDIAN AND FRONTIERSMAN

In analysis and use the mesquite beans are much like those of the old carob in the Old World (see analysis, Appendix). They have chiefly provided food for beast, but food for man also to a considerable extent. Some Indian tribes have had mesquite bread as a staple food for an unknown period of time.

The Pima and Papago Indians in Arizona have always made use of mesquite beans as food for themselves and their stock, particularly horses; and I think they are yet quite an important article of subsistence among the Papagos. (Levi Chubbuck.)

The Bulletin of the Biological Survey, Vol. V, No. 2, 1937, says that the mesquite harvest was a tribal event of these

Indians, and that their language and mythology indicate the mesquite's importance. When the pods are beaten in a mortar and sifted, the mass hardens and keeps indefinitely. It is eaten without cooking.

A caravan of Forty-niners seeking the golden sands of California lost some of their oxen as they toiled under terrible privation down the Gila River valley in southern Arizona; but when they reached the Colorado, near the present city of Yuma, they came upon a "grove of hundreds of acres of Mosquit Beans. These trees were full of beans and hundreds of bushels lay on the ground. These beans were reputed to be excellent feed for the cattle. . . . We remained at this place seven days, and our cattle gained strength and flesh remarkably fast, and with the two hundred bushels of beans we had loaded into our wagons, we felt warranted in making our start across the desert."

I continue to quote from the journal of Charles Pancoast, Salem, New Jersey. (See *A Quaker Forty-niner,* University of Pennsylvania Press.)

These Yuma Indians had a bad feeling towards the white people, and their hostility had lately been increased, in consequence of the acts of a lot of Texas emigrants, who, being too indolent to gather mesquite beans from the trees, broke open a number of Indian caches where they had stored their winter supply of the best screw beans, and loaded them in their wagons, for feed for their cattle. We did not do this, but picked up two hundred bushels or more from under the trees, and our cattle ate as much more, which did not please them very well, for it helped to diminish the supply they relied upon for their winter's bread. The soldiers had some of the bread made by the Indians from these beans. It looked like rich cake made from the yolk of eggs or nice corn bread. I ate a little of it and found it sweet and palatable, having, however, a little of the astringent tang of the acorn. . . .

As the mesquite aided the Forty-niners, so it has aided many a prospector, by feeding the beast that bore his equipment, but its chief use has been for animals on the range.

The mesquite is another bush or tree which is very abundant in the section lying southwest of the southern border of Utah, extending southwest over nearly to the coast. The mesquite bean is used in that section of the country quite extensively by the prospector and miner as food for their burros. (L. M. Winsor, San Luis Valley, Alamosa, Colorado.)

It was very noticeable during my work in Sulphur Spring Valley that the cattle were always in the mesquite bushes from the time they began to leaf out until the rainy season began, very few animals being found on the prairie; while as soon as the rains began, they transferred their grazing ground to the prairie. (R. W. Clothier, University of Arizona, Tucson, Arizona.

USE OF MESQUITE AS FORAGE FOR PASTURING LIVESTOCK

Mesquite beans are especially valuable because they ripen in August at the very time when drought may be expected to reach its worst. The beans are greedily eaten by cattle, horses, and goats. As C. W. Underwood, a rancher of Chillicothe, Texas, put it, "I have mesquite in my pasture and value a crop of beans very highly. I let the stock eat the beans on the trees, and a good bean crop means fat stock."

Similar testimony on the importance of the mesquite as range fodder comes from many parts of the Southwest. Some persons of responsibility report that the working horse does well on mesquite and that hogs have fattened satisfactorily on grass and mesquite beans.

The following information is furnished by Mr. N. R. Powell, Pettus, Bee County, Texas:

About six years ago one thousand pounds of mesquite beans were gathered and ground with the hulls and were then pressed by the Beeville Oil Mill into cakes the same size as cottonseed cake. Some of these cakes were kept two years and fed to cattle, which seemed to do them as much good as cottonseed cake. The difference between mesquite beans and cottonseed is that the former does not have to be ground as it breaks up easily.

Some seasons, as much as a trainload of mesquite beans are produced by him [Mr. Powell]. This is when a dry spring and summer occur. A good bearing mesquite tree will produce from fifty to one hundred and fifty pounds. Mr. Powell states that he thinks the value of mesquite beans in southwest Texas, if properly cared for, would be more than one million dollars annually. (Letter, Rex E. Willard, Assistant Agriculturist, Brownsville, Texas, February 11, 1914.)

However, the mesquite grows more on the flats than it does in the mountains; but it does grow to some extent in the foothills; and when there is a good crop of beans, these furnish a great amount of feed for hogs and also cattle and horses. When the mesquite beans get ripe along in August, and where they are very thick, you can see the cattle and horses grazing on them a great deal all over the ranges; and especially if the grass is short, which occurs in a dry year, a great many animals get their sustenance from these beans. Mesquite beans have an exceptionally high value as feed for all classes of stock, and they are looked upon as very fattening. I know of several cases where swine were turned out on the sand hills during August and September, and though they secured practically nothing else but the beans and what little grass they could get, they have been fattened sufficiently for market. There are other places in the mountains where they turn the hogs out just as they would cattle, and give them absolutely no feed, other than what they get themselves. The hogs range through the canyons and in the brush. I have seen a great many hogs gathered along before Christmas from places like this, and they were in as high finish as any you would find that had been fed in the lot. These are usually fattened on acorns.

In some places where the mesquite bushes are exceptionally thick the people, especially the natives, gather them and feed to their horses and cattle. I have used mesquite beans to feed to horses myself on trips over the country when we had no grain, and I find that they are not only relished by the animals, but they are very good feed. (Letter from W. H. Simpson, Professor of Animal Husbandry, New Mexico College of Agriculture and Mechanic Arts, State College, New Mexico, June 3, 1913.)

In the western and southwestern sections of the State the mesquite bean affords during some years a large proportion of the feed supply of the horses and cattle of those sections. (Letter from John C. Burns, Professor of Animal Husbandry, Agricultural and Mechanical College of Texas, College Station, Texas, June 6, 1912.)

The wild mesquite appears to be a crop plant of great promise if scientifically used and improved. It covers a wide territory and grows under very adverse conditions and in the unimproved state contains many good productive specimens.

THE NATURAL HABITAT OF THE MESQUITES

Robert C. Forbes, Director of Arizona Agricultural Experiment Station, says, in Bulletin No. 13, Arizona Station:

The mesquite tree (*Prosopis juliflora*), known in some localities as the algaroba, honey locust, or honey pod, is found, roughly speaking, from the Colorado and Brazos Rivers in Texas, on the east, to the western edge of the Colorado desert in California on the west, and from the northern boundaries of Arizona and New Mexico southward as far as Chile and the Argentine Republic, and eastward to Brazil, according to Bulletin of Biological Survey, Vol. V, No. 2, 1937.

The American Naturalist, Vol. XVIII, May 1884, "The Mesquite," by Dr. V. Havard, U. S. Army:

It flourishes in the southwestern territory of the United States, especially in Texas, New Mexico, and Arizona, being by far the most common tree or shrub of the immense desert tracts drained by the Rio Grande, Gila, and Lower Colorado.

The plant endures in all kinds of soil except that which is wet, resists great drought by means of small water consumption and a root system of great depth.

University of Arizona, Agricultural Experiment Station, Tucson, Arizona, letter, October 22, 1913, from Robert C. Forbes, Director:

Although mesquite grows abundantly in regions of the lowest rainfall, it is found for the most part in the washes, where occasional flood waters undoubtedly penetrate to considerable depths and thus afford a water supply far in excess of that which would be available on elevations. Standing on a mountainside and looking off across the country, one can easily trace the drainage by the long lines of dusty green mesquite which thus occupy the drainage lines.

The mesquite, as well as some other desert plants that I have observed, has two distinct root systems—one spreading laterally in every direction from the tree and evidently availing itself of the occasional supplies of moisture coming from heavy penetrating rains, and the other striking straight down to great depths and, presumably, feeding upon deep ground water supplies. I once, personally, dug up a lateral root running along a ditch, which at rare intervals carried flood water, a distance of exactly fifty feet from the trunk of the tree. This root was not as large as my little finger at its base and tapered out to a small filament at the end, where I lost it.

Mr. Forbes sent me the following item, February 23, 1914:

W. M. Riggs says that in boring a well in the San Simon Valley, using a drop auger that brought up a core, they found fresh living roots at a depth of eighty feet. There were growing at that point greasewood, sagebrush, and scrubby mesquite. The roots must have been mesquite. There was an earthquake crack near by which may have facilitated the penetration of the roots. Earthquake, 1886. Well bored, 1909.

The American Naturalist, Vol. XVIII, May 1884, "The Mesquite," Dr. V. Havard, U. S. Army:

Sometimes in the Southwest tents are pitched on claims where no timber or fuel of any sort is visible. It is then that the frontiersman, armed with spade and ax, goes "digging for wood." He notices a low mound on whose summit lie a few dead mesquite twigs; within it, he finds large creeping roots, which afford an ample supply of excellent fuel. These roots can be pulled out in pieces fifteen or twenty feet long with a yoke of oxen, as practiced by the natives in the sandy deserts of New Mexico and Arizona, where no other fuel can be had.

Of the vertical roots, the taproot is often the only large and conspicuous one. It plunges down to a prodigious depth, varying with
that at which moisture is obtainable. On the sides of the gulches one
can track these roots down thirty or forty feet. They branch off and
decrease in size if water is nearby; otherwise they, even at that
depth, retain about the same diameter, giving off but few important
filaments. How much farther they sink can only be conjectured.

Between these heaps of shifting sand are sometimes found large,
vigorous mesquite shrubs, the only arborescent vegetation there.
The inference would be that water, although too deep for the ordinary shrubs of the country, is accessible to the mesquite and should
be reached at a depth of about sixty feet, a conclusion practically
verified by the digging of wells along the Texas-Pacific Railroad.

Mesquite posts, much used in fencing, are said to be indestructible,
whether under or above ground.

As fuel, the wood from both root and stem is unsurpassed. It is
the most commonly used from San Antonio, Texas, to San Diego,
California.

It is a mistaken localism to think that the mesquites are purely
North American. Argentina has eleven species of mesquite,
while the United States has but six. (See pp. 87-89.)

Dr. Walter S. Tower, geographer, late of the University of
Chicago, reporting on his explorations in South America, says:

I have ridden all day through northern Patagonia with a temperature of 10° F. above zero, and have walked all the next day through
continuous forests of algaroba. I think their species would grow in
our own arid wastes of the Southwest because I found with it identical species of cactus growing in New Mexico and Arizona.

Dr. Tower further writes:

As I recall conditions, the algaroba grows pretty generally in the
dry fringe along the western and southern margins of the Pampas.

To the best of my recollection, the algaroba is common at least
as far south as latitude 40°, and at least as far north as latitude 30°.

The trees are small; as I recall them, few were more than ten or twelve feet high, rather bushy, and pretty well protected with long, sharp thorns. In the more northerly sections of its distribution, I think the size of the trees was rather larger than toward the south, where the growth was more in the nature of scrub than of what one commonly thinks of as real trees.

Speaking of the Pampas of northern Patagonia, Bailey Willis says in *Northern Patagonia*, p. 109:

The shrubs which are present in the flora throughout the entire range of the bushes are the algorrobo or algorrobilla (*Prosopis chilensis*), and the jarilla *larrea divaricata*. The algorrobilla is an acacia, a bush of strong growth, characterized by the delicate foliage, strong brown thorns, and large beanpods of the family. Sheep and cattle eat the beans when ripe, and the large roots are dug for firewood.

This was in an area (latitude 40° 45′ S. and longitude 65° W. and on to westward) entirely too dry for agriculture without irrigation; it had had recently observed temperatures of 106° F. and 12° F.

Through the kindness of Mr. Tracy Lay, American Consul at Buenos Aires, I have received a communication from the Argentinian Minister of Agriculture quoting from the book entitled *Contribución al Conocimiento de los Árboles de la Argentina* (*Contribution to the Knowledge of Trees in Argentina*) by Miguel Lillo, to the effect that Argentina has eleven species of Prosopis, that one of these, the *juliflora* (*Chilensis*), exists in eleven provinces, including Buenos Aires and Corrientes on the extreme east and all the western arid provinces from Patagonia in the south to Salta and Tucumán in the north; in fact, almost the whole of that vast country. The Minister further reports that the beans are very valuable for live stock, especially horses and mules, in place of green grass in times of drought.

The Minister of Agriculture further reports that one species, *Prosopis alba*, called in translation the white carob, analyzes twenty-five percent sugar, sixteen percent starch, and ten percent protein. This species was reported growing in nine provinces.

According to the Argentinian Minister of Agriculture, the mesquites grow in nearly all parts of Argentina.

Dr. Clarence F. Jones of Clark University reports bean-bearing mesquites which he called *Prosopis juliflora* (synonym for *Prosopis chilensis*) in widely scattered locations in arid and semiarid sections of South America. That portion of his letter follows:

1. In scattered patches in the dry coast of northern Colombia.

2. In the dry section of western Ecuador.

3. Along stream courses on the western flank of the Andes in Peru and Chile.

4. In the eastern lowlands and savannas of Bolivia.

5. In the Chaco of northern Argentina and Paraguay. A closely related tree, hymenaea courbaril—known also as algaroba—is found near northern Uruguay. . . .

It bears numerous straight or sickle-shaped pods about six inches long. . . . In Ecuador, the Chaco, and the savannas of Bolivia, it is prized as an article of food, being prepared in a number of ways. The leaves and tender shoots are grazed by cattle. . . .

The pods, which are very saccharine, are greedily eaten by cattle. In the coastal desert of Peru both pods and beans may be gathered and fed to cattle, especially in years of scant pasturage. In northern Colombia, in Ecuador, and in Peru the beans, being rich in tannin (sometimes containing 45 percent), are gathered and made into tanning materials for domestic use; also there is quite a trade developing in the beans for the manufacture of tannin. . . .

Algaroba beans are also used in the manufacture of dyestuffs and coloring materials.

THE ASTONISHING POSSIBILITIES OF THE MESQUITE GROUP OF SHRUBS AND TREES

This group of bean bearers holds out interesting possibilities of increased productivity for the arid lands of our Southwest,

of Mexico, and of similar lands in each of the other five continents. (See Climate Regions Map, Fig. 138.)

The possibility of further and useful adaptation to particular
places and needs lies pregnantly in the statement that the trop-
ically tender keawe of Hawaii is of the same species, *Prosopis
chilensis* (synonym *juliflora*), as many frost-resistant strains
that are scattered from Texas to Patagonia. The botanists have
had (and perhaps still are having) a glorious game of their
favorite indoor sport, "How many species there are and which
is what." I do not attempt to referee this game.

*A Chemical and Structural Study of Mesquite, Carob, and
Honey Locust Beans,* by G. P. Walton, Assistant Chemist, Cattle
Food and Grain Investigation Laboratory, Bureau of Chemistry,
U. S. Department of Agriculture, Department Bulletin No.
1194:

Mesquite grows over a wide range of territory and will flourish
where the more valuable carob cannot exist. It has been introduced
into India and South Africa, where it is attracting favorable attention. (Brown, W. R., "The Mesquite *Prosopis juliflora,* a Famine
Fodder for the Karroo." In *Journal* of the Department of Agriculture [Union of South Africa, 1923], Vol. VI, pp. 62-67.)

According to C. V. Piper, U. S. Department of Agriculture
(letter, May 1923):

The mesquites belong to the botanical genus Prosopis, in which
there are about thirty valid species, although many more than this
have been proposed. One species occurs in Persia and India, one in
the eastern Mediterranean region, two in Africa, and the rest in
America. Argentina is richest in species, fifteen occurring in that
country. In one group of species, the pods are coiled and hence
called screw beans. According to some botanists these constitute a
distinct genus, Strombocarpa. Two species of screw beans occur
in the United States and four in Argentina. . . . *P. juliflora* is
apparently the same as the older *P. chilensis,* which ranges from
Patagonia to Texas. The species so abundantly introduced into the

Hawaiian Islands is *P. chilensis* and is there known as kiawe or algaroba. The common species in the United States *P. glandulosa Torr.* occurs from southern California to Texas and Oklahoma. In modern times it has spread greatly and now occupies extensive areas formerly prairie. This is probably due to the seeds being carried by horses and cattle and not being injured in passing through the intestinal tract.

According to Frederick V. Coville, Principal Botanist, U. S. Department of Agriculture, letter, August 17, 1927:

The range of each of the American species of mesquite and screwbean covers a wide area in our Southwest.

Strombocarpa pubescens (Benth. A. Gray) ranges from western Texas to California and northern Mexico.

Strombocarpa cinerascens A. Gray, ranges from southwestern Texas to Nuevo León, Mexico.

Prosopis glandulosa Torr.

Prosopis juliflora glandulosa Cockerell. The range of this species is given as "Louisiana to southern California" and Mexico.

Prosopis velutina (Wooten).

Prosopis juliflora velutina (Wooten Sarg.) ranges from Arizona to Lower California and Michoacán, Mexico.

Prosopis palmeri S. Wats. Lower California.

Prosopis juliflora (Swartz) D. C. West Indies and Central America, including Mexico.

Dr. Coville stated further that "All the species are tropical or subtropical. Within our borders, they grow in the creosote-bush belt of our southwestern desert region. Standley, in *The Shrubs and Trees of Mexico*, refers all the species of Strombocarpa to the genus Prosopis."

Mr. Caledonia V. Perada, Ave. Quintana 93, Buenos Aires, claims 17 species in Argentina.

With all this American desert experience, the plant has wide areas awaiting in Old World deserts.

THE PRODUCTIVITY OF THE MESQUITE

J. J. Thornber, Botanist, Arizona Agricultural Experiment Station, Tucson, Arizona, says:

I will state that the yield of the mesquite tree ranges anywhere from fifty to one hundred pounds per tree of good size. This is in deep rich soil of our valleys with, of course, the rainfall and no irrigation. These trees are anywhere from twelve to eighteen inches in diameter and perhaps fifty years old. Some of them stand as tall as forty feet. They are not a tall-growing tree, but they have very wide-spreading branches, so that a single tree may cover a diameter of fifty to seventy feet. Smaller trees will bear less in proportion. Such trees commonly grow anywhere from forty to one hundred feet apart over the ground under native conditions, and I have seen the crop of mesquite beans so abundant under them as to cover the ground everywhere over considerable area as much as one or two inches in depth. (Letter, October 16, 1913.)

A space ten feet square would hold thirteen bushels or more than two hundred and fifty pounds, if the beans were two inches deep.

Robert C. Forbes, Director of the Arizona Station, wrote a bulletin (No. 13) suggesting the preservation of mesquite as a crop because of the two qualities of good food and valuable productivity of the tree. He went on to state:

It is difficult to state the yield, more than that it is usually very abundant, often amounting to from one to several bushels on a small tree. It should also be noticed in this connection that the beans are quite bulky. One bushel weighs about twenty-one pounds.

Analyses in the Appendix of this book show the high content of sugar and other nutrients and explain why the animals are so fond of the beans. In *Arizona Bulletin No. 13*, Mr. Forbes says:

The air-dry fruit, entire, was found to contain from 17.53 to 17.67 percent of cane sugar, all of which was in the pods. Further examination of another sample of pods showed them to contain 2.4 percent of grape sugar, and 21.5 percent of cane sugar, no starch or tannic acid being present.

G. P. Walton, U. S. Department of Agriculture, states:

After a favorable season the quantities of mesquite beans available over large areas of Southwestern United States are limited only by the facilities for gathering the ripe fruit.

C. P. Wilson, writing in the *New Mexico Farm Courier,* Vol. 4, pp. 7 and 8, concerning the use of the beans as pig feed, reports that in southern New Mexico it is not uncommon to see a medium-sized bush, with a spread of not more than fourteen to eighteen feet, bearing from one to one and one-half bushels of beans. Although the process of gathering the fruit is tedious, during the 1917 season the beans could be secured for from twenty to thirty cents per one hundred pounds. A native worker at the New Mexico Agricultural Experiment Station gathered about one hundred and seventy-five pounds of dried beans in a day. Since the pods weigh but twenty-one pounds to the bushel, however, the man gathered only eight and one-third bushels, not a very strenuous day's work. In a northwestern province of India, a good tree may yield more than two hundred pounds of ripe fruit a year.

In 1917 mesquite beans were gathered and shipped by the carload in Texas.

The yield of fruit, of course, varies with the type and size of the tree or bush. It has been stated in U. S. Department of Agriculture bulletins, Agrostology, No. 2 and No. 10, that one acre of land well covered with the trees may produce one hundred bushels of fruit per year. Two crops a year have been produced in Arizona and in Texas, the early crop ripening during the first half of July and the second during the first half of September.

As to the value of the beans, Professor Robert C. Forbes (Bulletin 13, Arizona Experiment Station) says that according to analyses the entire beans, weight for weight, compare favorably with alfalfa hay, are of slightly less value than wheat bran, and contain more protein but less fat and carbohydrate than shelled corn. It must be remembered, however, that these ingredients are partly contained in the hard kernels and that grinding is necessary as it is in carob, keawe, and honey locust.

THE SCREW BEANS

Mesquite has a kind of first cousin in the screw bean (Strombocarpa), or tornillo, which is so greatly like it in both botanical and economic aspects that they commonly are classed together, and should be. D. E. Merrill, Biologist, New Mexico Agricultural Experiment Station, wrote in a letter:

The tornillo grows extensively at lower levels in the southern part of New Mexico on the flood plains, as a rule. The trees grow from fifteen to twenty feet high. Posts from the larger trees are very durable. The stems and roots make excellent fire wood. The crop of "screw beans" is usually very prolific, though badly infested with bruchids. These beans have a large amount of sugar in the substance surrounding the seeds, and are, for this reason, eagerly eaten by stock. I have no particular data on the productivity, except that last fall one tree fifteen feet high on the campus here yielded one bushel and one-half of the beans measured.

Perhaps adequate testing will show that both mesquite and screw bean have their places in a reasonably scientific agriculture for the semiarid lands. John C. Burns, Professor of Animal Husbandry, College Station, Texas, stated it thus in a letter:

In the western and southwestern sections of the State the mesquite bean affords during some years a large proportion of the feed supply of the horses and cattle of those sections. As far as I am aware no investigations have been made in regard to the actual feeding values of these products. Neither has anything been done towards the develment of more productive strains of trees. It seems to me that the field offers considerable opportunity for investigation.

THE NEED FOR SCIENTIFIC WORK

This combination of the above-mentioned qualities, productivity of trees, ability to stand drought and frost, and good analysis of beans and their appetizing quality, certainly makes reasonable the statements of the scientists and ranchmen of the Southwestern plateaus that the mesquites are worthy of experimentation and gives reason for Mr. Forbes's belief that gathering beans on a commercial scale "seems to be practicable in some parts of the country." In considering all these statements it should be remembered these are wild plants, quite unimproved either by propagation of the best strains or by breeding. Suppose we should improve them as much as we have improved the apple, peach, English walnut, and many other food plants!

CHAPTER IX

The Real Sugar Tree

Several generations of Caucasian Americans have called the sugar maple the "sugar tree." It had been done before by countless generations of American Indians. Rare indeed is the person who will not say that maple syrup and maple sugar are delicious.

The sugar maple is a fine tree. Its spring sap has from 3 to 6 percent of sugar. It grows over a wide area of cold, rough, upland country with a poor agricultural surface and in some cases a poorer agricultural climate. Possibly plant breeding could do wonders with the maple similar to those it has already done with the sugar beet—namely, raise its sugar content several fold in a century and a quarter.

But why wait? Behold the honey locust! Look at Figs. 34, 35, 36! There is a wild tree, native, hardy, prolific, and yielding beans more than a foot long.

The beans from some of these unimproved and unappreciated wildlings carry more than 30 percent of sugar. This is equal to the best sugar beets and more than the yield of the richest crops of sugar cane. This, too, after man has been struggling with the sugar cane for centuries.

And Secretary of Agriculture Jardine told me that his department has no time for such new things as honey locusts, that they were busy with the bugs and bites and blights of crops already established. Such is the scientific side of this democracy!

Who will apply science and horse sense to this wonderful

95

bean tree, which may hold a hundred thousand gullying hills with its roots while its tops manufacture the world's sugar, if need arises?

Consider the history of the sugar beet, and it seems perfectly reasonable to picture, fifty years hence, a thousand mountain farm wagons hauling locust beans down to the sugar factory in some Carolina valley.

This sugar factory should also sell thousands of tons of cow feed, rich in protein and having enough molasses left in it to make the cows fight for it.

But this requires *constructive intellect* and *time*. Time appears to be more common than intellect, and neither is used very much.

CHAPTER X

A Summer Pasture Tree for
Swine and Poultry—The Mulberry

For a large section of the United States the mulberry is easily the king of tree crops when considered from the standpoint of this book; namely, the establishment of new crops which are easily and quickly grown and reasonably certain to produce crops for which there is a secure and steady market for a large and increasing output.

A KING (OF CROPS) WITHOUT A THRONE

The mulberry is excellent food for pigs. To harvest mulberries costs nothing, because the pigs gladly pick up the fruit themselves. Therefore, mulberries fit especially well into American farm economics because labor cost is high.

The mulberry tree is no new wildling just in from the woods and strange to the ways of man. It is one of the old cultivated plants. It has resisted centuries of abuse. It has been tried and found to be good and enduring.

It can perhaps be called the potential king of tree crops for the Cotton Belt and part of the Corn Belt. In actual production today the pecan is far ahead of the mulberry, but the potential market for mulberries is far ahead of that of pecans. Pigs eat mulberries, and feed for farm animals is the major objective of the American farmer. The honey locust, oak, and chestnut probably have greater promise, because their crop can be stored; but the mulberry has already arrived and has proved its adaptability and its worth.

The mulberry is a tree with good varieties already established and waiting to be used.

THE ADVANTAGES OF THE MULBERRY TREE

1. The trees are cheap because they are so easy of propagation.

2. The tree is very easy to transplant.

3. It grows rapidly.

4. It bears as early as any other fruiting tree now grown in the United States, perhaps earliest of all, for they often bear in the nursery row.

5. The fruit is nutritious and may be harvested without cost.

6. The tree bears with great regularity as far north as the Middle Atlantic States and New England, also through the Cotton Belt and much of the Corn Belt, and even beyond it into the drier lands.

7. It has a long fruiting season.

8. It bears fruit in the shady parts of the tree as well as in the sunshine, and thus has unusual fruiting powers.

9. It has the unusual power of recovery from frost to the extent of making a partial crop the same season that one crop is destroyed. This results from the remarkable habit of putting forth secondary buds and producing some fruit after a frost kills the first set of buds.

10. The fruit has a ready and stable market since swine and other animals turn it into meat, a product for which there is no prospect of a really glutted market, such as that which haunts the growers of so many crops.

11. The trunk of the tree is excellent for posts, and the branches make fair firewood for the farm stove. It is doubtless worth growing in many sections for wood alone.

12. While attacked to some extent by caterpillars, it prospers at present in most parts of its area without spraying, and

seems to have fewer enemies than most other valuable trees. Unfortunately, according to a letter from the Fruitland Nursery, Augusta, Georgia, the San José and India scale in that locality require one good dormant spray per year to keep the tree in good health. This, however, is not a serious burden, especially as it is a winter job. For many years I have had annual crops and no pests on the everbearing trees in my yard in a climate like that of Philadelphia.

As the mulberry tree has been cultivated for ages over a wide region, this comparative pest immunity is probably more dependable than it would be on a tree that has just come in from the forest and has not been subjected to the crowded conditions of artificial plantings. It is also probably a much safer tree than some fresh importation would be.

13. Growth of the mulberry for forage has gone forward in the United States, so that for a large area the experimental stage for the *tree* is past (but not for the crop). In localities where the mulberry is not well established, experiments are aided by the incomparable boons of low-cost trees, rapid growth, and ease of transplanting.

THE MULBERRY CROP IN THE UNITED STATES

Every claim that I have made for the mulberry has been backed up by correspondence with persons interested in mulberries or with interviews that I have had. In most cases my information comes from the statements of people who grow mulberries or are closely associated with those who did.

G. Harold Hume of the Glen Saint Mary Nurseries Company, Glen Saint Mary, Florida, wrote April 12, 1913:

All through the Southern States, mulberries are commonly used as feed for pigs and poultry. In North and South Carolina and Georgia nearly every pig lot is planted with these trees, and the mulberries form a very important addition to the pig's diet. There is one variety, Hicks, which will give fruit for about sixty days, in some seasons even for longer.

In eastern North Carolina it is the common practice to plant orchards of mulberry trees for hogs to run in. (W. F. Massey, Associate Editor, *The Progressive Farmer*, Raleigh, North Carolina, letter, March 11, 1913.)

The everbearing mulberry in this country is so common as to occasion very little comment; in fact, they become unpopular on account of their profuse bearing, especially if there are not pigs and chickens enough to pick them up. (Letter, John S. Kerr, Texas Nursery Company, Sherman, Texas, November 19, 1913.)

The mulberry grows to perfection, fruits abundantly, and is used both for hogs and for poultry. (Professor C. C. Newman, Clemson College, South Carolina.)

With a proper selection of varieties, this season might be extended.

In 1927 Mr. Hume, formerly a professor in horticulture, reported that there was little change in the situation.

I found a very general belief in the Cotton Belt that one "everbearing" mulberry tree is enough to support one pig (presumably a spring pig) during the fruiting season of two months or more. Professor J. C. C. Price, Horticulturist, Agricultural and Mechanical College, Mississippi, said:

The everbearing varieties will continue to bear from early May to late July, a period of nearly three months. I believe that a single tree would support two hogs weighing 100 pounds each and keep them in a thrifty condition for the time that they are producing fruit. They could be planted about 35 trees to the acre.

F. A. Cochran, breeder of Berkshire swine at Derita, North Carolina, reported:

I only have a few trees, but they are large ones, 100 feet apart. . . . I would not take $25 per tree for the old trees. I have three hogs to the tree. They are doing fine, in good flesh. . . . I have not weighed any hogs that were fed on mulberries, but estimate that they gained one pound and over per day. My hogs have a feed of two small ears of corn twice a day.

James C. Moore, farmer of Auburn, Alabama, wrote:

I never weighed my pigs at the beginning and close of the mulberry season, but think I can safely say that a pig weighing 100 pounds at the start would weigh 200 pounds at the close. . . . Three-fourths to the mulberries is safe calculation of the gain. I have had the patch about 18 years bearing. I planted my trees just 32 feet apart, and now the branches are meeting, and I have about 40 trees. I have carried 30 head of hogs through from May 1 to August 1, with no food but the gleanings of the barn and what slops came from the kitchen of a small family.

J. C. Calhoun, farmer of Ruston, Louisiana, said:

The variety is the "Hicks." I set them out 30 x 30 feet apart. I have 50 trees. They were mere switches when I set them out about three feet high. They began bearing the second year and made rapid growth. The fourth year after putting them out the trees would nearly touch, and they are abundant bearers—ripening from the last of April to the last of July. There is nothing that a hog seems to enjoy better than mulberries. I always feed my hogs at least once a day, but I find that it takes considerably less feed for them to thrive and do well during mulberry season.

In the course of much correspondence and a long journey through the Cotton Belt in 1913, I met many such enthusiastic statements.

The large mulberry tree of which I spoke is in the south part of Wake County, North Carolina. I stopped at the place to get some water and spoke of the large mulberry trees and made the remark that it was wisely said that one tree fed one pig during fruiting season. The owner said it certainly would, for this tree was feeding fifteen hogs at that time and was probably one hundred years old. The place is near Fuquay Spring, North Carolina. (J. W. Green, Green Nursery Company, Garner, North Carolina.)

It may properly be objected that this man is a partisan, being a nurseryman with trees to sell, but many farmers without any axes to grind are of the same opinion.

R. H. Ricks, a farmer specializing in cottonseed at Rocky Mount, North Carolina, gives the following testimony:

I planted two hundred mulberry trees of the everbearing variety thirty-three years ago on what I regarded as waste land. They commenced having some fruit at once, but they did not have profitable crops until the fifth year. I have since planted another orchard of fifty trees. I have carried fifty to sixty hogs on the fruit ten to thirteen weeks every year for the last twenty-seven years without other food, and in the main bearing season the two hundred and fifty trees would carry twice the number of hogs. Nearly every farmer has a small orchard in this section, eastern Carolina. I regard the mulberry for hog food with much favor.

I wouldn't take a pretty for my mulberry orchard [said one of these Carolinians]. "It's funny to me to see how soon a hawg kin learn that wind blows down the mulberries. Soon as the wind starts up, Mr. Hawg strikes a trot out of the woods fer the mulberry grove. Turn yer pigs into mulberries, and they shed off and slick up nice. It puts 'em in fine shape—conditions 'em like turnin' 'em in on wheat. They eats mulberries and goes down to the branch and cools off, and comes back and eats more—don't need any grain in mulberry time.

"They begin to beah right away," one Carolina farmer declared of his trees. "Why, I've seen 'em beah in the nursery row, and they begin to beah some as soon as they get any growth at all."

I have myself seen wild mulberries bearing in the Virginia woods when only five feet high. As to their hardiness, another Carolina mulberry grower testified as follows: "Yeh can't kill the things. Yeh kin plant yeh mulberries jest like yeh would cane—cut it off in joints and graft from the one yeh want to. I moved a tree last yeah. Jest put a man cuttin' roots off—and he cut 'em scandalous—and I hooked two mules to it and hauled it ovah heah. I didn't 'spect it would live, but it did." ("A Georgia Tree Farmer," J. Russell Smith, *The Country Gentleman*, December 4, 1915, pp. 1921, 1922.)

In that east-central part of North Carolina where the mulberry orchard is a very common part of farm equipment, a veteran of the Civil War (a captain) declared, "When I lived ovah the rivah we had a lot of mulberry trees—300 to 400 mammoth big ones. We had fully 200 hawgs, but we had to send fer the neighbors' hawgs to help out and to keep the mulberries from smellin'."

Where the captain then lived he had a bunch of thirty-five hogs of various sizes running in mulberry and persimmon pasture, all of which I saw. He estimated that one-third of their weight, or one thousand pounds of pork, live weight, was due to the mulberries from eighty trees set twenty-four by thirty-three feet. That runs out about six hundred and twenty-five pounds of pork, live weight, to an acre of rather thin, sandy land with little care and no cultivation. A big yarn, you say? I'll willingly take it back just as soon as any experiment station makes a real test and disproves it.

The trouble is that no station, so far as I can find out, is in a position to disprove it, because none of them has any facts. This is a reproach to Station staffs and also a really interesting piece of human psychology.

Blush, O ye Caucasians!

The attempt to obtain scientifically determined facts from southern Stations supported by State appropriations brought nothing but much favorable opinion and a suggestion that something might have been done at Tuskegee Institute. Dr. G. W. Carver, a Negro, Director of Research and Experiment Station at Tuskegee Institute (Alabama), not supported by State appropriation, wrote (November 2, 1927), "The small amount of analytical data that I have been able to find on the mulberry shows it to be higher in carbohydrates than pumpkins, being fourteen percent carbohydrates, a rather convincing evidence that it is really worth while as a fattening food." Dr. Carver also told of their own use of it in growing their pork supply.

A letter to the Experiment Stations in all Southern States in 1948 has brought out no new results.

THE NEED OF SCIENTIFIC TESTING AND EXPERIMENT

As nearly as I can learn through trusted correspondents in several Southern States, there was little change in the mulberry situation between 1913 and 1927. The neglect of the mulberry as a crop in the face of such evidence seems to require some explanation. However, this psychological and economic phenomenon becomes easier to understand when one recalls the slavish dependence of the southern farmer on the one crop of cotton. By tens of thousands they have resisted the temptations of clover and cowpeas and soybeans and vetch. They still buy hay for the mule. Nor had they planted pecan trees in their door yards. They grow no fruit, and some do not even have anything worth the name of garden. So the mulberry is, after all, in good company with the things they haven't done. Occasionally one finds a man who has tried mulberries and does not like them because of caterpillars, but in the main I have found enthusiasm among those pioneer farmers who were trying out the crop.

It is still true that in the principal cotton sections, particularly the "black belts," anything that can really be called a garden is quite scarce.

You are also undoubtedly correct in saying that there are still not only thousands but tens of thousands of cotton farms which make practically no hay and depend almost entirely upon buying hay if any is used, and this in spite of the fact that there has been a tremendous increase in the acreage in alfalfa, soybean, cowpea, and other hays in the past few years. You could even go further and say that thousands of such farms have no milch cows, practically no poultry and no hogs.

A real home orchard is yet a comparative scarcity in the Cotton Belt. (Letter, J. A. Evans, Assistant Chief, Office of Cooperative

Extension Work, U. S. Department of Agriculture, September 7, 1927.)

Take a journey through the South in the summer, and you will agree that there has not been very much change since the above paragraph was written.

THE POSSIBLE RANGE OF THE MULBERRY
IN THE UNITED STATES

As with any other little-used crop, the exact range over which the mulberry can eventually spread is today unknown. For example: Kansas seems to have some mulberry territory and some that is not mulberry territory.

State Forester Albert C. Dickens makes some interesting observations in Kansas Bulletin 165 (pp. 324-326). It should be remembered that as a forester he naturally is dealing primarily with wood trees rather than fruit trees, and that therefore his statements concerning fruits might tend to be weak rather than strong:

The success of the Russian mulberry has been quite varied. In northern Kansas it has been injured very frequently in severe winters.

In the southern counties of the State, Russian mulberry seems much less liable to winter injury. At the fair grounds at Anthony, Kansas, the rate of growth has been especially good, trees set four years ago having attained a height of fourteen feet and a diameter of four inches.

The fruit is not of high quality but is often used when other fruits are scarce; and as it ripens with the cherries and raspberries, it seems to attract many birds from the more valuable fruits, and it is frequently planted in the windbreaks about fruit plantations with this end in view. The fruiting season lasts a month or more. The need of some careful selection and breeding of this species is clearly indicated. The species is quite readily grown from cuttings, and the better individuals may be propagated and the uncertainty which attends the planting of seedlings be avoided.

There is little doubt that at least a million square miles of the United States, and in the most populated parts of the country, are now capable of producing crops of fruit from the everbearing strains of this remarkable tree. Fortunately, two of the commoner varieties, the Downing and the New American, originated in New York.

With chickens, ducks, birds, and pigs clamoring for the fruit, the everbearing mulberry is certainly a candidate for experiment in your poultry yard or pig lot; but if you live north of Mason and Dixon's line, make some investigation as to the hardiness of varieties that are offered you. ("A Georgia Tree Farmer," J. Russell Smith, *The Country Gentleman,* December 4, 1915, p. 1822.)

STABLE PRICE AND EASY EXPERIMENT

The price stability of the mulberry should be emphasized in a country where so many commodities find markets that are often glutted in normal times. Sometimes peaches are not worth the picking. Apples and oranges occasionally rot on the ground, as do beans, peas, and all the truck crops. In contrast to this the mulberry is in the class with corn. We have not had much trouble about a glut in the corn market, or in that of meat, its great derivative, except as measured by the low prices of the Great Depression, 1929-1939. A stable market is a fact of great importance in considering any crop. The fact that the mulberry has no harvesting costs and needs no special machinery minimizes the risk in experimenting. We need to get the facts on the feeding value of the mulberry from State experiment stations. However, any landowner can try it now. Commercial nurseries have the trees ready.

For the actual use of the mulberry as a farm crop, see Chapter XXIII.

The farm yard, the rocky slope, the gullied hill, the sandy waste invite you to try this automatic crop for which there is a world market at a stable price, and probably a very good price when one considers cost of production.

OLD WORLD EXPERIENCE WITH THE MULBERRY

Perhaps someone thinks that I should mention the silkworm. That classic domesticated worm has for centuries made the mulberry leaf worth its millions of dollars yearly and thus renders its great service to humanity by enabling hundreds of thousands of hard-worked Orientals to eke out a hungry existence. We have the climatic and soil resources for the mulberry trees, but the silk crop is not for us—not in this next hundred years. It requires the labor of human fingers, lots of it, and in this we have no present prospect of competing with China, Japan, or other very populous countries. (See *Industrial and Commercial Geography*, J. Russell Smith and M. O. Phillips, Henry Holt and Company.)

Nor is there much likelihood that in this century we shall ever be sufficiently lunatic to put on a tariff that would drive us to such a crop. But if we do ever want to feed silkworms, we have the resources, because the humid summer of our Corn and Cotton belts keeps the trees growing and producing leaves.

The mulberry has another great use among the Asiatics. It is a food of value for a dense population pressing upon resources more heavily than we do. This fruit has long been an important food in many parts of Western Asia.

Dried white mulberries, practically but not quite seedless, and extremely palatable, form almost the exclusive food of hundreds of thousands of Afghans for many months of the year. This use of dried mulberries suggests a new tree food crop. This particular variety, if needed in America, should be expected to thrive in the irrigated lands of our West and Southwest, where dry summers and frosty winters somewhat like those of Afghanistan are found.

Analysis of these dried mulberries (see p. 108) shows them to have about the food value of dried figs, and the fig is one of the great nutritive fruits. (See table, Appendix.)

Ellsworth Huntington of Yale, geographer and explorer, says that in Syria the troubles of the beggar and the dog are over for a time when mulberries are ripe, for both of these mendicants move under the mulberry tree and pick up their living. "Not only do the people eat large quantities of the fruit, but they also dry it and make a flour out of which a sort of sweetmeat is made."

But I am not urging diet reform for people—only for pigs. They are much more amenable to reason, much more easily pressed by necessity. But, nevertheless, this Afghan dried mulberry seems to be a remarkable food, according to the explorer's record of its use in that country, where it is more important than bread is to many people of the United States.

The dried mulberries form the principal food of the poor people of the mountain districts of "Koistan." In the valleys of Koistan and around Kabul there are extensive orchards of this mulberry, all irrigated, and the yield seems to be heavy. There is a howl if you have cut down a mulberry tree. When the mulberries are ripe, they sweep under the trees and let the fruit fall down and dry them just as they do the plums in California. For eight months the people live entirely on these mulberries. They grind them and make a flour and mix it with ground almonds. The men come month after month with their shirts filled with them. They can carry in their shirt enough of these dried mulberries for five days' rations. These men are commandeered and they bring their food with them. They get no other food whatever; mulberries and water are the whole diet. They sit down on the rocks and lunch and dine on nothing but these dried mulberries (Jewett). Here is the analysis of the dried mulberry, thirteen ounces in pulp, from Afghanistan made by F. T. Anderson of this Bureau. (Courtesy of Mr. Peter Bissett.)

Total solids...........................	94.81%
Ash..................................	2.75%
Alkalinity of ash as $K_2 CO_3$..............	.414%
Ether extract.........................	1.60%
Protein (N × 6.25).....................	2.59%
Acid as malic.........................	.30%
Sucrose..............................	1.20%
Invert sugar..........................	70.01%

Starch.....................................Absent
Crude fiber............................ 2.65%

(From records of Division of Seed and Plant Introduction, United States Department of Agriculture. Copy of Inventory Card. 40215 (F. H. B. No. 3445) *Morus alba.* Mulberry. From Afghanistan. Presented by his Majesty Habibulah Khan, Amir of Afghanistan, Kabul, through Mr. A. C. Jewett. Received February 23, 1915. See Plant Immigrants, 1916, Bureau Plant Industry, U. S. Department of Agriculture, 1916.)

Please note that ground almonds are high in protein; the mulberry, in carbohydrate. These mulberries were dried. Those analyzed by Dr. Carver of Tuskegee were doubtless fresh and 75 percent water.

CHAPTER XI

The Persimmon: A Pasture Tree for the Beasts and a Kingly Fruit for Man

One of the remarkable things about the human mind is its power of resistance to new ideas. If you have not been convinced of this by the facts about the honey locust and the mulberry, consider the present status of the persimmon in American agriculture. The persimmon has been praised, and its bright future has been predicted by the earliest explorers and the latest horticulturists. Captain John Smith, first explorer of Virginia, declared that the persimmon was as delicious as an "apricock." Said he:

Plumbs there be of three sorts. The red and white are like our hedge plumbs; but the other which they call Putchamins, grow as high as a Palmeta; the fruit is like a medlar; it is first green, then yellow, and red when it is ripe; if it be not ripe it will draw a man's mouth awrie with much torment; but when it is ripe it is as delicious as an apricock.

I am convinced that the persimmon is destined to be one of the most important fruits grown in the United States. (Walter T. Swingle, Physiologist in Charge, United States Department of Agriculture, letter, August 25, 1927.)

The persimmon is gradually being recognized as an important food for hogs. (C. C. Newman, Horticulturist, Clemson College, South Carolina, letter, May 27, 1913.)

Persons interested in the persimmon as human food should know that the well-known puckering astringency can sometimes be removed by simple processes. (See U. S. Department of Agriculture, Bureau of Chemistry, Bulletins 141 and 155.)

Ten generations of Americans, myself included, have spent some thrilling autumn nights pulling fat opossums out of the persimmon trees, where they so love to feed. Every animal on the American farm eats persimmons greedily. Millions of our people eat them occasionally with relish. Nevertheless, the persimmon has not become an important crop in America.

Throughout the region where persimmons are found in abundance the fruit is considered as being "good for dogs, hogs, and 'possums." Occasionally a family is mentioned as having lived for several months upon the fruit from a single large tree. (U. S. Department of Agriculture, Farmers' Bulletin 685. *The Native Persimmon*, W. F. Fletcher.)

The persimmon grows on a million square miles of the southeastern part of our country. It bears fruit profusely, often as much as the tree can physically support, and many trees bear with great regularity. Yet the persimmon as a crop in American agriculture has not arrived, despite its two great chances, one as a forage crop and one as human food. To make the case even stronger, it is a good-looking tree for your lawn, and the ripe fruit will lie clean on your grass for a month or two in autumn and early winter. Some trees hold the half-dry fruit until spring. It is delicious, and some are seedless.

THE ORIENTAL PERSIMMON INDUSTRY

Our failure to appreciate the persimmon becomes the more conspicuous because persimmons have been a major fruit crop and a standard food in the Orient for many centuries. I am one

of many thousands of American erstwhile travelers who hunger
for the persimmons of East Asia. How I would like to chew the
firm flesh of the persimmons such as I had in Korea, and still
more do I crave the soft, luscious golden saucerful such as I ate
week after week through the autumn in Peking.

The *Yearbook* of the U. S. Department of Agriculture (1910,
p. 435) quotes plant explorer Meyer, who had spent years in
China:

> The fruit of this particular variety (now called in America the
> tamopan) has a bright orange-red color, grows to a large size, meas-
> uring three to five inches in diameter, and sometimes weighs more
> than a pound. It is perfectly seedless, is not astringent, and can be
> eaten even when green and hard. It stands shipping remarkably
> well.

The persimmon is a fruit of great climatic range. It is the
major autumn fruit of both North and South China. I have seen
the rich orchards bending down with the big fruits in the
shadow of the mountain range north of Peking that bears
the Great Wall, northern boundary of China. Peking has the
latitude of Philadelphia and the climate of Omaha (almost
precisely), save possibly some spring changes from hot to cold.
I have also seen the Chinese persimmon trees growing abundantly and rendering important food service in the hills of
Fukien back of Foochow in the latitude of Palm Beach, Florida.

In certain localities of China, Plant Explorer Frank Meyer
reported (see *Yearbook*, U. S. Department of Agriculture,
1915, pp. 212-214) that the valleys are entirely given over
to the cultivation of persimmon. "Hundreds of varieties exist
there, and the trade in dried as well as fresh persimmons compares in importance with our trade in peaches."

The fruit is eaten raw, kept through most of the winter in a
fresh condition. It is also extensively used over wide areas dried,
as we use figs and prunes. Meyer goes on to say (U. S. Department of Agriculture *Yearbook*, 1915) :

In certain rections of the provinces of Shantung, Shansi, Honan, Shensi, and Kansu one finds that strains of persimmons are being grown for drying purposes only. These strains are quite different— not as juicy as those which have been so far cultivated in this country.

A dried persimmon in looks and taste resembles a dried fig, with the exception that it is devoid of small seeds and is coated with a heavy layer of fine grape sugar.

Dried persimmons of different varieties differ both in taste and in appearance. This difference is due not to the variety alone, but to the greater or less care employed in their preparation. The coarser sorts, upon the preparation of which little care has been bestowed, taste very much like cooked pumpkin, but those of finer quality are as fine as dried figs, being even juicier and more palatable because of the absence of objectionable small seeds.

I noticed that the Chinese used a stock which was entirely different from the American persimmon and also was not merely a seedling stock.

At last, in a valley north of Peking, near the Nanku Pass, I was shown wild trees of this stock. I recognized it at once as a species of persimmon, *Diospyros lotus,* which is also found in northern India, Persia, the Crimea, and the Caucasus. In the last-mentioned country it is known by the Turkish name of "ghoorma."

This ghoorma when found in its native haunts seems to be able to withstand drought and neglect to a remarkable degree, and it is for that reason, no doubt, that the Chinese have selected it as a stock. It has already proved to be better adapted to our American semiarid Southwest than our native persimmon, *Diospyros virginiana,* which has been the only one heretofore used. These varieties for drying purposes budded upon the ghoorma as a stock will probably be very well adapted to large areas of land in the Southwest. Americans heretofore have never realized what an important food product the Oriental persimmon is in its native country.

It is an interesting contrast to see near the Ming tombs north of Peiping native wild persimmons three-fourths of an inch in diameter standing near orchards with fruit four inches in diameter.

THE ORIENTAL PERSIMMON AS HUMAN FOOD
IN AMERICA

It would appear to be a simple process to establish the Oriental persimmon in America as a commercial orchard industry sending its products to city markets for human consumption in large quantities. It bears close analogy to the peach industry —the transfer across the ocean of the improved strains of a productive species. There are already available and growing here and there in the United States a few of the hundreds of varieties of the Chinese, Japanese, and Korean persimmons which have resulted from centuries of plant improvement by the patient Orientals. As the peach came across the eastern ocean to be a staple food, so the persimmon came across the western ocean. At present it is grown commercially only in California and is merely a food novelty in most markets. Any new fruit must fight its way to acquaintance, and in the beginning it moves in such small quantities that it usually must pay high freight and sell at unreasonable prices. Such fruits must move in carload lots before they can be sold at prices within the reach of large numbers of people. Thus any new fruit industry is at a kind of impasse for a time.

In addition to all this Oriental success, the excellence of the native persimmon is widely recognized, and in the year 1915 our Department of Agriculture published Farmer's Bulletin No. 685, which tells about thirteen named varieties of native persimmons and gives fourteen recipes for the use of persimmons as food, including recipes for bread and fudge.

After all, when the persimmon, native or Oriental, comes to the American market for human food, it finds two great drawbacks; first, conservatism: we are creatures of habit. *We like what we eat.* And then the American stomach is already full, and the markets are often overloaded with our old familiar fruits. Introducing a new food is usually a process of slow and expensive education. Who is going to pay for the introduction of the persimmon as human food? If a corporation like the

United Fruit Company, which is back of bananas and has most of them to sell, could handle the persimmon, it would have a good chance. As it is, the whole thing stands awaiting an educational or marketing program; or perhaps it would be better to say the development of new food habits.

The cost of production of persimmons is much lower than in the case of the citrus fruits, and eventually the fruit will undoubtedly sell for much lower prices than have been obtained up to the present. I am quite certain that persimmon growers can make good money at prices of two or three cents per pound for the best grades. (Letter, January 4, 1928, Robert W. Hodgson, Associate Professor of Subtropical Horticulture, University of California.)

The Oriental persimmon acreage in California has not increased as much in the last decade as was expected, because of lack of knowledge as to friendly relations between choice varieties and different stocks, *Diospyros kaki*, *D. lotus*, and *D. virginiana*.

When traveling in China in the autumn of 1925, I obtained cions of nine varieties of persimmons growing in central Korea, in North China, near Taiyuan, the capital of Shansi, and along the base of the Great Wall, beyond which no persimmons grow. These cions were sent to a nursery in Yokohama, Japan, and were grafted on established trees, which came to the United States in 1927. Cions from these were grafted on wild trees on an abandoned farm on a slope of the Blue Ridge Mountain, northern Virginia (latitude 39°, altitude 800 to 1,300 feet), and to my surprise nearly all have survived.

They did so well that the Department of Agriculture sent me many more varieties for tests, and in 1946, twenty-nine varieties bore fruit on my grounds. The same number fruited in 1947.

Some of these trees made perfect unions on the native persimmon stocks; some did not. They vary greatly in appearance of tree, productivity of tree, and also in size, shape, and flavor of fruit. Some varieties bear so heavily that the fruits, two and one half inches in diameter, perhaps more, touch each other as do grapes in a cluster, but such unreasonable loads cannot be expected every year.

Trees propagated by me have lived and have done well as far north as Reading, Pennsylvania, and some have survived in south-central Indiana. They should of course be regarded as experimental here in the eastern United States for years to come, but if I were a young man, I should certainly put out an orchard of them in some good fruit location, near latitude 39°, where I have had them growing for more than a dozen years.

These trees should have a great appeal to the home gardener and to the landscape architect. Their foliage is dark and glossy, almost like that of the orange, and in autumn they develop shades of red the like of which I have never seen except in some Chinese lacquer. Many of the leaves will have some of these various reds splotched with yellowish green. The tree is worth having purely as an ornamental, for its leaves only, but many trees hold a crop of orange-colored fruit for days or even weeks after the leaves have dropped. This is both interesting and beautiful, and—don't forget it—the fruit is good to eat, *very good*.

In some cases the fruit is without astringency and can be eaten as one would eat an apple. In others, the fruit must get soft so that it can be eaten with a spoon. Merely writing this makes my mouth water!

These trees have been subject to all the botanic and entomo-logic barbarities and fungus attacks of the long, hot, humid summer of southeastern Asia. Thus far, my trees have been attacked by no fungus and no insect save the Japanese beetle (to some extent), which fortunately we know how to handle (with DDT). I have little doubt that this fruit has a large area for commercial production somewhere between Pennsylvania, the Deep South, the Atlantic Ocean, and the Appalachians, where I have tried it. I do not know how much farther west its safe range may be. Its enemy is the cold wave which follows a warm spell. It should be noted that, while the average temperatures of Peiping and Omaha are essentially identical in January and July, the Chinese winters, and especially the Chinese autumn and spring, are much less subject to cold and warm waves than those of Omaha.

My trees have been through the hot and cold waves of the three devilish spring seasons 1945, 1946, and 1947. They suffered less than did apples, peaches, and cherries, alongside. Only highbush blueberries did better.

I should be quite satisfied to plant an orchard in Virginia or in Maryland, of the best varieties I have, if only I had the luck to be forty-five years old. What might be done to improve the Oriental persimmon, *Diospyros kaki*, if scientific horticulturists worked at it for a few decades, is purely speculative. The Chinese may have improved them to the limit. The fact that some of the varieties bear seeds may open the road to plant breeding and possible development of hardier strains. They have thus far refused to hybridize with the American persimmon, *D. virginiana,* and most of the seeds refuse to grow.

THE PERSIMMON AS A CROP TREE

Meanwhile the opportunity, the real opening, and the great need for the persimmon is for forage. Here and there a Negro mammy sits over a few native persimmons in some town or city market, and a few men in the Middle West have grown and sold a few bushels of named varieties of natives, but forage —pig feed—is the big outlet for the native American persimmon. Let the pigs pick up persimmons as they do mulberries.

Letter, A. D. McNari, Agriculturist, Box 316, Little Rock, Arkansas, March 24, 1913, says:

In this connection I will state that Mr. S. A. Jackson, of Monticello, Arkansas, has some very poor land on which persimmon trees are growing and which he thinks furnish more hog feed than if the same land were planted to cultivated crops. These are wild persimmons, but there is a great difference among the trees in the time of ripening the fruit. Some are ready for hogs to eat in September, while others are fit only after frost or as late as November.

Letter, University of Tennessee Experiment Station, Department of Horticulture and Forestry, Knoxville, Tennessee, May 26, 1913, says:

Everybody in Tennessee considers the persimmon a good pig feed, but nobody so far as I know has attempted to grow persimmons specifically for this purpose, nor has any attempt been made in the improvement of fruits of forest trees as a forage for domestic animals. (Charles A. Keffer, Acting Director.)

Letter, Joseph H. Kastle, Director of Kentucky Agricultural Experiment Station, State University, Lexington, Kentucky, September 27, 1913, says:

I have also heard it stated many times that hogs fatten rapidly on persimmons.

The persimmon tree has magnificent natural qualities of great aid as a crop maker:

1. Its extreme catholicity as to soil. It thrives in the white sand of the coastal plain, the clay of the Piedmont hills and the Blue Ridge, in the muck of the Mississippi alluvium, and on the cherty hills of the Ozarks. This chert is a covering of flints which remain when certain limestones dissolve and pass away. In some cases it covers the Ozark hills for several inches in depth, making tillage almost impossible but permitting full growth for forest or crop trees. It is a blessing for the future, because it prevents erosion.

2. Another soil aspect needs to be emphasized—the ability of the American persimmon to grow in poor soil. I have seen them grow and produce fruit in the raw subsoil clay of Carolina roadsides and in the bald places ("galls") in the hilly cotton fields where all the top soil had been washed away and there was neither crop nor weeds—save the persimmon, which is one of the great weeds of the South.

3. The persimmon is remarkable in the length of its fruiting season. With the persimmon, nature unaided has rivaled the

careful results of man with the peach and apple, for the wild persimmons ripen often in the same locality continuously from August or September until February, dropping their fruit where animals can go and pick it up through this long season of automatic feeding. In this respect it is ahead of the mulberry. Dr. John E. Cannaday of Charleston, West Virginia, reports one tree in his neighborhood (letter, September 27, 1924) that ripened its fruit in July.

4. Furthermore, it should be pointed out at once that this long season combines the added virtue, the great virtue, of automatic storage. It is true that nearly half of the season of persimmon dropping occurs after frost has stopped all growth, and farm beasts are usually eating food stored in barns. Truly these are two great virtues for a crop tree.

5. The fruit of the persimmon is very nutritious. It is said to be the most nutritious fruit (analysis, Appendix) grown in the eastern United States.

It is too much to expect the persimmon tree to have the complete and amazing collection of virtues cited for the mulberry. Compared to the mulberry, the persimmon (a deep rooter) is not easy to transplant. Therefore it is produced in the nursery at greater expense. It does not grow so rapidly as the mulberry. It does not even grow quite so rapidly as the apple. Benjamin Buckman, a private experimenter at Farmingdale, Illinois, said:

A persimmon, stem grafted in a two-year-old and vigorous stock, may make anywhere from one to four feet the first and second years, but later the new growth will be shorter, especially in years of heavy fruiting.

"Early Golden" should have a "gallon to the tree" in three or four years after planting.

F. T. Ramsey of the Austin Nursery, Austin, Texas, reported:

I budded about forty to fifty trees in four-foot rows—only a small part of an acre—three years ago last March. They are now bending with their third crop.

I happened to bring in a leaf this morning of Shingler 10 x 5¾ inches on a thrifty graft.

Albert Dickens, Horticulturist, Kansas State Agricultural College, Manhattan, Kansas, reported:

The seedling persimmons we have grown here have in the fifteen years they have been growing attained a height of from twenty-five to thirty feet and a diameter averaging about six and a half inches. These trees have been growing in good ground and have been given good care, and I think their size is considerably less than that of standard varieties of apples would have been under similar conditions. We consider from four to five bushels a fairly good yield for these trees.

E. A. Riehl, Alton, Illinois, a farmer with mental curiosity, put it thus:

Persimmon, not usually considered a timber tree, but the most valuable as food for stock of any forest tree that I know of, as it is a most profuse bearer, seldom failing to bear a crop and very nutritious.

The persimmon tree does not grow so rapidly as the mulberry tree, but it is a much more common tree in America than the mulberry, growing wild in much greater abundance. This is partly due to the fact that the leaves are shunned by most pasturing animals, including the sheep and goat. I have proved this in my own experience. This is a point of great importance, because seeds can be planted in pasture fields, pasturing can continue without interruption, and the trees can be grafted to better varieties when they reach a suitable size.

6. Native persimmons in my native locality, northern Virginia, bloom late in June, when wheat is ripe. This almost sure escape from frost injury would seem to be a great advantage in favor of growing these trees.

THE PERSIMMON AS A FORAGE CROP IN AMERICA

While making a journey of investigation through the South, I found that Dr. Lamartine Hardman of Commerce, Georgia, later governor of that State and owner of sixty farms, was one of the most enthusiastic tree farmers and persimmon growers anywhere to be found.

Cows keep the persimmons picked up clean [he said]. Hogs, cattle, cows, and horses are fond of them. I couldn't tell which likes them better—and mule colts! Just turn them out and they go to the persimmon tree first thing. They just love persimmons better than anything. Yes, sir!

Then telling of his farm practice he continued:

I found that I had a lot of land; it was just washing away, and here was natural produce in these good persimmons. I just put my men to looking for the best persimmon they could find, and we planted the seed.

To my question about the time of bearing, Dr. Hardman replied, "The trees are full before they get as high as your head. Another thing is, they do not seem to need any rich land, but just grow right out of the side of a gully. They don't seem to have any disease except a girdler, and he doesn't hurt them much—they live on. Crops grow right up to a persimmon tree." My own observation of many wild persimmons in Virginia confirms Dr. Hardman's statement as to precocity.

The facts which caused Dr. Hardman to plant persimmon seed in his fields seem to be widely known throughout the South and parts of the Ohio Valley. However, this knowledge appears to have produced little result other than the fairly common practice of leaving the wild persimmon standing in pastures when clearing the wood and thickets.

In North Carolina I have seen boys getting persimmons from the trees where the hogs were pasturing, and it required three boys to get the persimmons—one to keep the hogs back, one to knock the fruit from the tree, and one to pick it up. And the boys had to be quick if the hogs did not get a share. (D. S. Harris, of Roseland, Capital Landing Road, Williamsburg, Pennsylvania, in a letter to H. P. Gould, Pomologist in Charge, Fruit District Investigations, Washington, D. C., March 26, 1913.

This situation is fairly well proved and described by the following excerpts from some letters kindly secured for me by Messrs. W. A. Taylor and H. P. Gould of the United States Department of Agriculture:

In regard to the use of the native persimmon as a hog crop, will say that I do not know of any parties in the State who have planted the persimmon with this idea in view, yet it is very common in the making of hog pastures to retain any native persimmons that may be growing there with the idea of the hogs gathering the fruit. I believe that this is a most valuable fruit for hogs, and I have been collecting a number of our native persimmons that ripen their fruit at different stages with the idea of having them come on in succession from early fall until midwinter. (C. C. Newman, Horticulturist, Agricultural Experiment Station, Clemson College, South Carolina.)

A number of native persimmon trees are found in almost every pasture in this section of the State, and the hogs consume the fruit freely; however, very little attention is paid to the trees. (Letter, April 10, 1913, H. P. Stuckey, Horticulturist, Georgia Experiment Station, Experiment, Georgia.)

I have noticed this, that hogs seem to like them very much and make paths through the woods and fields going to the trees to pick them up as fast as they fall and seem to relish them very much, and they are generally in fine condition during persimmon season, though the acorn and other nut crops generally come off at the same time. Owing to the fact that the persimmon is very full of sugar, I think it very fattening and would be a fine thing to include in hog pastures following the everbearing mulberry, which lasts through the summer. (Letter, April 22, 1913, John F. Sneed, proprietor, the Sneed Wholesale and Retail Nurseries, Tyler, Texas.)

As a matter of fact a great many hog lots have persimmon trees in them, some of which may have been left because of the liking of the hogs for the fruit of the trees. Only one man of whom I know has any considerable quantity of the persimmon trees in his hog lot. It seems that he has chosen this lot in part because of the presence of the trees. He also has mulberries growing in the same area for the same purpose. I refer to Mr. Sam Wilder, proprietor of the Trinity Dairy, whose post-office address is Cary, North Carolina. (Letter, April 29, 1913, J. P. Pillsbury, Agricultural Experiment Station, West Raleigh, North Carolina.)

Here in southern Indiana where native persimmons abound, nearly everyone appreciates the value of persimmons for hogs. I had a hog pasture on one of my farms in a native persimmon orchard, and their value is almost equal to corn, after they get ripe, but cannot be utilized at any stage of growth like corn. But after they begin to ripen and lose their astringency, they turn rapidly to sugar that has a decided fattening property. I had a lot of hogs in a pen that contained two wild or native persimmon trees; as I was fattening these hogs I kept plenty of corn by them, but it was very interesting to watch these porkers ever on the alert for the familiar sound of persimmons, following and vying with each other in trying to be first to reach the fruit, leaving any kind of feed. I know a party near me who planted a considerable orchard of Golden Gem persimmons on purpose for hogs and paid (I think) $1.00 each for the trees. (Letter, March 25, 1913, Alvia G. Gray, Salem, Indiana, R. No. 4.)

I have found one other farmer, the late R. O. Lombard, of Augusta, Georgia, who was enthusiastically grafting the native persimmons that stood in his fields. He did this to get hog feed as one of the crops in his systematic series of tree-crops-for-hogs-forage system. I saw the trees standing in the white sand of his coast-plain cowpea fields and bending down with fruit.

In 1934 the *Southern Agriculturist,* stimulated by the TVA's tree-crop interest, located and published this interesting information: "A farmer in Arkansas pastured on a 3-acre grove of native persimmons as many as 35 hogs, 2 mules, and 2 calves from September to January 1st. The mules were worked every day during the fall season and kept slick on persimmons and grass." (Facts from John W. Hershey.)

It should be a nice element of farm management to let the pigs that picked up their own living on mulberries in June and July continue the process with persimmons from September until Christmas or snowfall.

CREATIVE WORK WITH PERSIMMONS

All this promise in the American fields and woods comes from one species, *Diospyrus virginiana.* Its range is at least a million square miles of area.

In this wide variety of climates, soils, elevations, and exposures, nature has made this species into an almost infinite variety of forms. These offer great promise alike to the searcher for trees fit to propagate and for the plant breeder. W. F. Fletcher of the United States Department of Agriculture, who has spent much time studying persimmons, states it thus in *Farmers Bulletin 685:*

The wide variations shown by the fruit in size, color, season of maturity, and tendency to seedlessness, and by the trees in size, shape, and vegetative vigor, indicate the possibility of greatly improving the native persimmon.

Analyses of persimmon pulp expressed in percent pulp show one specimen of *Diospyros virginiana* to have 29 percent solids while another had 48 percent solids. The low one had 26.30 percent nitrogen-free extract, the other 43.88—suggestive variations.

Now here is the stupendous fact. There are two hundred

species of persimmons scattered about the world. A veritable gold mine, first for the plant hunter, and then for the plant breeder.

There should be a corps of men at work right now upon the persimmon. Think of the work involved in finding the dozen best wild parent persimmon trees suited to make a crop series for East Texas, the Ozarks, southern Iowa, southern Indiana, central Tennessee, north Florida, central Georgia, eastern North Carolina, western North Carolina, central Virginia, southeastern Pennsylvania, Connecticut, and Rhode Island, and for extending the range to places where it does not now grow!

Concerning one of the few American experiment station plantings, Professor Albert Dickens writes:

Our planting of persimmons has been very satisfactory, but the named varieties are not much more desirable than a number of our seedlings. As a matter of fact, I think that two of the best we have are seedlings.

In spite of the fact that they are particularly tempting to the student body, we have harvested considerable quantities and have had a fairly ready sale for them when marketed in strawberry boxes or small baskets.

The trees that have borne five bushels of fruit were seedlings of the 1901 crop, being now twenty-six years old. They are about thirty-five feet in height and six to eight inches in diameter at breast height. They are almost as large as an apple tree of that age. For fruit production, I think the trees should not be spaced closer than thirty feet. (Letter, September 2, 1927, Professor Albert Dickens, Department of Horticulture, Kansas State Agricultural College.)

Professor James Neilson, Horticulturist at Port Hope Station, Ontario, reports a seedling persimmon from southern Missouri stock bearing fruit on the farm of Lloyd Vanderburg, Simcoe, Ontario, ninety miles from Niagara Falls and seven miles from Lake Erie. This place is somewhat protected from spring frosts by the lake, but the late-sleeping persimmon **is** not supposed to be a spring frost victim. Professor George L. Slate, Experiment Station, Geneva, N. Y. reports wild persimmons thriving in his locality.

The adequate search of American fence rows, fields, and woods for parent trees is a heavy task. It is a work of years. Testing these trees is another work of years. Breeding better ones is yet another work of years. And no one is doing it. And then there are the one hundred and ninety-nine foreign species, of which one, *Diospyros kaki,* has been made into the glorious golden fruit known to the Peking travelers, important to the Chinese in so many forms, and now coming into our markets.

Walter T. Swingle of the U. S. Department of Agriculture says:

There are at least six species of *Diospyros* (temperate and subtropical)—*D. virginiana,* U. S. A.; *D. kaki,* China; *D. chinensis,* China; *D. conazotti,* Mexico; *D. sonorae,* Mexico; *D. rosei,* Mexico; and probably more that yield edible fruits and are hardy enough to be grown in the limits of the United States. I think a careful study of the species of this very large genus might bring to light two or three times this many. They should be studied exhaustively.

At least four species are now used for stocks: *D. lotus, D. kaki, D. chinensis,* and *D. virginiana.* Many more should be tested and probably some would be valuable.

The persimmon alone could occupy profitably for many decades the time and resources of an institution with a staff of twenty to forty persons. They could probably produce crops that might rival corn, as well as the apple and the orange.

CHAPTER XII

A Corn Tree—The Chestnut

THE CHESTNUT SYSTEM OF EUROPEAN MOUNTAIN FARMS

Three men sat chatting at leisure under a chestnut tree on the little common of a Corsican village. It was a beautiful day in June. As the chestnut trees were only now blooming, it would be two full months before these men, one-crop farmers and owners of chestnut orchards, would have to go to work.

For miles I had ridden along a good stone road that wound in and out along the face of the mountainside, a mountainside that was much like my own Blue Ridge in Virginia, with one chief difference—this mountainside was higher and more prosperous than its Virginia prototype. The road at about two thousand feet above sea level went for miles along the mountain through a zone of chestnut orchards. In and out it went, in and out through coves and around headlands. Down near the sea level it was too dry for the chestnut, for Corsica is a land of Mediterranean climate, with little summer rain at low altitudes. Up near the top of the mountain it was too cool for the chestnut; but throughout a middle zone one thousand feet or more in elevation from bottom to top, the chestnut was at home. I think that in fifteen miles I had not been more than one hundred yards from a chestnut tree. If I was correctly informed, every chestnut tree in the wide area that stretched up the slope and down the slope was a grafted tree. At frequent intervals I had

passed substantial, comfortable-looking stone villages, villages
that looked older than the gnarly chestnut trees that shaded
them.

I wanted to learn the system by which these trees made a
living for the village folk, so I joined the three men who sat
chatting on the little common. It was a simple story that they
told me. The men of the village were farmers. Their chief estate and sustenance were tracts of the grafted chestnut orchards
that surrounded the village in all directions. In addition they
owned little terraced vegetable gardens and alfalfa patches near
the village. Every morning a flock of milch goats, attended by
some member of the family, and perhaps accompanied by a
donkey or two or by a mule, went out to browse beneath the
trees. Goat's milk, goat's-milk cheese, and goat flesh were important articles of diet in the village. I found that a meal of
goat's-milk cheese and cakes made of chestnut flour was good.

The men told me that the year's work begins in August or
September. A few weeks before chestnuts are ripe, the orchards are scythed to remove the things that goats and mules
cannot eat. Then in September comes the chestnut harvest. At
that season there is no school. Even the children help the men
and women to pick up chestnuts. The nuts are carried upon the
backs of donkeys, mules, and humans to the village. After harvest, pigs are loosed to turn over the leaves and find the nuts that
have escaped the human eye. For the pigs, that is going back to
nature.

Paris restaurateurs and butchers say that chestnut-fed pork
is as good as any pork and is usually considered to be of superior quality. (Letter from American Vice-consul General, Paris,
August 12, 1913.)

Smoked pork coming from pigs raised in a chestnut district is regarded as a great specialty, and its superiority to ordinary pork is so
marked that the pigs are fed almost exclusively on chestnuts from
October to the end of March. (Letter, Wesley Frost, American Consul General in Charge, Marseilles, August 4, 1927.)

The American Consul at Seville, Spain, wrote, September 29, 1914:

Frequently the chestnut crop in the northern part of this district is sufficiently plentiful to go far toward fattening the famous Estremaduran hams.

Some of the Corsican nuts are shipped fresh to market, but the main crop is dried for local use. Upon the slatted floor of a stone dry-house, the nuts are spread to a depth of two or three feet. From a slow fire in the basement, smoke and heat arise through the slatted floor to dry the nuts that are spread upon it. This kills all worms and cures the nuts so that they will keep as well as any other grain. The air space between the shell and the shrunken meats makes good ventilating space.

The chestnut is to the Corsican mountaineer what corn is to the Appalachian mountaineer in the fastnesses of Carolina, Kentucky, or Tennessee, except that Corsica grows more of chestnuts than Appalachia does of corn.

The Corsican crop of 1925 was 95,000 metric tons, worth $1,650,000. (Letter, September 23, 1927, Lucien Memminger, American Consul, Bordeaux.) The production of chestnuts that year in Corsica per square mile was more than that of wheat in Kansas (28 tons to 25). The next year Kansas doubled its wheat crop. Consul Memminger states that the Corsican crop was reported to be 90,000 tons in 1924 and 95,000 in 1927.

In 1925 Corsica grew 28 metric tons of chestnuts per square mile.

The corresponding figures for corn for 1924 were as follows:

Bledsoe Co., Tenn.	29 tons
Yancy Co., N. C.	21 "
Buncombe Co., N. C.	17 "
Mitchell Co., N. C.	15 "
Harlan Co., Ky.	5 "
Bell Co., Ky.	8 "

On the per capita basis, the figures were 740 lbs. of chestnuts for Corsica, and of corn for the American counties:

Bledsoe Co., Tenn.	2,040	lbs.
Yancy Co., N. C.	1,630	"
Buncombe Co., N. C.	400	"
Mitchell Co., N. C.	800	"
Harlan Co., Ky.	200	"
Bell Co., Ky.	300	"

(Information from Office of Farm Management, U. S. Dept. of Agriculture, and from Consul Memminger, Bordeaux, letter, September 23, 1927.)

In place of the Appalachian corn bread, the Corsican has chestnut bread; in place of corn to feed the animals, the Corsican uses dried chestnuts. One of my informants—the mayor of the village—took me around to the barn and showed me how his horse relished a feed of dried chestnuts. She crunched them, shells and all, exactly as my horses crunched corn.

While my new-found friends were telling me about their system of agriculture, a woman and a little girl came out to show me cakes made of chestnut meal. The cakes were to be used at a feast in honor of the marriage of the priest's sister. For these festive cakes the chestnut meal was wrapped in chestnut leaves for baking. In Palermo, Sicily, I was told that a laborer's breakfast often consisted of leaf greens, bread, and chestnuts.

The leaves from chestnut trees also furnished bedding for the animals of Corsica, in lieu of the American straw; dead branches from the trees furnished firewood. They had a regular system of selling the old trees to the factories that manufacture tannin from chestnut wood. The trees stood about irregularly almost as nature would place them, as she had doubtless placed the first trees at the beginning of chestnut orcharding some centuries ago. As a tree approached old age, a young tree was planted as near to the old tree as possible. The younger tree lived for ten, fifteen, perhaps for twenty years a stunted and suppressed life, but it grew a little and got its roots well established.

The moment the big tree was taken away, sunshine, light, and free fertility made the erstwhile starveling grow rapidly to fill the place of the old giant that had made its final grand cash contribution. This regular system of retirements and replacements kept these orchards continually replenished tree by tree—generation after generation—century after century.

To encourage heavy crops of nuts the trees are kept far enough apart for the light of the sky fully to reach the ends of all branches.

The annual increase of wood in the chestnut orchards of Italy is reported to be very low because the situation is somewhat akin to the production of wood in an apple orchard. The chestnut trees are kept for their nuts long after they have passed the maximum of wood making. It is another way to say that the nuts are more valuable than the wood. (Raphael Zon, United States Bureau of Forestry, letter, March 23, 1923.)

I asked one of my Corsican informants how long these orchards had been established. This man happened to be a government official from the nearby city who spent his summers in the chestnut village of his nativity.

"Oh," said he, "a hundred years, five hundred years, a thousand years—always!"

In English phrase he might have said, "The memory of man runneth not to the contrary." It seems to be a matter of record that the chestnut was introduced into Corsica by the soldiers of the Roman occupation in the second century A.D., and the gentleman was right in maintaining that the chestnut business of his mountainside had been going on uninterruptedly for many centuries.

The date of the earliest grafting on chestnut trees cannot be determined, but there are in existence trees over one thousand years old which have been grafted. (Wesley Frost, American Consul General in Charge, Marseilles, letter, August 4, 1927.)

Grafting trees is no new art. The Romans did it by a number of methods, nineteen methods according to one writer. And that the art of grafting was well established in the Old World two thousand years ago is certified by the Apostle Paul's writings to the Romans from Corinth, prior to his visit to Italy. In the 11th chapter of his Epistle to the Romans, he uses as a simile the art of tree grafting, likening to it the progress of the Jews and the Gentiles. (Epistle to the Romans, Chapter 11, verses 16-24.)

This Corsican chestnut farming is typical of that which covers many thousands of steep and rocky acres in central France, some of the slopes of the Alps, of the mountains of Spain, of Italy from end to end, and of parts of the Balkans. Especially do I recall when crossing the Apennines from Bologna to Florence, the marked and sudden increase of population that occurred at about two thousand feet elevation. The slopes below two thousand feet were treeless and on them were few evidences of people. At two thousand feet, where the chestnut forests begin, the villages were numerous, large, and substantial.

Compare this age-old and permanent European mountain farming with the perishing corn farms of our own Appalachian mountains. The farmer of Carolina, Tennessee, or Kentucky mountains has the cornfield as his main standby. He has a garden, perhaps in the woods some pigs—largely acorn-fed, some cows and sheep which range the glades and hills and pick such living as they can. The corn crop is the main standby. Corn bread is the chief food of the family. If there be enough, the pig or sheep or cow may get a little, or again they may not. The part that corn whiskey has played in the history of this region need not be expanded here.

The economic contrast between the Corsican and Appalachian mountaineers is striking. In Corsica the stone house in contrast to the log cabin of Appalachia; in Corsica the good stone road going on a horizontal plane along the mountainside in contrast to the miserable trails running up and down the American mountain; the Corsican mountain covered with majestic trees whose roots hold the soil in place, in contrast to the American mountainside deforested, gashed with gullies, gutted, and soon abandoned.

When the Corsican starts a crop, he does it by planting beautiful trees whose crops he and his children and his children's children will later pick up from year to year, decade to decade, generation to generation. When the American mountaineer wants to sow a crop, he must fight for it, a fight without quarter, a fight to the death of the mountain. First he cuts and burns forests, then he must struggle with the roots and stones in the rough ground of a new field. The sprouting shoots of the trees and tree roots must be cut with a hoe. This is the most expensive form of cultivation, but often the steep and stony ground can be tilled in no other way. In a few seasons the mountainside cornfield is gullied to ruin, and the mountaineer—the raper of the mountain—must laboriously make another field. No race of savages, past or present, has been so destructive of soil as have been the farmers of the southeastern part of the United States during the past century. How long can the United States last at present rates of destruction?

There is one argument for corn. It is a great and destructive argument. The plant is annual. The labor of the husbandman is quickly rewarded. The ruin of his farm comes later.

As between corn and chestnuts as types of mountain agriculture, the labor cost appears to be plainly in favor of the chestnut, but there is that pesky time element.

The chestnut also seems to be more productive than corn. Much sifting of facts among the chestnut growers of Corsica and France seems to show that the chestnut is a better yielder of food in the mountains of those countries than corn and oats are in the mountains of Carolina and Kentucky.

An authoritative book on chestnut culture in France is *Le Chataignier*, by Jean-Baptiste Lavaille, Paris, Vigot Frères, 1906. This book says, "A good French chestnut orchard yields on the average thirty-two hectoliters per hectare," or about two thousand pounds per acre.

United States Daily Consular Report, July 20, 1912, p.

343, reports that "148,000 acres of chestnuts in the reporter's district in Spain yielded, 1910, 2,534 pounds of chestnuts to the acre."

The average yield of corn in seven mountain counties of North Carolina, Tennessee, and Kentucky for 1919 was 1,124 pounds per acre; for 1924 it was 1,145 pounds. For the same counties the yield of oats per acre was, 1919, 363 pounds; 1924, 524 pounds. Mr. Raphael Zon of the Bureau of Forestry tells me that the 1,600,000 acres of chestnut orchards of Italy (good, bad, and indifferent) yield on the average about 1,000 pounds of nuts per acre. The American Consul at Marseilles, France, reported in 1912, for the 190,000 acres in his district, a yield of 1,320 pounds to the acre, worth 0.8 cents per pound, or $10.46 per acre.

Professor Grand, Professor of Agriculture at Grenoble, said in 1913 that matured chestnut trees 70 years old, 25 to 30 meters apart, 12 to a hectare or 4 to an acre, would bear an average crop of 150 to 200 kilos per tree. This is 1,320 to 1,760 pounds per acre. He insisted on this as an average, and said that the yield at times would be 4,000 to 5,000 kilos of nuts per hectare in a year of big crop (3,520 to 4,400 pounds per acre).

A big tall tree near the village of Pedicroce in Corsica had a girth of 4.60 meters, a spread of 60 feet; it stood on a terrace with nothing on three sides of it, and beside it was alfalfa, on which its roots could feed. The owner stoutly held that it yielded 1,000 liters of nuts on the average, that the tree varied in production but little from year to year, and that he gathered nuts himself and therefore was sure of his facts.

The yields of corn in Appalachia and of chestnuts on the European mountainsides are about the same in quantity, but the corn yield can be made only *occasionally* and for a short period of time before erosion destroys the field. In contrast to this, the chestnut yields on and on and holds in place the ground it feeds on.

Mr. Pierri, a wealthy proprietor and merchant of the Corsican

village of Stazzona, valued chestnut orchards in his vicinity at
$230 per acre in 1913. For some orchards the price was more,
for others it was less. At $230 per acre an orchard should have
thirty-five to forty trees per acre, which would give a value of
about six dollars per tree. A tree with a girth of one meter was
worth six dollars, but a big tree was worth fifteen to twenty
dollars because it bore more nuts. This land valuation was based
upon an earning of sixteen to twenty dollars per acre net at that
time, 1913. This income in turn was based upon an average
production, according to Mr. Pierri, of 3,100 pounds per acre,
with fluctuations ranging between 1,700 and more than 4,000
pounds per year.

I saw this land. It was as steep as a house roof and is shown
in Figs. 48 and 49. Similar land without chestnuts had almost
no value. Note the number and value of trees in Mr. Pierri's
statement above.

THE CHESTNUT FIELD IN THE SYSTEMATIC FRENCH FARM

The chestnut is a regular crop on systematized farms in at
least one section of south central France. I saw it near the towns
of Jouillac and Pompadour in the department of Corrèze. Under this system the farmer plants about one-third of the farm
land to grafted chestnut trees. The crop function is almost identical with that of corn on a farm in Pennsylvania, Kentucky, or
Wisconsin. As the corn is used in these states for forage, so is
the chestnut used in France. When the French farms are rented,
the agreements usually contain a provision similar to that found
in many American leases with regard to corn, the French provision requiring that the chestnut shall be fed on the farm so
that the land may benefit by the fertilizing value of the crop.
Sometimes this land that is in chestnuts is good arable land,
sometimes the trees are planted in rows and cultivated—true
tree-corn indeed.

The following interesting prices were reported from the
chestnut-growing provinces Corrèze and Aveyron by American

Consul Lucien Memminger, Bordeaux, letter, September 23, 1927:

	Per 100 kilos
Dry chestnuts.........................	230 francs
Wheat...............................	155 francs
Indian corn..........................	170 francs

I think it is accurate to say that in this district of systematic chestnut farming, which covers many square miles of gently rolling country, one-third of the area is in trees, ninety-nine percent of the trees are chestnut, and virtually all the chestnuts are grafted. It is the regular rule of the country that one-third of a man's farm is in chestnut for nut crop with a by-product of wood; one-third of the farm is in tilled fields; one-third is in pasture and hay meadows.

I have seen other French localities in which the fields were small and every fence row or boundary was bordered by a solid row of great chestnut trees, which thus covered a substantial percentage of the area.

The area of cultivated chestnuts seems to be declining in most of the French districts, especially Corsica. The following reasons are cited:

1. Such a large income is to be derived from sending the trees to tannin factories, a comparatively new industry.

2. The ravages of a disease called "maladie de l'encre" (*Blepharospora cambivora*).

3. A great increase in cost of gathering them, which now amounts to fifty percent of their value. This is due to the increasing scarcity of hired labor.

A chestnut grove of, say, four hundred trees (about twenty-five acres) costs as follows to harvest:

Labor for 45 days to clear away shrubs, etc....	600 francs
40 loads of wood as fuel for drying..........	250 francs
200 days' labor, mostly women and children, for harvesting chestnuts.....................	2,000 francs
Total.............................	2,850 francs

The crop would amount to 1,200 decaliters (or 12,000 liters) of dried chestnuts, worth at four francs per decaliter a total of 4,800 francs. (Letter, Wesley Frost, American Consul General in Charge, Marseilles, August 4, 1927.)

The franc here referred to was worth about 19.3 cents of a dollar that is no more.

These facts of decline should be considered in connection with the following facts. There has been recent decline of rural population in all the chestnut districts of France as well as in nearly all the other districts of France. This is accompanied by a decline of acreage of nearly all other crops in the chestnut localities and also the closing down of mines in Corsica. It should also be remembered that during the period 1920-1926 there was a very sharp decline in rural population in nearly every American state, and many farms were abandoned, as much, for example, as five hundred thousand acres of land in the state of Ohio alone. (Information on chestnuts from Lucien Memminger, American Consul, Bordeaux, letter, September 23, 1927; and from Hugh H. Watson, American Consul, Lyons, France, letter, October 12, 1927.)

CHESTNUTS IN JAPAN AND CHINA

Japan has a species of chestnuts, *Castanea crenata,* different from those of America, *C. dentata,* or those of Europe, *C. sativa.* They are larger than any of the American or European chestnuts, are less sweet, but like the sweet potato they are full of starch and nourishment. The chestnuts of Japan, like those of Europe, are used for both forage and human food, almost invariably cooked. The Japanese laugh at us for eating them raw. Japanese government bulletins recommend the use of chestnuts for hillside planting and grafting as it is done in Europe and in America. In one of the best-known chestnut localities in the Japanese mountains, which I visited, about forty miles northwest of Kyoto, the value of the poorest chestnut land (30 to 35 yen per tan) was more than that of the poorest rice land, while the best (irrigated) rice land was twice as valuable as the best mountainside chestnut land above it.

Japan's mountain chestnut orchards do not differ greatly from those of Europe. As I observed the mountainsides of Japan, the conspicuous thing about them seemed to be the small area given to tree crops. This seems unfortunate when one considers the great proportion of Japanese land that is not tillable and the great need for food in that crowded land. Perhaps the chief reason for the small extension of hillside chestnut growing in Japan is to be found in the widespread practice of cutting grass and herbage from the hillsides annually and carrying it down to fertilize the rice fields in the flat lands. The mountainside cannot yield fertilizer and wood and also nuts, as the Koreans have so sadly proved. It erodes.

Orchards of either nuts or fruit are not common in China, but scattered trees for fruit and nuts are widespread in many hilly localities. The Chinese chestnuts are more like the American nut in flavor, and many are larger in size. Chinese chestnuts hold great promise as a basis for an American crop that is now in its very early but unexpectedly promising youth.

A Chinese chestnut tree, leaving a stump 6 inches in diameter, cut in the spring of 1949, showed in 12 months 23 suckers more than five feet long—several of them were eight feet. It grew in fair circumstances, with no cultivation or fertilization. I incline to the belief that the Chinese chestnut is as keen a sprouter as was the old American.

THE AMERICAN CHESTNUT

The American chestnut is a fine tree for timber, and also a good producer of tannin. It has the good timber qualities of swift growth and ability to throw up shoots or suckers from the stump, if cut in the dormant season. The value of this is seen by comparison with pine. Cut down a pine, and it dies; cut down a chestnut and at the end of the first year the stump will have twenty or fifty shoots, some of them six feet high. At the end of the second year the suckers may well be eight to ten feet in height. In a very few years they will have attained a size sufficient to make them useful as poles.

This quality of rapid growth of the suckers is also a great advantage in the production of nuts. When the trees are cut for lumber, the resulting shoots can be grafted in a year. Fruit can be had as quickly as from the apple tree, or even more quickly in many varieties of chestnuts. I cannot definitely compare the American with the Chinese chestnut, but my experience with *molissima* shows it also to be a vigorous maker of suckers, and certainly not far behind the American.

The native American chestnut has a delicious flavor, but very little use was made of these nuts, considering the fact that they once grew wild to the extent of millions of bushels scattered over hundreds of thousands of square miles of the eastern United States. The Indians mixed chestnut and acorn meal to make bread baked in cornhusks, but for the European population of this country, the small size of the chestnuts was against them, in addition to the weevil worms. The nuts were a source of income for the Appalachian mountaineer in many sections and for boys on farms from Maine to Georgia. Looking for the beautiful brown nuts under the trees in the woods, on farmsteads and in fence rows is a lure to the hunting instincts of man.

Only a few million pounds were sent to American markets. These nuts were eaten along the street, at Hallowe'en parties, and beside the open fire after supper; perhaps we should not omit their service in school to alleviate for country boys the tedium of lessons. American wild chestnuts were important to the Indian, the squirrel, the opossum, the bear, and the frontiersman's hog; but a century and a quarter after the Declaration of Independence, they rotted by the million bushels in the forests from Vermont to Alabama.

The great drawback of the American chestnut was its small size and the added disadvantage that many nuts stuck fast in the bur and had to be removed by force. These disadvantages helped to make the Indian's corn preferable as the frontiersman's chief crop. For the same reason the large nuts of Europe appealed to the first experimenters with grafted chestnut trees and chestnut orchards.

Thomas Jefferson grafted some European varieties of chestnut on his Monticello estate in 1775, but an extensive introduction by Irenée duPont de Nemours of Wilmington, Delaware, about the beginning of the 19th century, seems to have been responsible for the rather wide distribution of these nuts and their accidental hybrids, by the year 1900, over southeastern Pennsylvania and the adjacent parts of Delaware and New Jersey. As early as 1891 the late Edwin Satterthwaite at Jenkintown, Pa., ten miles north of Philadelphia, had the roadsides and fence rows of his truck farm lined with a great assortment of chestnut trees. He also had them in regular orchard form, probably several dozen varieties. They were grafted and produced nuts of many different sizes and shapes, some nearly two inches in their largest dimensions; these nuts were sold in Philadelphia markets.

This rich collection of trees seems to have escaped the attention of professional horticulturists. I saw them only with the undiscriminating eyes of a schoolboy. My most vivid memory of them is the delightful speed with which they filled my surreptitious pockets; but I am sure that they were of many sizes and shapes and mostly of European origin. Some, however, seemed to be natives of small size that ripened nearly a month before other natives in the same locality. It is possible that *that* chestnut planting (and stealing) is responsible for this book. Virtue is not the only thing that has rewards.

The Paragon variety, undoubtedly a hybrid American x European, became the favorite of a young and promising American industry in the Nineties of the last century. This Paragon was very vigorous. I have seen grafts make six feet the first year, when grafted on suckers from a stump. It was not uncommon for the grafts to yield good nuts the second year. Not unnaturally, there was quite a boom for orchards of grafted chestnuts in the Nineties. For example, Mr. John G. Reist, of Mt. Joy, Pennsylvania, together with some associates, had eight hundred acres of hill land near the Susquehanna River in Lancaster County, Pennsylvania, stump grafted to Paragons. In the year 1908, before this orchard was mature and after the blight had begun to kill trees, it produced thirteen hundred bushels, which netted five and one-half cents per pound. C. K. Sober, of Lewisburg, Pennsylvania, had three hundred or four hundred acres on a nearby mountain ridge. I had twenty-five acres on the Blue Ridge Mountains of Virginia about fifteen miles southwest of Harpers Ferry, West Virginia. These are only a few of the plantings.

Then came the chestnut blight. It came with an importation of some Oriental plants. It spread concentrically from the Brooklyn Botanical Garden, where it first broke out in 1904. In a few years all these commercial orchards were gone. I turned the planting over to the U. S. Department of Agriculture for experimental spraying. They did everything they could think of, but by 1925 every tree was dead on my twenty-five acres save one Japanese tree which sprang up from a seed dropped from a Japanese variety. It is still there.

This blight is a fungus which gets through the bark, lives in the cambium, spreads concentrically, girdles the tree, and kills it. The spores seem to spread by means of birds, winds, and probably commerce in many forms. All attempts to stop it have failed. All attempts to kill it by sprays have failed. It has spread to the outer limit of the chestnut area in the East, the North, the West, and the South and is now finishing off the chestnuts in the southern Appalachians.

No tree has been found completely proof to the blight, but trees differ greatly in their capacity to resist it. A nice chestnut tree that shaded my Virginia mountain porch began to blight in 1908. In five years it was dead, but some trees in the nearby mountain still survive, now in 1950, through the process of throwing up suckers when the tree is girdled. The sucker thrives until it gets to be the size of a baseball bat; then the smooth bark begins to crack, the spores enter, and the blight starts. These shoots often bear a crop of nuts the year they die.

Sometimes there is vigor to send up a second crop of shoots, and sometimes a third and a fourth. Thus, these trees that began to blight thirty-five or even forty years ago still live. What is much more important, some of them produce seed.

If I could be granted an extra century or two of life and enough income to employ three or four helpers, I strongly suspect that, with the necessary land and equipment, I could (if I stuck at it) produce an effective blight-resistant strain of the American chestnut. I would do this by planting, generation after generation, the seed of the most resistant descendants of these tough specimens that have already resisted the blight enough to live with it for a third of a century.

Some people think that the American chinkapin, *Castanea pumila*, a diminutive (bush) species of chestnut, is the survivor of a larger species that was ravaged by some blight in past epochs yet managed to survive in the dwarf form. Chinkapins blight to some degree, but they keep on sending up enough suckers to be prosperously productive, and they keep this up.

This blight came from eastern Asia, where the Chinese and the Japanese chestnuts have been exposed to it for an unknown period of time and have developed varying degrees of blight resistance.

The spectacular Japanese chestnut had been introduced to this country by private individuals several decades before the blight came. However, upon the appearance of the blight, the U. S. Department of Agriculture acted with wisdom and promptness. It imported seed of the Chinese chestnut, raised seedlings, and distributed them far and wide for tests; and now, after nearly forty years of this testing, these trees are growing and producing nuts from southern Alabama to northeastern Massachusetts and in hundreds of localities between, also in many places west of the Alleghenies in scattered locations as far west as Iowa, Missouri, Arkansas, and Texas.

It is but natural that all this should produce a rapidly rising interest in planting these trees. It is also but natural to expect that this species, like almost every other species of tree, varies greatly. The Chinese trees have been frozen out in some places, and in other plantings nearby they have not. Local conditions often combine with the natural variation of individual trees to cause this variation in survival. For example, the chestnut is easily injured by wet ground in places where the pecan would rejoice and wax fat and green. In my more than twenty years' experimentation with the Chinese species I have, on two occasions, planted rows of them in places where a portion of the row was in well drained ground and another portion stood in ground only two or three feet lower yet not well drained. Blight ravaged the trees with the wet feet. The trees forty feet away on the well-drained soil missed it almost entirely.

We are now in a rapid process of testing individual trees, any one of which may become a variety such as the Abundance, the Connecticut Yankee, the U. S. Department of Agriculture No. 7930, now christened "Nanking," also the Meiling and Kuling and many others that will follow.

The variation of individual trees within the species is attested to in New Jersey Experiment Station Bulletin No. 717, 1945, which reports the planting of 150 Chinese chestnut seedlings in orchard form, in 1926. The bulletin reads as follows:

A number of distinct tree forms developed, and a few of the trees developed true central leaders and have formed a distinctly upright type. . . . After several crops had been produced, marked differences were noted in the regularity of bearing and in the productiveness of individual trees. . . . The variations were at least as great as among an equal number of apple seedlings. . . . Since the trees have come into bearing, a few produce annually, many biennially, and still others are uncertain as to regularity of bearing and quantity of nuts produced. It thus appears that an orchard of unselected Chinese chestnut seedlings may be very variable in production and consequently may be disappointing to the grower. . . .

The nuts from the different seedling trees in the College Farm plantation vary also in quality. Some are of good quality . . . having thin skins that are readily removed from the kernels. The nuts of other seedlings are more starchy and coarse in texture and have thick pellicles or skins that do not separate readily from the kernels. Such a natural variation in productiveness and quality is to be expected in seedling trees of any species or variety.

You see here the reason for the universal use of budded or grafted trees in the American fruit industry. However, I do not fully agree with the New Jersey bulletin as a description of the whole species, which indeed it does not claim to be. There is evidence that some *strains* of Chinese chestnut seedlings vary less than do others which may not be strains at all, merely mixtures.

Some plantations of seedlings are producing enough nuts to encourage their owners mightily and to cause the vendors of seedlings perhaps to become overenthusiastic about their wares and to mislead the public. There is no evidence that any hundred seedlings are on the average as good as an orchard of grafted trees.

So much for the Chinese chestnut, which seems securely started toward a rapidly increasing industry, with at least half a million well-watered square miles in the eastern United States as its field, and also a good little corner of the Pacific Coast.

We now have good varieties of Chinese chestnuts, good enough for an industry that is starting. Better trees may be discovered any year among the thousands of seedlings now scattered over more than twenty states. Further than this, there is no reason why we might not easily produce still better trees of the Chinese species by a very simple process of selection. The fact that each chestnut tree is almost self-sterile indicates that we can take two of the best trees we have, plant them side by side a quarter of a mile from any other chestnut trees, and be reasonably sure that almost all of the nuts are produced by a fusion of the strains of these two trees. This is very easy plant breeding, if we may so use the term. Plant these nuts, fruit them, take the best two trees of this generation, and repeat. By this process, we can be on the road to chestnut trees that will regularly bear excellent nuts, probably up to the physical limit of the tree to hold them up.

THE CHESTNUT AS A FORAGE CROP

If we could just get a new idea into the minds of the American farmers and of the creative leaders of American agriculture! That new idea is this: The chestnut offers an opportunity for an American forage crop, especially pig feed, productively covering hundreds of thousands of acres in addition to the few tens of thousands of acres needed to supply the human food market. I can see no reason why the best Chinese varieties we now have should not replace corn on many an Appalachian hillside, and perhaps on level land.

There is no reason to think them inferior to the European in productiveness. Here is a record made with care on U. S. Department of Agriculture grounds:

The seed of the parent tree of Nanking variety was planted in 1936.

In 1943 it bore 2.3 pounds.
In 1944 it bore 34.4 pounds.
In 1945 it bore 37.8 pounds.
In 1946 it bore 1.0 pound (due to big freeze).
In 1947 it bore 87.7 pounds.
In these 5 years it bore 163.2 pounds.

If that tree is given room, air and food, and no freezes, it will probably bear somewhere between 500 and 750 pounds in the next 5 years. A row of such trees would almost support a family at the present price of 30¢ to 50¢ a pound.

For the forage crop, we probably need new varieties of chestnuts, and the attempt to get them may easily give us good new varieties for the table. Breeding possibilities appear to be very great. The Chinese chestnut, the Japanese chestnut, and the chinkapin all interbreed with ease, and they offer rich possibilities to the plant breeder.

Within these three species there is a great variety of quality.

The Japanese chestnut is a weaker tree than the Chinese and somewhat more susceptible to blight, but there are many trees now living in the United States after forty years of exposure to the blight. This species has an enormous productivity, great precocity, and an astounding variation in the flavor and size of the nuts, which range from the size of your little finger up to nuts that rival eggs in size.

A Japanese chestnut called the Japan Mammoth, grown years ago by one Julius Schnadelbach of Grand Bay, Alabama, was photographed life size, and the photograph showed a profile 2.5 inches by 1.9 inches. A large egg measures 2.16 by 1.72 inches. Multiply the length by the breadth, and the Schnadelbach chestnut gives 4.75 square inches, the egg 3.715 square inches. Harvesting chestnuts like the Schnadelbach would therefore be like gathering eggs or potatoes, and could perhaps be done by hand for forage, perhaps even in high-wage America, because almost the only labor cost about it would be for picking them up. But pigs will pick them up without charge!

The flavor of the Japanese chestnut is not high, but the nut is regarded as a good food in Japan, when cooked. Certainly it is good enough in its natural condition for the pigs and other farm animals.

The American chinkapin is relatively blight resistant, tremendously productive, and very precocious, producing its burrs in strings. The very small, shiny nuts are easily the king of all chestnuts for sweetness. Turkeys might pick them up and fatten themselves on them and get a rare flavor in so doing.

BREEDING CHESTNUTS

The theory of plant breeding depends upon (1) the fact of variations of individuals within the species or within the crossing range, and (2) crossing to get new combinations of qualities, and what is even more significant, *getting qualities not in either parent.*

The amount of variation among trees of the same species is a surprise to most laymen. One of the qualities in which trees differ is that of precocity. Those who think of all nut trees as being so very slow in coming into bearing may be amazed at the precocity shown by some trees as reported by Dr. Walter Van Fleet, at one time experimentalist for the *Rural New Yorker* and later with the Department of Agriculture. Dr. Van Fleet wrote to me on April 13, 1914, giving the following surprising results of a series of experiments in getting a precocious strain of Japanese chestnuts:

No. 1. 1898. Japanese chestnut seeds planted.
No. 2. 1902. Fruit produced by trees from No. 1.
No. 3. 1903. Cross pollination of earliest ripening trees of No. 2.
No. 4. 1904. Cross pollinated nuts of No. 3 planted.
No. 5. 1906. No. 4 bore fruit, immediately planted.
No. 6. 1909. No. 5 bore fruit profusely (in 3 years)—immediately planted.
No. 7. 1911. No. 6 bore fruit (as much as 32 nuts per tree, in two years after seed ripened)—fruit planted.
No. 8. 1913. Trees from No. 7 bore in two years.

In each generation of those crossed he selected the earliest ripening nuts for seed. Note the increasing precocity of the generations.

The above experiment was started before the blight came. After the blight appeared, Dr. Van Fleet started to breed blight-proof chestnuts by crossing Chinese and Japanese chestnuts with the chinkapin.

Unfortunately, Dr. Van Fleet is dead, and this work with Japanese trees, continued for a time by Dr. Gravatt, was discontinued because of the excellence of the Chinese chestnut as revealed by the seedlings previously distributed by the Department. This is regrettable, for there is no reason to think that he made more than a beginning at the possibilities.

Dr. Gravatt began again on another objective. Here is a report which he sent to me in February 1948:

CHESTNUT BREEDING AND TESTING WORK BY
the Division of Forest Pathology at Beltsville, Maryland.

Present breeding program, under G. F. Gravatt and R. B. Clapper, has been conducted for 23 years; several thousand hybrids have been produced, using varieties and the strains of Chinese and Japanese chestnuts, the *henryi* chinkapin and sequin chestnut, both of China, also the American chestnut and the several species of native chinkapins. The principal objects of breeding are to obtain forest types of hybrids and hybrids as food source for game. Some hybrids and selections are being tested for nut production. Various strains and varieties of Chinese chestnut are being investigated for suitability for forest plantings. Most important results are: (1) Some Chinese chestnuts show promise as forest trees when growing under favorable situations; (2) Some first-generation hybrids of Chinese and American show promise as fast-growing forest trees with sufficient blight resistance when grown under favorable conditions. Back crossing to secure more resistance is being continued. . . .

About 7 acres in chestnut selections and hybrids at Bell, Maryland, and 4 acres at Beltsville, Maryland.

At *Chico, California,* U. S. Plant Introduction Garden: Testing strains of Chinese chestnut for nut quality and cropping ability. About 2 acres in chestnuts.

At *Savannah, Georgia,* U. S. Plant Introduction Garden: Testing strains of Chinese chestnut for nut quality, resistance to seed decay; progeny testing of selected trees for orchard and forest plantings. About 2 acres in chestnuts.

The thing that bothers me is—when will hybridizing open-pollinated trees produce seedlings that come anywhere near true? Rather than revert to ancestral types?

Dr. Arthur H. Graves has done much chestnut breeding. He began the work in 1930 under the auspices of the Brooklyn Botanical Garden. Hurrah for a botanical garden that follows an economic trail. Other gardens, and especially arboretum managers, please take notice!

Dr. Graves started after timber trees. He crossed American *Castanea sativa* with Japanese *C. crenata.* They started beautifully and then died of blight.

I now quote from a letter that Dr. Graves kindly sent to me:

. . . some of our hybrids of Chinese crossed with Japanese-American, now ten years old, give ample promise of fulfilling the required specifications of timber quality plus resistance to the blight.

Hybrids of Chinese and American chestnut also show great promise. These and many other combinations have been made, including further breeding together of the F 1 Japanese-American, amounting altogether to more than fifty hybrids new to science. Nearly all known species of chestnuts (twelve species) have been utilized in this breeding work—even many species of the chinkapins or dwarf chestnuts, which presumably might have desirable characteristics recessively.

Chestnut Blight in Italy. Professor Smith has given us a graphic picture of the dependence of the Italians on the chestnut for food and even for a means of earning their livelihood. Unfortunately, the dread blight has penetrated deep into Italy (having been discovered there in 1938) and now all trees in the country, said to form fifteen percent of the forests there, are threatened with extinction. Not only does the country use the nuts as a source of food and income, approximately sixty million pounds being exported annually in former years, but the young coppice shoots are used for the weaving of baskets, older ones for poles for vineyards, still older for staves of wine casks, and the oldest for telephone and telegraph poles. Before the war, chestnut flour was the principal food in many localities, but during the war a serious food shortage forced the people in many other areas to rely solely upon chestnut flour for weeks at a time.

Professor Aldo Pavari, Director of the *Stazione Sperimentale di Selvicoltura* at Florence, Italy, visited this country in the summer and fall of 1946, and he is now receiving the full cooperation and exchange of material with Graves and other American breeders.

Forestry News, Washington, D. C., February 1948 (p. 6) reports that Professor Pavari has discovered in Spain a strain of the Spanish chestnuts which to all intent and purposes was immune to the blight. This is a discovery of great possible significance for the entire world if true, but I shall take my time before believing it.

I again quote Dr. Graves:

Natural Variation as Regards Blight Resistance. Throughout the range of the American chestnut there exist certain strains or races and here and there individuals which show more or less resistance to the blight. For more than a quarter of a century we have been searching for these individuals or groups with the result that many are now known to us. By grafting cions of these trees onto the more resistant Chinese stock they are being reproduced on our own plantations, and their resistant qualities can be incorporated with our hybrids by breeding.

Following this same angle of research, we are planting all nuts of American chestnut sent to us, since these nuts may contain valuable blight-resistant characters, coming as they do from a region where the blight has eliminated all but the most resistant trees. We are therefore calling on all patriotic Americans for nuts. They should not be allowed to dry out after harvesting, but should be mailed to me within a day or so, wrapped in moist sphagnum, commercial peat moss, moist cotton, paper napkins, or something of the sort. If the nuts become dry and hard, the embryo is killed. All nuts sent in are planted immediately and labeled with the name of the donor and their source. Address: A. H. Graves, Chestnut Plantations, Wallingford, Connecticut.

Chestnuts for Eating. As to edible nuts, we have developed several hybrids where the nuts are equal in flavor, or possibly even superior, to those of the old American species. Work on these hybrids is still in progress. The Chinese chestnut, *Castanea mollissima,* which shows a high degree of resistance and in some individuals seems quite immune to the disease, offers a splendid solution of our nut problem. The nuts are usually much larger than the American chestnut, but often not as sweet, although there is much variation between different trees as regards flavor, some individuals being quite sweet.

With some cultivation, fertilizing and pruning, plenty of light, and rich, deep soil, the Chinese chestnut should begin to bear nuts in five or six years from seed. Just now this species as a nut bearer is enjoying a wave of popularity throughout this country, and except in northern New England and other northern regions of the country where it suffers from extreme cold, I feel that this popularity is entirely justified.

Fig. 6. Double ruin and quick. The hills are ruined by gullies. This Chinese valley is completely ruined by the wash from the hill, which was settled by Chinese farmers after the Atlantic seaboard of the United States was settled. The valley was then good land for farms. The ruins of the hill ruined the valley. (Facts and photograph Bailey Willis.)

FIG. 7. *Top*. Oriental Penance. With great labor, Japan reclaims her mountains denuded by past carelessness. Little trees planted by hand on terraces built by hand labor finally renew the forest and hold her mountainside. (Photo Shitaro Kawai. Courtesy U. S. Forest Service.)
FIG. 8. *Bottom*. By hand labor the Chinese near Nanking carry from a lake bottom back to the mountainside a small fraction of that which need never have left it. (Photo Prof. Joseph Bailie. Courtesy U. S. Forest Service.)

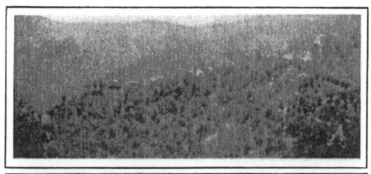

FIG. 9. *Top*. Zone of Corsican chestnut orchards (or forests) and the villages they support. Note village in left distance. FIG. 10. *Center*. Characteristic road and slope in Fig. 9. All trees are chestnuts and all are grafted. A stand centuries old. Note the different ages. FIG. 11. *Bottom*. Spanish Mediterranean island of Majorca. Limestone with fissures and pockets of earth in it. The man stands by grafted wild olive. At left, grafted carob. At right, acorn-yielding ilex. No earth in sight. (Photos J. Russell Smith.)

FIG. 12. *Top.* Olive and fig trees on the hills of Kabylia, foothills of the
Atlas Mountains, Algeria. Population twenty-five times as dense as on
the same hills where there are no tree crops. FIG. 13. *Center.* The pasture
year on the two-story Majorca farm. Producing figs, wheat, clover, and
beans—four-year rotation. FIG. 14. *Bottom.* Olive trees in Central Tunis
planted (without doubt) by the Romans before A.D. 648 and still bearing
—a long-lived property. (Photos J. Russell Smith.)

Fig. 15. *Top.* (Photo George F. Ransome.) Fig. 16. *Bottom.* (Photo Arthur Keith.) I am grateful to Dr. George Otis Smith and Messrs. Ransome and Keith of the U. S. Geological Survey for this remarkable pair of pictures. They show essentially similar geologic formations. Both are old granite. In the dry climate of Powers Gulch near Globe, Arizona, there is no vegetation to hold the rotting rock. At Roan Mountain Station, Carter County, Tennessee, the rain supports vegetation, and the vegetation holds the earth in which it lives, and covers up the bare bones of granite. If careful, very careful, man can keep his earth. Otherwise—?

FIG. 17. *Top.* This very thin layer of clay loam resting on solid granite in Chekiang province, China, supports the rapid-growing bamboo with its myriad uses, including edible shoots. How long would that slope last if treated as the land in Fig. 2 is treated? (Photo F. N. Meyer, U. S. D. A.) FIG. 17A. *Bottom.* Behold a ruined hill in Northern China, and a valley several hundred yards wide, covered by a layer of gravel and small stones, and also ruined. Not long ago the hill and valley were both good crop land. Man insists on a form of suicide that reduces the possibility of food production.

Fig. 18. Father, mother, and child, life-size. McCallister hiccan, Indiana, center, a chance natural hybrid almost certainly produced by crossing trees of low quality. A good Indiana pecan above, a good shellbark (laciniosa) below. (Photo E. R. Deats.)

Fɪɢ. 19, Fɪɢ. 20. *Top.* The trees in front of this hat, left, and behind the man, right (J. F. Jones), are hybrids (filbert x hazel) of same origin. Genetics is a useful science. Fɪɢ. 21. *Bottom.* Bottom row, nuts of parent trees. Progeny above. Compare size of parents and offspring. (Photos J. Russell Smith.)

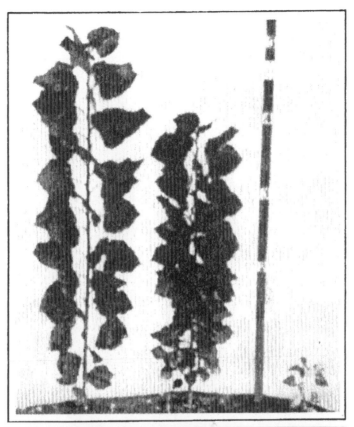

Fig. 22. *Top.* Three hybrid poplars from same parentage. Project by McKee, a chemist—quite a joke on the foresters. Some of these grow with great speed. (Courtesy Ralph McKee, A. B. Stout, and E. J. Schreiner.) Fig. 23. *Bottom.* Water-holding terraces made by machinery in clay hillside orchard of Lawrence Lee, inventor. This is one of the great inventions. The moisture substitute for cultivation is evident. The apple trees grew rapidly. (Courtesy J. R. Linter.)

FIG. 24. *Top*. The spreading keawe tree above the automobile was shown by Mr. Williams as one yielding from two to three tons of beans in a year. The mass of thick leaves near the horse is the tops of other trees in background. FIG. 25. *Bottom*. A fruitful branch of keawe, showing one of the two crops per year, Island of Maui, Hawaii. (Photos J. Russell Smith.)

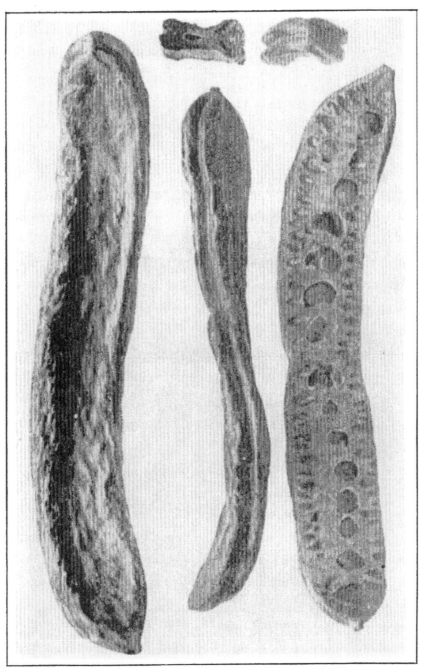

Fig. 26. Carob beans, showing their great pod development, and small seeds. Life size. A great gift to semiarid land. (Courtesy U. S. D. A.)

Fig. 27. *Top.* An expanse of young carobs on the eastern foothills of the Great Valley of California. There will be more in 50 years. (Photo H. J. Webber.) Fig. 28. *Bottom.* A seedling Carob street tree near campus of U. C. L. A., Los Angeles, Cal. Gets water by lawn irrigation, more than it needs. A very satisfactory street tree, with possibility of a large total income from that source alone. (Photo J. Eliot Coit.)

Fig. 29. *Left.* Young carob plants showing root development resulting from growth in slats saturated with nitrogen. This long root, in combination with ability to plant it uninjured in a vertical position with minimum of effort, is an invention of great importance for tree crops in dry-land agriculture. (Photo H. J. Webber.) Fig. 29A. *Right.* Road to top of Mt. Carmel, Palestine, showing newly grafted carob trees grown in the sandy soil above the naked limestone at the side of the road. Others in the distance above and below the road.

FIG. 30. *Top Left.* Limestone pasture in Portugal with scattered carobs. A good example of two-story agriculture. FIG. 31. *Center Left.* Cultivated vineyard interplanted in carob, Northern Algeria. Both make nearly 100 per cent crop. Details in text. FIG. 32. *Top Right.* Young carob tree in Northern Algeria, showing rather precocious burden of beans. FIG. 33. *Bottom.* Carob trees growing in the remnants of a wasted upland below a monastery near Valencia, Spain.

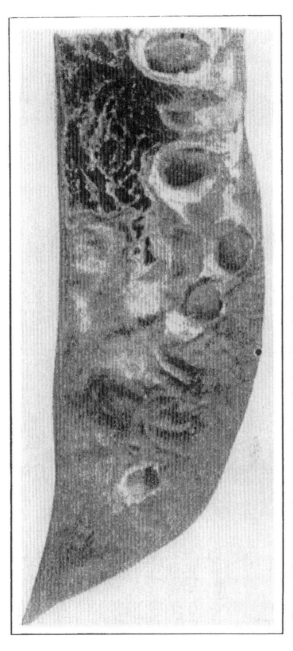

FIG. 34. Life-size part of honey-locust bean winning prize given by American Genetic Association, 1927. The seeds are grown in a thin edge of pod containing little nutriment. The sugar grows in black masses at the side of pod, shown by cutting away part of husk. (See Fig. 35.) This picture shows a little more than one third of the whole bean. (Courtesy American Genetic Association.)

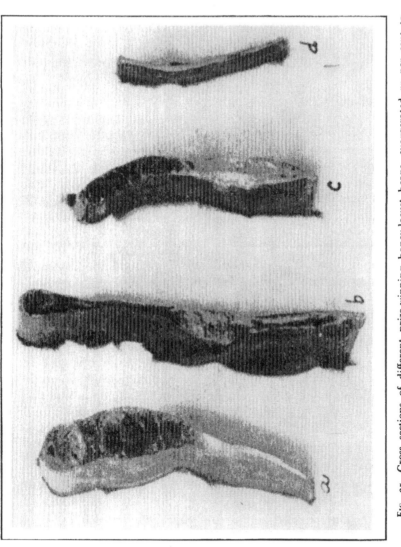

FIG. 35. Cross sections of different prize-winning honey-locust beans, exaggerated 50 per cent to display the variations in cross section, therefore variations in place where sugar is concentrated. That marked *a* is the prize-winning bean shown in Fig. 34. This tendency to vary suggests interesting possibilities to result from breeding experiments. Note the locations of sugar. (Courtesy American Genetic Association.)

Fig. 36. *Top.* Milwood honey locust tree bearing 273 lbs. air-dried pods. Auburn, Ala., Station. (Courtesy J. C. Moore.) Fig. 37. *Bottom.* Velvet mesquite beans and branch. Santa Rita Experimental Range, Arizona. (Photo Kenneth Parker, U. S. Forest Service.)

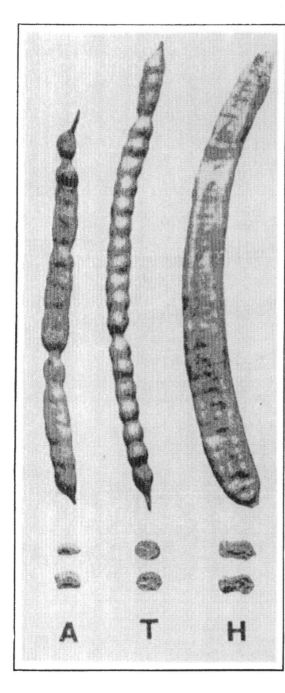

FIG. 38. Mesquite beans, cross section, reduced two-sevenths in diameter. *H* is from Hawaii (courtesy J. M. Westgate). *T* is from Texas (courtesy H. Ness). *A* is from Arizona (courtesy J. J. Thornber). Mr. Thornber says: "I am somewhat in doubt about the species. The pods are very similar to Prosopis Velutina, but the tree resembles quite closely that of Prosopis Glandulosa except the leaves are scarcely Glandulosa and the tree grows to a considerable height. We have never determined definitely what species this mesquite is." That quotation reflects the glorious but bloodless hundred years' war that has raged and still rages among the botanists as to how many species there are of mesquite. Note the fatness of the beans. (Photo E. R. Deats.)

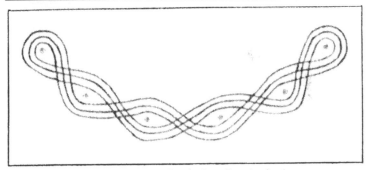

FIG. 39. *Top.* A systematic orchard of mulberries for hog pasture near Raleigh, N. C. The man stands on top of a mangum terrace. FIG. 40. *Center.* Hogs in a mulberry orchard planted for them near Fayetteville, N. C. FIG. 41. *Bottom.* Mulberry trees. The Southern Experiment Station staff member who said the use of mulberry was declining because of difficulty of sanitation also said (1) that nothing had been done about it and (2) that good disking as above would apparently solve the problem by burying the parasites. Comment unnecessary.

Fig. 42. *Top*. Characteristic burden of fruit produced by the wild American persimmon tree. Fruits nearly an inch in diameter, Augusta, Ga. Fig. 43. *Bottom*. Grove of persimmon trees, Diospyros, grafted on Diospyros Lotus. These are dry-meated varieties grown for drying, near Sian Fu, Shensi Province, China. Note the common Chinese method of growing them along the edges of fields. (Photos J. Russell Smith.)

Fig. 44. *Top.* Long strings of peeled persimmons hanging from a pole set up on the mud roof of a house at Siku Kansu, near Tibetan border of West China. This fruit is dried prune and dried fig for the Chinese. Generally relished by foreigners. (Courtesy F. N. Meyer, U. S. D. A.) Fig. 44A. *Bottom.* Mr. Li and some roadside persimmon trees near the Ming tombs. You can see here, by the difference in color, that they have been grafted.

FIG. 45. Fayette Etter's 7-year-old Sheng Chinese persimmon tree on American roots. Fruits nearly 3 inches in diameter. The author of this book has fruited 29 varieties of this species.

FIG. 46. Cross roads in North Carolina pasture. At the right is a poor mulberry tree, one of many connected by paths worn in July as the pigs made their mulberry-gathering rounds. Other paths develop in September as the pigs pass from persimmon tree to persimmon tree in the same enclosure. Pigs in background gathering persimmons. (Photo J. Russell Smith.)

Fig. 47. *Top.* South Central France. Nearly all trees in landscape, grafted chestnuts. A part of farm system—corn substitute. Fig. 48. *Center.* The Apennines near Florence, Italy. Terraced wheat fields, foreground. Grafted chestnuts, background. Value of terrace and orchard same as Illinois cornland. Fig. 49. *Bottom.* Corsican chestnut monarch. The man stands by its understudy and successor. (Photos J. Russell Smith.)

FIG. 50. *Top.* Systematic planting of chestnuts on level land, Southern France, to grow forage—regular farm system. (Photo J. Russell Smith.) FIG. 51. *Center.* Chestnut harvest from grafted trees (suckers) in Pennsylvania stump land, 1908. Before destruction by blight. (Courtesy J. G. Reist.) FIG. 52. *Bottom.* Chestnut suckers grafted to Paragon variety fifteen months before photographing. An important technique for tree crops. Perfectly applicable to the Chinese. (Photo J. G. Reist.)

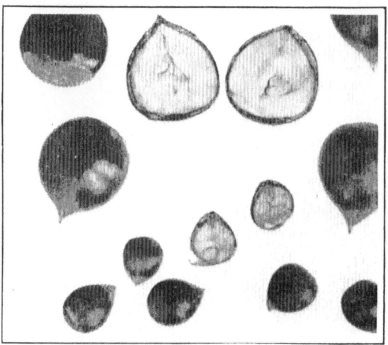

Fig. 53. *Top*. Below, American chinkapin (Castanea Pumila), natural size. Above, good-tasting hybrid, chinkapin x large tasteless Japanese. Fig. 54. *Bottom*. Fourth generation of selection in Japanese chestnuts. Tree twenty-three months from seed. (Courtesy U. S. D. A.)

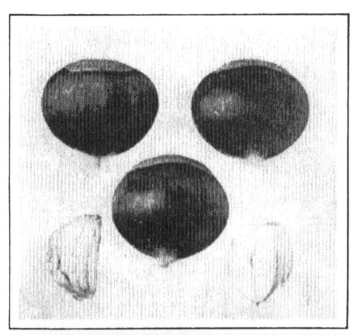

Fig. 55. *Top*. Chinese Chestnuts. Life size, grown in Massachusetts, 1949. Fig. 56. *Bottom*. Chinese Chestnut loaded with burrs in nursery row. The boy holding the sheet is 5 feet tall. (Photos J. Russell Smith.)

FIG. 57. *Top.* The Portuguese oak tree (ilex) that bore 1200 liters of acorns, close rival to an acre of American corn. FIG. 58. *Center.* Grafted oak trees standing in grainfields in Spanish island of Majorca. FIG. 59. *Bottom.* The Portuguese swineherd leading his wards from the earth-covered houses to the cork forest in the background where they harvest their own living. (Photos J. Russell Smith.)

FIG. 60. *Top.* Near Seville, Spain. At right, goat pasture scrubby with oaks and other brush. Trees allowed to grow up at left and yield acorns in the wheat field. (Photo J. Russell Smith.) FIG. 61. *Bottom.* Wheat field on naturally forested Andean slope. Venezuela. Altitude about 7500 feet. The gaping gully in wheat field at right was made after the planting of the crop, which is not yet harvested. Note ruined land above and stream bed at bottom right. How long will Venezuela last? Typical of wide-spread mountain farming in Latin America. (Photo Marjorie Vogt.)

Fig. 62. Helge Ness' hybrid oak, about 40 years old. Hybrid vigor. (Photo Bryson and Flory, Texas A. & M. College.)

FIG. 63. About 28 feet of the largest cork tree in the U. S., Napa, Cal. Height 75 feet, diameter 58 inches. Age 90 years. The much bifurcation increases the cork yield. The first 17 feet of this tree yielded 1,050 lbs. of good cork. (Photo Crown Cork and Seal Co., Balt. Md.)

FIG. 64. *Top*. Augusta, Ga. Mr. R. O. Lombard's plan of acorn-yielding oaks (for pig-feed) as successors to the fence post when the posts are gone. FIG. 65. *Center*. This hillside near Haifa in Palestine is on the way to becoming an orchard by the grafting of wild carobs. Anyone who has travelled the western U. S. or similar semiarid land in any of five other continents has seen wide stretches of such bush land. Are we smart enough to make the bushes become harvest producers? FIG. 65A. *Bottom*. Two-hundred-foot gullies (plow work) on naturally forested slopes of Sierra Nevada Mountains near Granada, Spain, a half mile from Fig. 66. (Photos J. Russell Smith.)

Beginning November 1, 1947, our chestnut breeding work has been sponsored by the Connecticut Agricultural Experiment Station at New Haven, Conn. It is being supported also by the Connecticut Geological and Natural History Survey and by the Division of Forest Pathology of the United States Department of Agriculture.

I wish to report that some grafted trees of varieties that I have grown begin to produce a few nuts the year after they are grafted, and keep it up. Not infrequently, they will bear the year after they are transplanted.

My friends Dr. Gravatt and Dr. Graves are professionals and experts. I cannot claim to be a botanist, a forester, or a plant breeder. I am only a professor of geography, and almost any businessman will tell you that professors are very impractical creatures. Nevertheless, I must (with apologies to the experts) call attention to what seems to me to be another and possibly a fruitful, possibly a quicker line of attack.

If Dr. Graves and Dr. Gravatt had been hybridizing to pro-duce nut-bearing trees, the process would have been simple: Get one (only one) good blight-resistant hybrid and graft a mil-lion of them on pure *Castanea mollissima*—and we are off to the harvest. But to get a hybrid strain that will reproduce itself true from seed is a very different, very much slower, and more difficult matter. Therefore, I call attention to the following:

1. All the *C. mollissima* in this country is grown from orchard or nut-bearing strains, the result of nobody knows how many generations of selection for this quality. Squat tops might be an advantage, not an objection.

2. Lack of evidence (is there any?) that the more upright *C. mollissima* are more squatty in habit than the American chestnut or the white oak, which make those beautiful hemi-spheres of top when allowed to grow in the open.

3. There is great variation in the form of *C. mollissima* seed-lings as we grow them, also in their speed of growth. I have one that grows twice as fast as any near to it, and some make poles very naturally if crowded as in a forest.

4. There must surely be wild *mollissima* on the mountains somewhere in China, but they have not received much attention from Americans who report in scientific circles. Presumably, they have forest form and produce the small nuts from which the orchard forms originated, as has been the case with so many other crop trees.

I am told that the United States Government has not had an agricultural explorer in China since 1929. When one compares that neglect with the amount of attention we understand the Russians are giving, it makes us appear disadvantageously, to say the least. Think of the ineptitude! Hundreds of American planes soaring over the Chinese mountains for several years past, and not one plant explorer dropped off to bring us out a pocketful of precious good timber-tree (*C. mollissima*) seed! And think of the dozens of competent men who would be delighted to go! How blind is bureaucracy! Certainly its left hand knoweth not what its right hand does!

5. The Zimmerman Chinese chestnut, of which I have distributed a good many hundreds through a little retail nursery, has proved, thus far, to be completely blight resistant. I have seen dozens of *grafted* Zimmerman trees die in a place where the ground was too wet and the blight attacked the stock but never the Zimmerman *wood*. The Zimmerman tree is surprisingly erect.

6. Here is a plan: Plant a thousand Zimmerman chestnut seed twelve inches apart, in rows eight feet apart; give liberal applications of good, complete fertilizer such as Vigoro, or Davison Company's Complete (there may be others, but I know these two) ; sow soybeans August 1, and repeat for three years. At the end of the fourth year there should be trees from eight to twelve feet high, if the planting is done on good soil. There will be great variation in shape, but take cions from the five most promising as timber trees. Graft these cions into the tops of well-established seedlings, to fruit them as soon as possible. Plant the resulting seed, and repeat the process for two or three cycles, and I think the result will be a blight-resistant,

fast-growing tree as erect as the American chestnut, which, like the white oak, often made an umbrella of itself when it got a chance.

You might grow a good timber tree out of your first thousand Zimmerman seedlings, which would require an eighth of an acre!

THE CHESTNUT WORM

The bane of the chestnut industry in the past has been the weevil. Nearly everyone who has eaten a chestnut has met a chestnut worm under conditions of extreme and unpleasant intimacy. Neither party was pleased, but despite the fact, let's look at the life history of this insect.

The worm, which comes out of the chestnut in the autumn, goes into the ground, pupates, and comes out the next summer as a very tough, armor-coated weevil. It has a very long proboscis with a little pair of scissors on the end. It flies enough to get up on the chestnut burr, then it reaches down between the spines, cuts a hole in the burr with the scissors, and deposits an egg in the hole. This egg hatches shortly before the nut ripens, remains quiescent until the nut is about ripe, then eats and grows rapidly, comes out, goes into the ground, and the cycle is complete.

In the past, the only cures have been as follows: First, pick up the nuts every day before the worms come out; that gets most of them. Secondly, plant the trees in the chicken yard, where the chickens will do a fairly good job of collecting the worms or the weevils, or both. Why cannot the ornithologists turn constructive and work out some technique whereby we learn to grow insect-catching birds, which would do for bothersome insects what the cat does for the mice?

But, bravo! We have discovered that three or four applications of DDT spray make a fairly clean sweep of the weevil, about 90 percent, apparently. This really opens a new era in chestnuts, for it has been known for several years that the real enemy is the weevil, not the blight.

TWO PIPE DREAMS

This paves the way for two pipe dreams, which seem to me not so very "pipey." One, the farm with a field of forage chestnuts of specially bred Japanese and Chinese varieties that ripen from July until late October. Here the pigs can gather the crop and fatten themselves, and produce pork of high quality (if the French example amounts to anything). A recently discovered fact that the turkey does well on pasture raises the suggestion, not proved, that the turkeys might forage a chestnut orchard during weevil and harvest season, catch the adult beetles and also the escaping worms, and perhaps pick up the crumbs of nuts that the pigs dropped. (Pigs are not tidy eaters.)

The other pipe dream is of a turkey farm with a field of chinkapins. These nuts are small enough for the turkey to eat, and this probably would dispose automatically of the worms, if the beetle had not been caught a few weeks earlier. Then too, the habit of a flock of turkeys of spreading out four or five feet apart and moving across the field in far-reaching phalanxes in search of grasshoppers and other food, suggests that turkeys might be very effective beetle destroyers, indeed might turn them into marketable turkey.

If the turkeys will be kind enough to do this, the saving of DDT cost and spraying would be a handsome return on the orchard investment.

Will the United States Department of Agriculture please breed these trees for us? Or will some State station, or some botanic garden, or some arboretum? Or will it be necessary, after all the appropriations and the institutions, for individuals to do it? One trouble with us as individuals is that we die so soon, and so often. The institution lives on.

When this work is done, we shall be ready to clothe hundreds of thousands and even millions of acres of our hill lands with a crop like that of Corsica. We can make an Appalachian or an Ozarkian Illinois which will yield corn substitute for centuries, preserve the lands, and support generation after generation of people in greater physical and intellectual comfort than is the lot of the present generation of hill destroyers. There is good reason to believe that the present most productive strains of Chinese chestnut may, in a term of years, out-yield corn. I'm not talking about the future. I'm talking about now.

You say this is too slow for the average American. There is the rub! But perhaps there is a way around this difficulty if we can focus a small fraction of our brains and money on it. One moderate-sized philanthropy can produce the new varieties and the farm technique. Then we need some philanthropy or some State or some promotor to make a lot of real demonstrations that will teach our boneheaded brethren to save themselves and their country.

CHAPTER XIII

A Corn Tree—The Oak
as a Forage Crop

THE SURPRISING AND NEGLECTED AGRICULTURAL
POSSIBILITIES OF THE OAKS

The oak tree should sue poets for damages. Poets have used the oak tree as the symbol for slowness—sturdy and strong, yes, but so slow, so slow! The reiterations of poetry may be responsible for the fact that most people think of this tree as impossibly slow when one suggests it as the basis of an agricultural crop. On the contrary the facts about the oak are quite otherwise. I am sure no poet ever grew a grove of the faster growing varieties, for he would have put speed into his oak poetry.

The genus of oak trees holds possibility, one might almost say promise, of being one of the greatest of all food and forage producers in the lands of frost. Why has it not already become a great crop? That is one of the puzzles of history, in view of its remarkable qualities.

1. Some oaks are precocious in bearing nuts (acorns).

2. Some grow swiftly.

3. Some are very productive.

4. Some acorns are good to eat in the natural state, and most can be made good to eat by removing the tannin which makes some acorns bitter to the taste. However, tannin is a useful, commercial product easily removed and in steady demand.

5. The acorn has been used as a standard food for ages.

6. The food value of the acorn (pp. 180, 189) shows that it stands well in the class of nutrients. Historically, it has been a food for ages of the squirrel, opossum, raccoon, and bear. Among the four-footed brethren the hog, above all, might almost be called an acorn animal. For untold ages he has lived in the forests, from Korea to Spain. In the autumns he has larded himself up with a layer of fat to carry him through the winter. A food for man, the acorn has certainly been used for unknown centuries in many lands. It is still being used as food by man, his beasts, and the wildlings.

As the pioneer farmers of Pennsylvania pushed aside the flowing stream of oil from their springs so that animals might drink water, so the modern world has pushed aside this good food plant, the oak tree.

There is a strip of hills from New England to Minnesota, from New England to Alabama, from Alabama to Ohio, from Ohio to Missouri, and from Missouri down to Texas. On these hills men have been making their living by growing wheat, corn, oats, clover, and grass. Yet I am confident that in every county there are oak trees of such productivity that if made into orchards they would in any decade yield more food for beast and possibly for man than has been obtained on the average in any country in any similar period on the hill farms of this wide region.

THE PRODUCTIVITY OF THE OAK TREE

The oak tree is productive. Down in Algarve, the southernmost province of Portugal, I heard of an ilex that bore 1,200 liters (35¼ liters equal 1 bushel) of acorns. The definiteness of the report was pleasing to the ear of one in search of facts, and the size of the yield was surprising. I went to see the tree. It stood alone in an unfenced outer yard at the edge of the village of St. Bras. The branches had a reach of only fifty-one feet. By long and devious methods I cross-questioned the owner, a widow of sixty. She always told the same story, and the neighbors believed it. The woman said that the tree bore 1,200 liters on full years and 240 liters in the alternate years, and that the average yield was 720 liters per year. She said that she knew this because she had picked up the acorns and sold them in the village. The usual price was 260-300 reis per 20 liters. This made 15,600 milreis—18,000 milreis or $16.84 to $19.44 American gold for the one tree on the big crop year. Data 1913. This income was based on pork, which brought nearly double the American price because of Portuguese tariff.

Selling acorns is a common but not an extensive practice in many parts of Spain and Portugal. The swine usually do the harvesting. It is interesting to compare the yield of the ilex oak in Algarve with a statement in the *Journal of Heredity* for September 1915, to the effect that a certain valley oak tree in California bears a ton of acorns as a full crop. (The average yield of corn per acre in the United States, 1943-1945, was 1,830 lbs.; in North Carolina, 1,288 lbs.; in South Carolina, 907 lbs., and in Georgia, 700 lbs.)

The Portuguese manager of an English-owned estate who had spent his life raising and managing forests (cork oak mixed with ilex, or holly oak), whose chief products were cork and the pork of acorn-fed pigs, assured me that there were ilex oaks that bore 100 kilograms (220 lbs.) of acorns, 200 kilograms (440 lbs.), 300 kilograms (660 lbs.), and 400 kilograms (880 lbs.), but not 500 kilograms. He further stated that a *tract* of ilex oak was fairly constant in its yield, although the individual trees usually bore on alternate years.

An oak tree reported, according to family history, to be 100 years old stood alongside of a road in the village of Pedicroce, Corsica. Because it stood along the edge of a street, its acorns were picked up, and the owner stoutly maintained that it produced in good years 350 kgms. (770 lbs.); the next year 20 to 50 kgms.

The tree had a girth of 12 feet and a spread of 51 feet.

A similar report of heavy alternate yield for *trees* with the forest yielding annually was made by Forest Supervisor Louis Knowles, United States Forest Service, for bur oak, *Quercus macrocarpa,* in Sundance National Forest, Wyoming.

Mr. John L. Wilson, an Englishman with an engineering education, the managing director of the company owning the above-mentioned Portuguese cork estates, took the trouble to measure the crop of one particular cork tree which had a spread of 45 feet and which yielded 840 liters of acorns. This amount is enough to furnish a pig that weighs 100 pounds his standard Portuguese ration of 7 liters per day for 120 days. This, by Portuguese expectations, should increase the weight of the pig to 230 pounds, thereby getting 130 pounds of live pork by way of one pig from an exceptional cork tree for one season. The tree would be entitled to loaf the next year and would doubtless claim its rights.

Compare that permanent producer with the average corn yield of the gullying fields of our South. Is it any wonder that many Portuguese landowners live in town and loaf their lives away in the cafés?

The late Freeman Thorpe of Hubert, Minnesota, after some years of experimentation and actually measuring the acorns from test trees, was confident that the Minnesota black oak would average 100 bushels of acorns per year on sandy land of low fertility—land that would make not more than 30 bushels of corn. He also thought that he could harvest the nuts as cheaply as he could harvest corn. Perhaps Colonel Thorpe was overenthusiastic. He was a man with a flame in him, and he loved his trees. However, we can cut his production estimate in two and double the cost of harvesting and still have a sound business proposition and an astonishing production for *chance seedling trees.* That is virtually all the labor there is to producing the acorn crop.

I was told by landowners in Majorca that a woman could pick up 100 liters of wild acorns in the woods and 200 liters under the grafted trees.

The late R. O. Lombard, of Augusta, Georgia, had worked out a series of tree crops for an almost automatic production of pork. He experimented with oaks, deliberately planting them for acorn production. As a result of years of experimentation he was sure that on the sandy soils of the coast plain of Georgia the water oak was much more productive per acre of hog meat than was corn.

Through the much appreciated kindness of Mr. Raphael Zon and other members of the staff of the Forest Service, forest rangers gathered acorn-production data, published at considerable length in the first edition of this book, which may be seen in many libraries. I select one or two for reproduction here to show the trend of many.

Q. garryana

Trinity Forest. One tree noted produced five hundred pounds of acorns, and another six hundred pounds. A third, rather small, produced two bushels. Crops occur at intervals of from three to four years. Fruit is frequently killed by frosts.

Q. lobata

Sierra Forest. Maximum seed production about one hundred and seventy-five pounds per tree. Good crop one year out of three, and partial crop two years out of three.

Q. wislizeni

California Forest. Observations on twelve trees, average D. B. H. 26″, indicates that the maximum production is about two bushels per tree and that there is a crop every two years.

[Signed] C. S. Smith, Acting District Forester.

Apache, Arizona, October 21, 1913.

The emory oak bears nearly every year; in fact there have been some acorns every year since I have been in this part of the country. Last year (1912) I saw trees that I would judge bore at least four bushels of acorns.

[Signed] Murray Averett, Forest Ranger.

Bonita Canyon, Arizona, October 29, 1913.
The quantity produced by one certain tree*—the one standing in my yard, and under which my office tent is located—bears fruit nearly every year; a maximum crop it was considered to have had last year, 1912, would have amounted to at least one hundred pounds. This year it produced not more than fifty pounds on account of the few very cold nights, that froze many buds. The fruit is very sweet and relished by nearly all who have tasted it.

[Signed] Neil Erickson, Forest Ranger.

These reported yields of two bushels per tree, of two hundred pounds per tree, of three hundred pounds per tree, of four hundred pounds per tree, even of six hundred pounds of acorns per tree, seem to give a sense of verity to the claims of Messrs. Thorpe and Lombard, quoted above, and a sense of reasonableness to my own claim that an oak orchard of *selected* trees would be more productive than the existing agriculture of the American hills.

Even more astonishing is the report of a Michigan man traveling in Colombia, South America, telling of solid forests of oak in the mountains of that country, yielding acorns four times as large as those in Michigan (forty of them weighed a pound) and lying so thick upon the ground that he thought he could pick fifty bushels in a day.

ABILITY TO YIELD HEAVILY, AND ON POOR SOIL

The oak tree is productive in the quantity of fruit produced, and some species seem to display a well-nigh marvelous ability to produce heavily while still very small in size. Sometimes this fruit weighs more than the tree. Some dwarf forms bend over to the earth with the burden of their crop.

Some of the oaks can produce their burdens from soil of the greatest variety and great sterility. In this connection it should be remembered that the hopeless sand barrens of Long Island, New Jersey, North Carolina, Georgia, and of the Appalachian

* Cions from that tree would apparently start a commercial orchard.

summits are shunned by the farmer as though they were the Sahara Desert, yet these sand barrens are often productive of many acorns—good carbohydrate nutriment.

These are strong statements, but here are some of the reasons why I make them:

> U. S. Department of Agriculture,
> Bureau of Plant Industry,
> Washington, D. C., August 6, 1914.

A letter received from Mr. W. O. Wolcott, Bucaramanga, Colombia, says, "By this mail I am mailing you one hundred acorns of a species of oak that grows in this state of Colombia. I traveled for two days on mule, coming from Ocana through these oaks all the way. You will notice these acorns are about four times the size of our acorns in the United States, and as the trees are wonderful bearers these nuts should be of use as a stock and hog feed. The trees grow in the mountains at about four thousand to seven thousand feet altitude, and so close together that they grow to thirty to sixty feet high and not over four to six inches in diameter. Where the trees are in the open some of them are from ten to twenty inches in diameter. In many places the trees were not over two feet apart, like the cedar swamps in Michigan when I was a boy. The natives fatten their hogs on the acorns. I could have gathered fifty bushels in a day, I should judge. . . ."

> Very sincerely yours,
> [Signed] David Fairchild,
> Agricultural Explorer in Charge.

In a letter addressed direct to me Mr. Wolcott said:

A man told me that, wherever there is a forest of these, the natives never clear off the forest, as the land is so poor it won't grow anything; and now I remember that all through that section what few ranches I saw were awful miserable shanty ranches, and yet all of them had fat hogs and goats. I believe goats can be fattened on these acorns the same as hogs.

They tell me the acorns begin to fall in February. I gathered those sent on the second and third of June. Then there were piles of them in the very trail itself. They tell me they bear only once a year but every year.

Look at Fig. 61 in this book to see what happens when Andean forests are cleared and the land is put to wheat.

Office of Experiment Stations

Honolulu, Hawaii, July 18, 1916.

. . . a small oak tree growing near Manhattan, Kansas, which is very peculiar, its small size being ordinarily less than two feet in height. I have taken specimens of an entire tree, shrub, or bush, which had acorns on it weighing in the aggregate more than the tree from which they came.

John M. Westgate, Agronomist in Charge.

Quercus primus pumila, chinquapin or dwarf chestnut oaks, one of the smallest genus of two—four feet high. The acorns are of middle size and very sweet. Nature seems to have sought to compensate for the diminutive size of this shrub by the abundance of its fruit; the stem, which is sometimes no bigger than a quill, is stretched at full length upon the ground by the weight of its thickly clustering acorns. This shrub grows most abundantly in the northern and middle States of America and is usually found in particular districts of very poor soil where alone or mingled with the bear oak it sometimes covers tracts of more than one hundred acres in extent. (Loudoun, *Arboretum et Fruticetum,* Vol. III, p. 1876.)

Britton, *North American Trees,* p. 327, emphasizes its great fruitfulness and wide distribution, Maine to Minnesota, North Carolina, Alabama, and Texas.

Q. prinoides, chinquapin oak. The acorns are produced in the greatest profusion, covering the branches in favorable seasons with abundant crops one-half to three-fourths of an inch in length and from one-third to nearly one-half an inch in breadth, with a sweet seed. Massachusetts to North Carolina, southeast Nebraska to Texas, rocky slopes and hillsides. (Sargent, *Silva,* Vol. VIII, 59.)

I would like to have the opportunity to demonstrate whether the California scrub oak, *Quercus dumosa,* Nuttall, could be made to yield a return such as you have described for *Quercus ilex.* This little oak grows in a country that has no further value. It produces acorns freely and with great variation.* (Letter, William E. Lawrence, Oregon Agricultural College, Corvallis, September 17, 1916.)

Q. catesbaei. The barren scrub oak. Carolina and Georgia. It grows on soils too meager to sustain any other vegetation, where the light movable sand is entirely destitute of vegetable mold. Old trees only productive and only a few handfuls. (Loudoun, Vol. III, p. 1889.)

It can be seen growing on the sand dunes in pure sand, exposed to the sea winds, on the marshes in strong marsh clay, and on the upland; it seems to thrive equally well on gravel, chalk, or sand; good soil or bad makes little apparent difference to its well-being.

From its earliest stages it is not difficult to raise, but is rather slow of growth. Most years there is an abundant crop of acorns. [When was it that Anglo-Saxon man became economically blind?]

THE CORK AND ACORN AS A COMMERCIAL CROP

It is not unnatural that the oak tree should render its greatest service to commercial agriculture in the modern world through the cork oak, the species which has the greatest number and variety of products.

The oak tree is the source of one of the chief exports and two of the important industries of Portugal. If I wanted to secure permanent and comfortable financial ease from agricultural land, few more secure bases need be sought than the undisturbed possession of a large tract of Portuguese land having a good stand of cork-oak trees, *Quercus suber,* and evergreen-oak trees, *Quercus ilex.* If the stand of trees were good, it would make little difference even if the land happened to be rough untillable hillsides. Such land would still yield its crops of cork and pork (the pork made of acorns). The virtues of the Portu-

---

* Plant breeders! Please take note—two prime qualities for you—*"produces freely* and with great variation." What an interesting job to make a real crop out of that combination!

guese cork forests are quadruple, and the forests are almost perpetual if given a little intelligent care.

With reasonable care, which consists of occasional cutting out of an undergrowth of bushes inedible to the goat, and occasional thinnings, the forest will live and reproduce itself for centuries and yield four kinds of income. The trees yield best of cork and acorns when the stand is not thick enough for the trees to crowd each other. Thinning to attain this end permits some forage for sheep and goats to grow beneath the trees. It is the common expectation that the pasturage income in a proper oak forest will pay for the labor of grubbing bushes, which, aside from fire protection, is the only maintenance charge. That leaves the other three sources of income—wood, cork, and pork—to offset interest and taxes and to make profit. The landlord's task is easy, as the pasture is rented by the tract, the cork is sold to contractors by the ton, the wood by the cubic meter, and the hog pasture on some lump-sum arrangement. Absentee ownership could scarcely be more providentially arranged. If you wish to see the Iberian owners of cork estates, go to Lisbon, Madrid, or Seville. Your chance of finding them on their estates is small indeed.

Every nine or ten, or eleven or twelve years, according to locality, the cork bark is stripped. An acre with a full stand of young trees should have seventy trees yielding fifteen kilograms per tree, or 2,300 pounds per acre per stripping.

Consul General Lowrie at Lisbon reported (1913) the annual average cork production of the 900,000 acres of Portuguese cork forests was 240 kilos per hectare, 214 pounds per acre per year.

The average annual yield of cork per acre in Portugal is put at 275 pounds per acre by the *Quarterly Journal of Forestry*, Vol. VII, No. 1, p. 57.

Mr. John L. Wilson, an Englishman educated as an engineer and for many years managing director of Bucknall Brothers, Lisbon (English owners of cork estates, cork merchants and manufacturers), reports one particular cork oak, the trunk twenty-four feet in circumference, seventy-two feet in spread, which yielded at one cutting 2,112 pounds of cork; twelve years previously 2,310 pounds of cork, giving this tree an income value at that time (1913) of about one pound sterling per year from cork alone. The current quotations on cork will give values at any particular time—$30-$600 per ton, depending on quality, 1931-1948.

As the trees grow older and are thinned out to prevent crowding, the increased yield per tree keeps the cork output up to the average which, with care, can be maintained indefinitely at over a ton per acre per stripping. When the trees come down, they make the precious charcoal for the domestic fires beside which the Portuguese nation shivers in winters while it cooks its simple meals. In a properly cared-for forest, every old cork tree has beneath its branches several half-starved understudies all leading a submerged life until it comes their turn to have space in which to spread.

I have seen a large cork tree, fifteen feet in girth and with a reach of fifty-six feet, that had yielded 1,980 pounds of cork at a stripping. An acre will easily hold eleven such trees, and I am told by competent authority (Mr. John L. Wilson, above mentioned) that there are many such acres in Portugal. It is an interesting and peculiar fact that the poorer the land the better the cork. This superior quality results from a finer texture due to slower growth.

So pronounced is this influence of slow growth, that cultivation of cork oak trees often injures the quality of cork. Therefore, some fine cork forests are on sandy or stony land, which in any other country would be called "barrens" and would find its highest productivity by growing a pine forest. In some districts the trees are made the sole basis of estimating the value of the land on which they stand. Near Evora, in south-central Portugal, the method of calculating the value of a cork forest is as follows: The twenty-year yield of cork (two strippings) is taken as the basis. The buyer pays (1913) 600 to 650 reis (1,000 reis = $1.08) per arroba (33 pounds) of cork capacity.

Thus a good acre of cork trees would bring $125 or more, and the man who sells throws in the land.

English owners of cork estates in Portugal estimate that acorns alone produce from a half to two-thirds of the total pork crop of that country.

The *Quarterly Journal of Forestry*, Vol. VII, No. 1, p. 37, estimated the annual acorn consumption of the Portuguese pig at 200,000 tons. This figure is interesting when compared with the one million tons of corn produced in the state of Virginia by cultivating 1.5 million acres to its serious injury by washing away of soil. Virginia corn also feeds cattle, cows, horses, poultry.

	Area Thousand Sq. Miles	Population Million	No. of Swine Million
Portugal............	35	8.1	1.1
Virginia...........	40	3.1	0.8

The Portuguese hog leads a lean and hungry life for a year and a half, or two years, and then comes an orgy during which he eats acorns all day and sleeps on them at night, as he lives in the open beneath the trees. In three months he doubles or triples his weight and passes on to the ceremonies at the abbattoir. Separate tracts are fenced off for the winter food of hogs that are to be kept until the next year.

Careful experiment in central Portugal has shown that 5.3 liters of acorns will make one pound of pork. Thirty-five and one-fourth liters make a bushel. Therefore, a bushel makes 6.6 pounds of pork. These figures were given me by Mr. John L. Wilson, mentioned above. An American finding is as follows: Mr. G. T. McNess, Superintendent, Sub-experiment station, Nacaᶜˣches, Texas, said (letter, February 11, 1914) to W. J. Spillmₐn, of U. S. Department of Agriculture, "One gallon of acorns is equal to ten good ears of corn."

As the acorn crop is not under much control of man, does not come regularly, and above all is not harvested and measured, it is difficult to make an exact approximation of pigs and acorns. This difficulty is met by the organization of fortnightly markets in the villages of the cork regions where pigs in any and all stages of fatness and leanness are bought and sold at any and all times. Two weeks before the pig is completely fatted, he may be sold by the man who has a shortage of acorns to the man who has a surplus. In estimating the number of pigs that he should buy, the Iberian oak tree farmer walks through his forest scanning the trees and noting the number of acorns lying on the ground. Every time he sees the fattening for one more pig, he puts another acorn into his pocket.

On the unfenced acres of a cork estate two kinds of herders are daily abroad: the swineherd, whose wards eat grass and acorns, and the shepherd, whose sheep and goats browse the bushes and grass and furnish wool and milk—the fifth and sixth products of the cork forest.

As I rode through one of these estates on an April day, I came upon a long lane full of sheep, seven hundred of them, crowded solidly between close fences. Two shepherds were busy at their regular afternoon job of milking the sheep. They started at one end of the long and motley mass and worked their way through it. One presided over the milk bucket, the other caught a ewe, backed her up to the bucket, and held her while the milker extracted a few spoonfuls of rich milk, after which the ewe was pushed aside to join the growing mass of the recently milked. That evening at the house of Joao Dias, estate foreman, I was offered rich cheese of sheep's milk, but my previously good appetite for it was diminished by my memory of the milking in the lane. (Sanitary inspection?)

The bark of the evergreen-oak, *Quercus ilex,* is of no value, but the tree yields more acorns than the cork oak. No Portuguese will cut down the one tree to make room for the other, so absolutely alike does he value the trees. Often the two species are mixed indiscriminately in the same tract, but in some localities the ilex forests are almost pure stand and are cared for exactly as are the cork forests. At Evora in south-central Portugal —a locality famed for its pork production—Estevao Oliveira Fernandes, a graduate of a German engineering school, told me that the local estimate of production for a ten-year period was as follows: For a cork forest, 34 kilos of pork per hectare (30 pounds per acre) and for an ilex forest, 68 kilos per hectare (60 pounds per acre). Compare these with pasture yields (p. 205) and then consider the low rainfall and dry summer of Portugal and its age-long exposure to a robber agriculture, and the oak tree stands forth as a crop plant most worthy of consideration.

TWO-STORY AGRICULTURE

In some sections of Spain and Portugal, the young ilex trees are allowed to grow where they have by chance sprung up in the fields. Around and under the trees the machineless cultivation of wheat and beans, barley and hay, goes on just the same. This combination of trees and crops gives a beautiful parklike landscape. The cultivation helps the oaks to make acorns, and after the grain and other crops are harvested, the hogs are turned in to gather the mast crop.

On the slopes of the Sierra Nevada in Spain, a few miles south of Granada, I saw the ilex rendering its supreme service—the oaks and their under-thicket holding the earth for man as no other crop could have done and at the same time giving him a living. In this locality the formation was unconsolidated clay and gravel outwash of such depth that bed rock was nowhere in sight. The slopes and height were such that the term "mountain" would be applied, perhaps even by the Swiss. On one of these slopes I examined an ilex orchard (or forest) that was giving a fair return as part of a farm. A part of this orchard was so steep as to be very difficult of ascent. When the tenant told me he let his hogs run there, I asked him how he kept them from falling out into the stream below.

"Oh," he said, "I don't let the big fat ones come here. I am afraid they would fall. I bring only the little ones into this part."

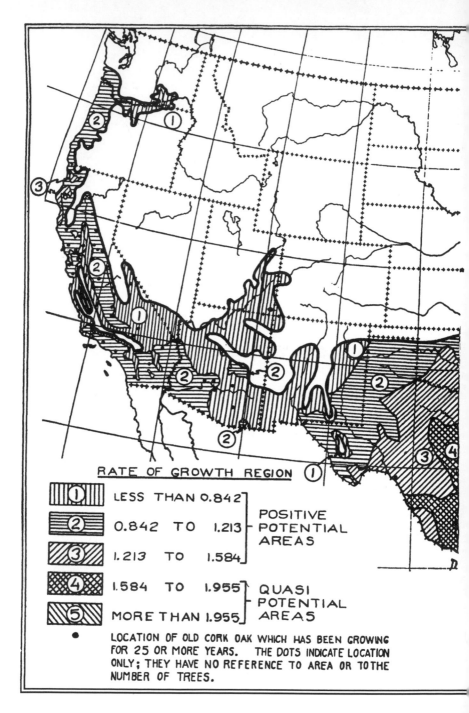

RATE OF GROWTH REGION ①

①	LESS THAN 0.842	POSITIVE POTENTIAL AREAS
②	0.842 TO 1.213	
③	1.213 TO 1.584	
④	1.584 TO 1.955	QUASI POTENTIAL AREAS
⑤	MORE THAN 1.955	

• LOCATION OF OLD CORK OAK WHICH HAS BEEN GROWING FOR 25 OR MORE YEARS. THE DOTS INDICATE LOCATION ONLY; THEY HAVE NO REFERENCE TO AREA OR TO THE NUMBER OF TREES.

AREAS IN THE UNITED STATES

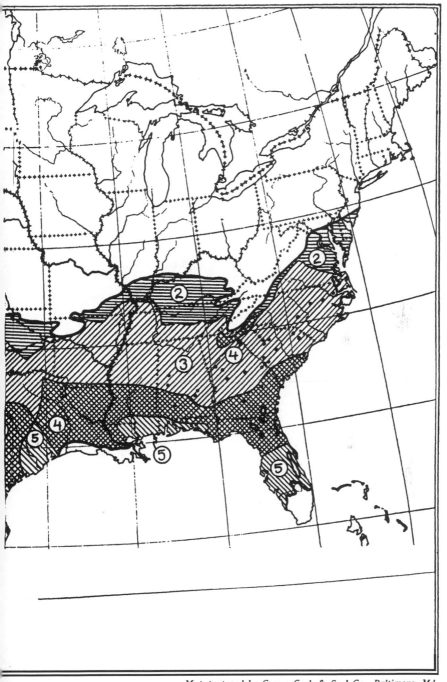

WHERE CORK MAY BE GROWN

A mile away, on a less steep part of the same slope, for centuries men had been supporting themselves by the agriculture of the plow, and nature had shown her resentment of this act of violence. In some places from half to three-fourths of the original land surface was gone, through the work of gullies that had become from fifty to two hundred feet deep. "After man the desert" was here demonstrated. Of the two slopes the ilex slope was much the more productive. It had been saved from the plow only by being so steep that it could tempt no plowman.

THE MIRACULOUS DISCOVERY OF CORK
IN THE UNITED STATES

After more than two thousand years of use in Europe, and four hundred forty-seven years after the discovery of America, cork trees were miraculously discovered growing in the United States. The late Charles E. MacManus (1881-1946), President of the Crown Cork and Seal Company, Incorporated, of Baltimore, Maryland, was vacationing in California and spied from his automobile something that looked like a cork tree. He jumped out, cut the bark with his pocket knife and discovered it was really cork. (See Fig. 63.)

Here is the miraculous part of the story. This tree that MacManus "discovered" was one of hundreds growing in California. One of them shortly yielded a thousand pounds of cork —the miracle is, they had been there for decades and nobody had done anything about them. They are a magnificent illustration of the impervious quality of the human mind to a new idea (relatively new, not really new).

About the middle of the last century the United States Government, through the Patent Office, I believe, imported some cork acorns and they seem to have been well scattered with fair success at growing. In 1854 the United States Commissioner of Patents published a "Report on the Seeds and Cuttings [of cork oak] Recently Introduced into the United States." The *Gardeners' Monthly* recommended its growth in 1876. The U. S.

Department of Agriculture reported on it in 1878. *Garden and Forest* published an article in 1890 about the California trees. The old Division of Forestry, U. S. Department of Agriculture, published nine pages about it in 1895. The *Southern Lumberman* recommended its growth in the South in 1904. *The American Forestry* had an article on American-grown cork in 1921. *The Pacific Rural Press* published an article in 1929 about "How California Might Grow Corks." The first edition of this book urged its feasibility in 1929, and there were others before MacManus made his great "discovery" and really started things happening. I suspect it is his influence that resulted in a publication, *Bibliography on Cork Oak*, compiled by Roberta C. Watrous and Helen V. Barnes, Assistants in the Division of Bibliography, Library of the U. S. Department of Agriculture in April 1946. It contains titles of approximately 570 references. Of these, 61 referred to American cork prior to the MacManus discovery, namely, from the beginning down to 1938, and 58 (at least 15 of them by Crown Cork and Seal Company employees or about Crown Cork and Seal Company projects) appeared concerning American cork between 1939 and 1946. Plainly MacManus increased the literary tempo.

As a matter of fact the history of cork trees in America is long and amazing in its apparently successful demonstration of rich possibilities, completely neglected.

The indefatigable Thomas Jefferson, perhaps the most talented of all the great men of America, was a diligent plant introducer, under the very bad conditions of sending plants from gentle climates to rougher ones and depending on sailing vessels for transportation. And there is no evidence that he had financial aid. In 1784 he saw cork-oak trees in France and realized the significance of what he saw. Along with many, many other European plants, he imported cork-oak acorns, but none of them grew. He might have planted one hundred of them in earth in a box twenty inches square and eight inches deep, nailed on the lid, bored a few holes by which to moisten it, brought it along home with him on the ship, and had a good chance of success. The real trouble was that acorns must not dry out, and many take root in a few days after they fall.

If Jefferson's acorns had grown, we *might* have had a cork industry by now. As it was, though disappointed by the "non-chalance" of his fellow citizens, he kept talking cork for forty years, but no one seems to have heard. For a quarter of a century after he passed away, his fellow Americans slept peacefully on with regard to cork.

About the middle of the century, the United States Government imported some acorns, as above mentioned. They have proved that the cork-oak is an amazing tree in its ability to survive a great variety of climates. For several decades the trees have been growing the whole length of California, also in Washington, Arizona, Texas, Alabama, Louisiana, Florida, Georgia, South Carolina, North Carolina, and Virginia. Some of these trees are of gigantic size.

In the late Eighties the Division of Forestry (now the Forest Service), U. S. Department of Agriculture, distributed some five thousand cork-oak seedlings in the southern tier of States. As stated above, they have been mentioned by frequent publications, and although we have had a Forest Service for fifty years and a galaxy of State foresters, and no end of publicly owned and privately owned forest land with proved cork climate, no one with constructive imagination and power saw the cork-oak trees until Mr. MacManus hopped out of his car.

MacManus did not have wooden eyes nor yet a wooden head, and when he saw that cork tree he promptly started in to *do something;* he started the MacManus Cork Project, and for the remaining seven years of his life he had a wonderful time.

The California cork trees produced respectable quantities of seed. In addition, MacManus imported tons from the Old World and began to disseminate and investigate cork. He distributed seeds and little trees wherever he thought they would grow and wherever people would promise to take care of them. And he had a good instinct for publicity. He put cork-oaks into Arbor Day exercises, with brass bands, school assemblies, and plenty of newspaper pictures of garden-club presidents and other promoters planting a cork-oak tree on Thomas Jefferson's lawn at Monticello, and pictures of governors planting cork-oak trees on their State Capitol grounds. He really knew how!

Mr. MacManus loaded a truck with cork acorns and publicity and ran it from the Pacific to the Atlantic, distributing seeds as it went, with plenty of newspaper reporting; and now there are probably hundreds of thousands, if not millions, of cork-oak trees started. In addition to those in the eleven States above mentioned, large plantings have been made in Maryland, New Jersey, Kentucky, Tennessee, Missouri, Oklahoma, Illinois, Indiana, Arkansas, Mississippi, and Oregon, with small experimental plantings in other States. In from thirty to forty years, good commercial crops may be expected.

The richness of the gold deposit that awaited exploiting by Mr. MacManus was revealed in the period 1940 to 1945, when five hundred American cork-oak trees were stripped and yielded 25,500 pounds of commercial cork. Four thousand mature cork-oak trees have been found in California alone.

From a tree at Chico, California, more cork was obtained in the six years of second growth than in 1940, when thirty-six years' growth of virgin cork was removed. The Chico tree (forty-two years old) had reproduction cork approximately one inch thick, while a younger tree at Davis (thirty-one years old), in a more favorable location, had produced one and one-half inches of cork in six years. (See December 1946 issue of *The Crown*, Crown Cork and Seal Company, Inc., Baltimore, Maryland; also the *Scientific Monthly*, May 1944 and February 1947.) Persons really interested should approach the above-mentioned company.

As this manuscript goes to the printer, I receive a book, *Some Geographic and Economic Aspects of the Cork Oak*, 116 pages, by Victor A. Ryan, Director of Research for the Crown Cork and Seal Company, Inc.

This carefully written and well-illustrated book was written to guide and locate our future cork industry. The established cork area in the United States, according to Ryan, is shown on pages 170 and 171 of this book. It is of vast extent, and it may increase by plant breeding.

To cap the climax, the research department established by Mr. MacManus has learned how *to grow rooted cuttings of cork-oak*. Cork-oak trees vary, of course, as to quality, quantity, and speed of cork and tree growth, but we are now in a position to propagate easily from the best cork-oak trees, and this greatly enhances possibilities. By breeding, it should be possible for the hybridizer to get good cork, fast-growing cork, on a tree that is also an extremely heavy acorn producer. Now, there's a piece of property for you! Cork, pork, sheep pasture, milk, wool, and wood, and good for a hundred years after it has become established.

This is a perfect example of the making of an industry. It awaits duplication by means of an indefinite number of varieties of trees awaiting the improvement and development now possible.

Hail to Mr. MacManus! May his tribe increase! Of course he could easily have wasted that much money on one steam yacht, one palatial showplace, or one playboy grandson. Why are most millionaires so unimaginative, so good at piling it up, yet so inefficient in using the pile? and, apparently, so bent on leaving playboys rather than men to follow them!

THE AMERICAN ACORN CROP

In America, our frontier farmer ancestors fed themselves in part for generations on mast- (chiefly acorn-) fattened pork. The late Dr. John Harvey Kellogg put it thus:

Battle Creek, Michigan, April 6, 1927.

My father was a pioneer in this country nearly one hundred years ago, and I understood from him and other pioneers that acorns constituted a considerable part of the sustenance of the hogs who found their living exclusively in the forest. These hogs were nearly as wild as wild bulls.

The raccoon, opossum, and bear also subsisted mainly on the acorn in the season of autumn fattening. Well fattened on acorns, the bear had nothing to do, so he hibernated till spring, thereby lengthening the season of acorn service. Yet no agriculturist seems to have applied constructive imagination to the very suggestive facts which have been thrusting themselves upon our attention over hundreds of thousands of square miles for several generations. Some agriculturists, however, are beginning to appreciate the possibilities of this neglected field. For example, Professor H. Ness, Chief Horticulturist, Texas Experiment Station, College Station, Texas, started hybridizing the oaks, and he thought that experiments with feeding mast might "give results which would be of the greatest value to the farmers throughout our southern region." In the explanation of this belief he said in letters of May 28, 1913, and April 18, 1923:

I wish to say that throughout the forest region of Texas which constitutes the eastern part of the State, the acorns furnish a very important part of the hog feed. As to the classes of this feed, there are two: namely, (1) from the white oaks, that is, the oaks that mature this fruit in one year and furnish what people call "sweet mast," which is considered equal to the best of our cultivated grains as hog food. Among the oaks of this kind may be mentioned the post oak, *Quercus minor,* because of its number and great fertility, and the white oak proper, *Quercus alba,* because of the excellent quality and size of the acorns, as well as their abundance. (2) The second class is furnished by the black oaks, *Quercus trilobata, rubra,* and *marylandica,* or those that mature the acorns in the second year from flowering. This is called the "bitter mast." It is very abundant, but is considered inferior because it gives inferior meat and lard of a dark color.

These five trees, when full grown, are all heavy yielders of acorns. Those (the white oaks) that produce what is called the "sweet mast" are especially abundant yielders in very nutritive food for hogs. Where the trees are properly thinned so as to develop freely and hence bear freely, an acre of land properly set with either the white oak or the post oak is almost equal to an acre of corn. The trouble in these forests is that the trees are prevented from producing fruit by being densely crowded. Many of them, therefore, develop no fruit at all, but act as a hindrance to others that are larger and would develop more fruit.

There is no attempt made to manage the oak forest for this purpose, although management to that end would be very effective in increasing the amount of hog feed, and would consist simply of thinning the trees to a proper stand for bearing. This could be done without decreasing the amount of wood produced, because each tree, if given a larger space, would not only produce a larger crop of fruit, but would grow to a larger size.

In the eastern part of Texas, or throughout that large forest region east of the Trinity River, I wish to say that the possibility by proper forest management of obtaining large quantities of feed for hogs is very great, and unappreciated by the people. It is merely incidentally taken advantage of, a good deal of the forest being unfenced and proper thinning and care of the trees utterly unknown.

The oaks of that region can easily furnish in the fall and throughout most of the winter the major part of the large amount of food necessary for raising and fattening hogs.

Proper thinning and judicious selection of good bearing oak trees would be a measure of high economic importance. The various individuals of the same species vary very much both in the amount of production and in size of the fruits. In many cases heavy-bearing trees can be selected, and the others of lesser value for the purpose might be cut out.

It is not only hogs that thrive and fatten on the "mast" of the forest, but also goats. During the early part of the season they feed on the underbrush and sprouts from the stumps of trees, and when fall comes they fatten readily on the acorns and other fruits. My personal experience is that in east Texas, hogs can be raised cheaper than anywhere else, provided advantage is taken of the forest. It also happens that much of the land covered with forest is of such nature that it cannot be readily put into cultivation, owing to the unevenness of the ground in some cases, to poor drainage in others.

Our Texas colleague here calls for what the European chestnut orchardists regard as a necessity—letting the light strike every branch of the tree.

John C. Whitten, long a horticulturist of Missouri, in discussing Missouri acorn-pork production concluded as follows:

I am convinced of this, however, that a considerable portion of the hilly, not easily tillable, regions of this State may more profitably be left in woods to produce hog feed than cleared off for other purposes.

And all of this on the basis of wild seedling trees unimproved by selection, by breeding, or by grafting! What possibilities are yet in store if and when we apply known and proved methods of ordinary horticulture?

Our forests are steadily being reduced in area and thinned out by the removal of good timber trees, but there still continues an amount of mast utilization that will probably be a surprise to many. It is still common in Texas and Minnesota, in Appalachia, Utah, and California, and it would doubtless be greater than it is but for the fact that the flesh of the acorn-fed hog is soft.

AMERICAN PRACTICE WITH THE ACORN

The first edition of this book contained many nuggets of acorn fact, of which I reproduce a part:

Jena, Louisiana, April 14, 1923.

The locust, walnut, and hickory are found almost altogether in the lower lands that are subject to overflow. These trees there together with the oaks supply the almost entire feed for the hogs for the people who live in or near this section of the country.

L. O. Summall, County Agent.

Kentucky Agricultural Experiment Station,
Lexington, Kentucky, September 8, 1916.

Quite a large number of hogs are annually fattened on acorns in this State, especially in the mountain regions. A number of these hogs find their way into the Blue Grass region where they are finished on corn, mainly for the purpose of hardening the fat. Lard of the acorn-fed hog will not, of course, congeal.

E. S. Good, Head of Department.

Covering the greater part of southern Missouri, Arkansas, and the eastern third of Oklahoma are low-lying hills, better known as the Ozark Mountains. This region is naturally a timbered country where

black and shellbark hickory, blackjack, red, post, and white oak, and
walnut grow abundantly. These trees never fail to produce part of
a crop of "mast" and usually are heavily loaded with nuts, especially
in the higher parts not so readily affected by the late frosts, which
in the low regions kill the blossoms of the young fruit. From a rough
estimate the bulk of the whole crop will run about ninety percent
acorns of the different varieties.

The general practice in the case of the hogs is to train them to
come to the barnyard once or twice a week for corn to keep them
tame and to enable the rightful owner to mark his pigs.

	Acorns Percent	Cornmeal Percent	Jap Clover (Hay) Percent	Red Clover (Hay) Percent	Cowpeas (Shelled) Percent
Water.........	55.3	15.0	11.0	15.3	14.6
Protein........	2.5	9.2	13.8	12.3	20.5
Carbohydrate..	34.8	68.7	39.0	38.1	56.3
Fat...........	1.9	3.8	3.7	3.3	1.5
Ash...........	1.0	1.4	8.5	6.2	3.2

(*The Country Gentleman*, December 13, 1913, p. 1821, "Ozark Nut-
fed Pork," by J. C. Holmes.)

This table of food values submitted by Mr. Holmes shows the
vital necessity of some protein food to make a balanced ration.
This is furnished by wild lespedeza (Japanese clover).

Another form of acorn analysis is as follows, furnished by
S. S. Buckley, Associate Animal Husbandman acting in charge,
Swine Investigations, U. S. Department of Agriculture, and sent
to me by Governor L. Hardman, of Georgia.

The digestible nutrients may be seen from the following
analyses:

	Acorns, Kernel and Shell	Acorn Kernel
Total dry matter in 100 lbs.......	72.1	65.6
Digestible crude protein in 100 lbs.	2.3	2.9
Digestible carbohydrates in 100 lbs..	36.2	27.3
Digestible fat in 100 lbs..........	3.8	4.7

University of Arizona, Agricultural Experiment
Station, Tucson, Arizona, September 4, 1913.

I will state that perhaps without exception the one known as
Emory's oak, *Quercus emoryii,* yields more abundantly than any
other oak tree. I have seen Indians gather as much as two or three
gunny sacks full of acorns from a single large tree. The acorns of this
particular oak are sweet and very agreeable to the taste; and if I am
not mistaken, some of them find their way to the market. We have
other good species of oak trees that produce heavily, among which
are the Arizona oak and the oblong-leaf oak. The acorns of this latter, however, are bitter to the taste, though they are excellent hog
feed.

J. J. Thornber, Botanist.

Speaking of mountain land near the divide between the Eel
and Sacramento rivers in Northern California, Will C. Barnes
in a personal interview and in an article in *Breeder's Gazette*
for May 29, 1912, said:

For forty miles we worked our way upward through the brush-
covered foothills into the zone of cedar and piñon, and finally
reached the great forest of yellow and sugar pine and firs. . . . This
is sheep and cattle country, but the most profitable animal these
men are raising is the hog. There are vast ranges of oaks all over
the mountains and the hogs which run at large the year round get
their living entirely on the acorns and other feed which they find.
Last year was an especially fine one for acorns, and we saw them
under trees where they lay thick on the ground to the depth of an
inch. Some of the oaks had very long acorns, almost two inches in
length, and there was a keen rivalry for them between the hogs and
the Indians. Sacked and packed on their ponies to the settlements
below, the acorns were bringing the Indians from 50 to 75 cents per
bushel for hog feed where the ranches were not near the forest. It
was no trouble at all to gather from a peck to a bushel of acorns
from beneath a good tree. Indians just scooped them up in armfuls.
Everything in the line of live stock seemed to be fond of acorns. One
saw the horses and cattle rooting among the leaves under the great
trees for them, and at times the crunching of them in their teeth
sounded like a lot of Missouri hogs eating corn.

I stopped to photograph a band of fine Shropshire ewes under an oak, all hunting for the toothsome acorns.

If acorns should be on one hundred square feet of ground to an average depth of one-half inch, the amount would be more than three bushels.

A newspaper report from California said:

Los Molinos, Cal., September 9—The crop of acorns is unusually abundant, both in the valley and the foothills. Many farmers are beginning to gather them for hogs. By September 10 or 15 the ground under many oaks will be covered with acorns.

Women and children take an active part in this work. Often one family will gather ten to fifteen bushels a day. The deer come down from the mountains during the latter part of October, and it is expected the venison taken this season will be of unusual quality, as the deer live largely on acorns for several weeks.

THE SOFT MEAT OF THE ACORN-FED HOG

In America, acorn-fed hogs bring lower prices in the whole-sale market because they have soft flesh. Is this a permanent handicap? I doubt it, if the problem is studied in a scientific way. In the first place, acorn-fed pork has fine (perhaps finer) flavor. For local consumption the meat (acorn-fed hogs) is satisfactory. If the lard is liquid instead of solid, what is the difference? One kind may go into a can while the other goes into a carton. Its meat drips; if so, the drip is good lard. Perhaps it needs to be subjected to some process such as 120° F. for a stated period to force and finish the dripping. This reduction of the fat might make bacon better. It is certainly no handicap to animals on a maintenance ration. "It might be stated his hogs would not eat corn in quantity until the acorns were gone." (J. C. Holmes, *The Country Gentleman,* December 13, 1913, p. 1822.)

There are a few hogs slaughtered in the smaller towns and on the ranches for local consumption. These animals are for the most part eaten within a short time, very little of the pork and meat being pickled or salted, and the animals which are used for local consumption in the region seem to be satisfactory to the consumers even though they are acorn fed. I believe that the packers desire as uniform a grade of meat and pork as is possible to secure, and the acorn-fed hogs would give a different quality and softer pork than the grain-fed animals. This I believe is the reason that the packers discriminate against the acorn-fed hogs. (C. F. Elwood, Director, Cooperative Extension Work in Agriculture and Home Economics, Berkeley, California, letter, August 1, 1927.)

We probably need to have packing-plant processes adjusted to the soft-meat hog. We probably need more of this kind of hog.

The present discounting of the acorn-fed hog may be largely a matter of psychology of the market. During the meat shortage of World War I, reindeer meat was brought forward to be used in place of beef. It was said to be as good, but the psychology of the process was that it was a substitute. This strange substitute sold at a discount in comparison to beef. It was withdrawn from the market, advertised a little, and sold as a novelty in exclusive clubs at a high price. Acorn-fed pork properly cured and properly sold might have a similar experience. We seem to have here a problem in market organization. The Acorn Pork Cooperative, with ten thousand acorn-fed hogs a year, might set a style and capture the fancy ham market.

Where are the men who start things?

The longer I live, the more amazed I become at the lack of constructive imagination, the lack of sheer curiosity, the desire to know. Why has no experiment station, no wide-awake farmer, given us some figures of pork yield per American acre of oak trees? The late Frank Vanderlip said we were a nation of economic illiterates. He was too polite. If we add resource destruction to neglected opportunity, is not "idiots" more nearly the word?

At last we have one figure, produced by enthusiasm, the kind of enthusiasm that has done so much to build civilization. John W. Hershey, now of Downingtown, Pennsylvania, then of the

T.V.A. Tree Crops Section, persuaded the county agricultural agent in Coffee County, Tennessee, in the fall of 1937, to fence off a two-acre farm wood lot of white oaks and turn in 20 Poland China shotes. The shotes averaged 30 pounds each, and in 42 days they had cleaned up the acorns and had gained 30 pounds each, or 600 pounds, a gain of 300 pounds of live pork per acre. You can figure what that is worth on your market today.

The shotes were then turned into an alfalfa pasture supplemented with 112 pounds of corn per day. After 50 days on this ration, they had gained 132 pounds each.

It should be noted that the corn and alfalfa give approximately a balanced ration, whereas the acorns alone, being rich in oil and other carbohydrate, are very distinctly an unbalanced ration. If the pigs in this test had been given a balancing ration of fish meal or tankage, or alfalfa while on acorns, the showing for the acorns would doubtless have been much better. The over-fat and unbalanced systems of the pigs on acorns perhaps rushed to balance themselves on alfalfa protein and added weight with unnatural speed.

On my farm in Loudoun County, Virginia, we hogged down a field of corn in 1947. The hogs were not happy. They roamed about; they tested the fences; they got out and made a path to the alfalfa field. They were then given tankage. They ate surprisingly little, but they ate and went to sleep. No more roaming! No more getting out of the field! A balanced ration had eased their spirits.

ACORN MEAL AS A DAIRY FOOD AND GAME FOOD

The question of breeding better oak trees is presented in the next chapter. However, I may point out here that it looks reasonable to expect scientific work to produce great improvement in acorns, improvement similar to that which has occurred in Persian walnut (p. 205), chestnut, and nearly every crop grown in our gardens and orchards.

There seems to be no reason why acorn meal should not be a

food of unsurpassed excellence for cows, stock hogs, all work animals, and probably for beef cattle.

If an acorn-meal industry were established and a person (man, woman, or child in early teens) could pick up five hundred pounds of acorns in a day (see p. 159), it seems a foregone conclusion that the Appalachian and Ozark Mountain regions of the United States could enter upon a new era of prosperity. Apparently present prices for cow feed would enable a good acorn crop to double the wages that thousands of American mountaineers now receive.

At a Pennsylvania meeting of experts concerned with preservation of game, a Mr. Freeborn recently pointed out that the acorn was one of the most important game foods in Pennsylvania for the following: quail, duck, dove, pheasant, grouse, turkey, bear (much of his food), beaver, fox, rabbit, muskrat, opossum, and raccoon.

Oak trees are not hard to graft. A farmer would soon have a valuable field if, in any one of five hundred counties in the eastern half of the United States, he would graft the young oaks and oak shoots on a tract of cut-over hardwood forest. He could do this by getting cions from the best trees in his own county. Now, however, having been to a school where he learned merely to repeat words out of a book, he is too blind to see such useful things near home, so he hunts a job in town. What proportion of the teachers of agriculture in American rural high schools know how to graft a tree, or do teach it to their pupils? Why is it that young men and women are the chief export of rural America? Going off to town.

John W. Hershey, Downingtown, Pennsylvania, one of the very few experimenters working with oaks, reports, 1948, a bur oak, producing half a dozen acorns, three years grafted on seedling and two years after transplanting; a chinkapin oak four years transplanted, a sprinkling of acorns; two bur oaks four years transplanted, a sprinkling.

He reports basket oak, bur oak, and chinkapin oak all doing

well on white-oak stock. I remark that the bur oak grows faster.

These findings of Mr. Hershey are very significant of crop possibilities.

ACORNS AS CHICKEN FEED

During World War II a letter was written to the *London Times* by a member of Parliament, Godfrey Nicholson, on "Acorns a Wartime Source of Food." It reads as follows:

I do not think that enough publicity has been given to the fact that acorns are an admirable substitute for the usual poultry foods. The ministry of agriculture, in their "Growmore" bulletin No. 5 published this year, give official blessing to this, saying that they "can be regarded as a substitute for cereals" and that "as much as two ounces per head per day of dried shelled acorns has been fed to laying fowls without affecting adversely either yolk color or egg production." In other words, practically half the diet of a laying fowl may consist of acorns.

A big bur oak stands in a Virginia farm yard. If one walks across the yard in autumn, the chickens follow, hoping that an acorn will be stepped on and broken so they may eat it. I have seedlings of this tree.

CHAPTER XIV

Some Bread-and-Butter Trees—
The Acorns as Human Food

ACORN FOOD AMONG PRIMITIVE PEOPLES

As to the oaks, the primary object of this book is to call attention to the excellence of acorns as a farm crop producing forage for pigs, sheep, goats, cows, horses, chickens, turkeys, ducks, and pigeons. Yet any balanced presentation of the economics of the acorn must point out its great nutritive value and its great use as human food. It may be possible that the human race has eaten more of acorns than it has of wheat, for wheat is the food of only one of the four large masses of humans, the European-North American group. The other three groups, the Chinese-Japanese, the Indian (Asiatic), and the tropical peoples, pay small attention to wheat; hundreds of millions of their people have never heard of it. Meanwhile those humans (and possibly prehumans) who dwelt in or near the oak forests in the middle latitudes— Japan, China, Himalaya Mountains, West Asia, Europe, North America—have probably lived in part on acorns for unknown hundreds of centuries, possibly for thousands of centuries. The prevailing style among anthropologists now is to lengthen the period of human prehistory to three hundred thousand, four hundred thousand, and even to five hundred thousand years or more. I'm in style.

It is almost certain that wheat has been of important use only in the era of man's *agriculture,* while the acorn was almost surely of importance during that very, very long period when man was only a food gatherer. (See Fig. 75.)

Persons interested in the possibilities of the oak will probably be surprised at the collection of thirty-eight references to edible acorns in many parts of the world published by Dr. Robert T. Morris, in the 1927 *Proceedings* of the Northern Nut Growers' Association, J. C. McDaniel, Secretary, Department of Agriculture, Nashville 3, Tennessee.

The excellence of acorns as food for man was forcibly called to our attention during the food hysteria of World War I by the well-known scientist C. Hart Merriam, in the *National Geographic Magazine* for August 1918. He pointed out that, for an unknown length of time, acorn bread had been the staff of life for the Indians from Oregon to Mexico, except those in the desert; that there were three hundred thousand of these Indians in California when it was discovered by the white man, and that acorns of several species were eaten by various eastern tribes from Canada to the Gulf of Mexico.

George B. Sudworth, Dendrologist of the United States Forest Service, told me (1909) that an Indian tribe in western Nevada made an annual autumn excursion over the Sierras into California. There they busied themselves in the national forests gathering acorns and carrying them out to the trails in sacks. Thence they were packed or hauled over the mountains to the tribal home, and the bread supply for a year was secure. It was a part of the squaws' daily routine to pound up a portion of acorn meal and soak it in water that the tannin might be dissolved and the bitter meal made sweet. The coarse, wet, farinaceous mass that remained was the oatmeal, shredded wheat biscuit, cornmeal mush, pone, wheat flour, potatoes, bananas, or boiled rice for the family.

In 1938 I was told in southern Arizona that Mexicans made an annual visitation in trucks to haul away the local crop of acorns for their own food.

ACORN BREAD AND ACORN BUTTER FOR MODERNS

Dr. Merriam stated in his article in the *National Geographic* that John Muir often carried the hard dry bread of the Indians during his arduous tramps in the mountains of California and deemed it the most complete strength-giving food he had ever used. Merriam gave the following analysis of California acorn and rival foods:

	Cornmeal	Wheat Flour	Leached Acorn	Cal. Valley White Oak, Unleached
Water............	12.5	11.5	11.34	8.7
Ash.............	1.	.5	.29	2.
Fat.............	1.9	1.0	19.81	18.6
Protein.........	9.2	11.4	4.48	5.7
Carbohydrate....	74.4	75.4	62.02	65.
Fiber...........	1.0	.2	2.06	
Tannin				6.63

In looking at this table, please note the very high percentage of fat, which makes it as nutritious as richly buttered bread. Dr. Merriam pointed out that one part of acorn and four parts of corn or wheat make palatable bread and muffins, adding to the cereal the value of a fat nut product. No wonder the Digger Indians, who feed upon acorn trees, are reported as being sleek, fat, and in good order.

The Missouri Botanical Gardens, one of the great institutions of the world for studying trees, is working on the problem of tree utility. It published a bulletin (1924) showing photographs (see Figs. 74 and 76) of acorns and muffins made from them. It says:

With modern kitchen equipment, acorn meal can easily be prepared at home. After husking the acorns they should be ground in a hand-grist mill or food-chopper. The meal is then mixed with hot water and poured into a jelly bag. The bitter tannin, being soluble, will be taken out by the water, but sometimes a second or even third washing may be necessary. After washing, the wet meal is spread out to dry and is then parched in an oven. If it has caked badly, it should be run through the mill again before using.

In cooking, acorn meal may be used in the same way as cornmeal. Its greatest fault is its color, muffins made from it being a dark chocolate brown. The taste suggests a mixture of cornmeal and peanut butter, and some people relish it at once, but others, it must be confessed, have to be educated to it. Because of the high oil and starch content of the acorn, it is very nutritious and is reported to be easily digested. Only acorns from white oaks should be gathered, as those from the black oaks are too bitter. Typical Missouri representatives of this group are the white oak, the swamp oak, the bur oak, and the chestnut oak. [The small pile of acorns shown in Fig. 76 made nearly two quarts of meal.]

Muffins, ⅔ acorn and ⅓ oatmeal, are reported to be good.

The acorns of *Quercus virens,* the green or live oak, were used by Indians to thicken their venison soups and to express an oil which was very much like the oil of sweet almonds. (Loudoun, Vol. III, p. 1919.)

C. S. Sargent, *Silva of North America,* Vol. VIII, p. 19, quotes Parkinson, 1640, first describer of the white oak: "The akorne likeweise is not only sweeter then others but boyling it long it giveth an oyle which they keepe to supple their joynts."

The simple process of straining by the jelly bag seems to be a good substitute for a much more cumbersome method used by original Americans and still in use by the Klickitat Indians of the Pacific Coast. This method is described by this same Missouri bulletin.

Dr. Merriam, in the *National Geographic Magazine,* said the California Indians boil their acorns with hot stones in baskets.

New York, November 4, 1926.

"The world do move." I have some perfectly good edible acorns without astringency and with real flavor which do not belong to the white oak group sent me by a doctor in Zenia, Ohio. (Robert T. Morris, M.D.)

The *Rural New Yorker*, p. 17, 1928, mentions edible acorns in Burnett County, Wisconsin. This prompted Mr. A. C. Innis (Connecticut) to write in the issue of February 11, 1928, about the great abundance of these acorns which grew on the bur oak. He had eaten them "much as we eat chestnuts here in the East." He also spoke of the rapid growth of these trees.

FOOD FACTORIES—THEIR SIGNIFICANCE IN RELATION TO THE ACORN

In this age more and more food is being prepared in factories and delivered to the consumer in packages ready to serve. Machine production is bringing many strangers to the shelves of grocery and pantry. Prepared wheat bran is now put into boxes in factories, and millions of people eat it with apparent relish.

The peanut is one of the most spectacular omens of others yet to come. It was unknown to Grandfather. Father cracked the shell and munched at the circus. Now, Milady often buys ready salted peanuts for card-table refreshment, puts up sandwiches of peanut butter (factory made) for Junior, while ten thousand drugstore counters have peanut candy bars and glass bowls of cracker and peanut sandwiches all ready for us to eat as we run (in circles, mostly).

Will the acorn be next, blended with some other cereals? The fact that the acorn carries its own butter is an attractive feature. Its amazing keeping qualities are also greatly in its favor. The acorn bread of the California Indian keeps indefinitely. This is a wonderful quality for factory foods that are to be distributed in packages.

Then there is that six percent of tannin. How easy for the chemical engineer to get it out if he had fifty thousand tons of acorns a year to deal with! Tannin is worth money. We scour the ends of the world for it. It is quite possible that income from tannin perfectly extracted might put a premium price on bitter acorns.

In praising the excellence of the acorn as food, Dr. Merriam showed that it can probably be kept in good order for a longer time and more easily than any other food product known. A common method of storage by the Indians was to bury acorns in mud kept cold by a spring of water. Dr. Merriam reports the discovery of such caches as these, that had lain for a period of thirty years, in which the acorn remained unsprouted and unspoiled. They were merely discolored. If there is another food product capable of such preservation I have not heard of it.

THE ACORN AS HUMAN FOOD IN EUROPE

Dr. Merriam states that "in Spain and Italy sometimes as much as twenty percent of the food of poorer people consists of sweet acorns." This statement chimes in well with a circular letter of a French bishop in the 9th century who called upon his priests to see that his people were supplied with acorns during a food shortage.

I have sat by the fire in Portugal and again in Majorca and eaten roasted acorns exactly as one eats roasted chestnuts in America. The variation among the fruit of the different trees of the evergreen oak (pages 162-186), like the variation among seedling apples, has caused some to be as good as chestnuts. The reference books show that this is no new discovery, nor is it limited to Spain and Portugal.

Loudoun, Vol. III, p. 1905, Oak. *Q. ballota.* The sweet acorn oak, *Chêne à glands doux, chêne ballota,* 20-30 ft. high, 3-6 circ., fruit 8-20 lines long, 4-6 wide.

Vast forests on the mountains of Algeria and Morocco, but only in small quantities on the plains.

The Moors eat the acorns raw or roasted in ashes; they are found very nourishing, and are not bitter. They are regularly sold in the market place, and in some places an oil is extracted from them which is nearly as good as that of the olive.

Loudoun, Vol. III, p. 1907, *Q. gramuntia.* Holly-leaved Grammont oak. Native of Spain, cultivated in England, 1730. Chêne de

Grammont (French). Wellenblattrige eich (German). Encina dulce and gouetta (Spanish).

Blossoms June, ripens fruit in autumn of next year. Fruits annually. Acorns edible, and when in perfection are as good as or superior to a chestnut. To give this sweetness they must be kept, as at first they have a considerable taste of the tannin like those of the other species.

These are the edible acorns of the ancients which they believed fattened the tuna fish on their passage from the ocean to the Mediterranean. A fable only, proving that they grew on the delicious shores and rocks of Andalusia, which unhappily is no longer the case. [They fattened the swine which produced the celebrated salt meats of Malaga and that vicinity.]

The wildest forests of it are now in Estremadura, where the best sausage and other salted meats are made from the vast herds of swine which are bred in them.

Produced by individuals and offered to the company as sweetmeats. Very hardy. Mountainsides in Castile and Aragón and in the wintry valley of Andorra.

Loudoun, Vol. III, p. 1902. *Q. ilex. 1.* Common evergreen or holm oak. *Chêne weit. Encina Span.* Deep rooter on well-drained soil. Wiffens Garcillasso of Spain writes:

> Hast thou forgotten, too,
> Childhood's sweet sports, whence first my passion grew,
> When from the bowery ilex I shook down
> Its autumn fruit which from the crag's high crown
> We tasted, sitting chattering side by side,
> Who climbed trees swinging o'er the hoarse deep tide,
> And poured into thy lap, or at thy feet,
> Their kernel's nuts, sweetest of the sweet.

In Spain and Portugal, the better trees have been grafted for some centuries exactly as we in America have grafted Baldwin apple trees, and these grafted trees (Figs. 58, 67, 111) are commercially grown to a small extent in Majorca and some parts of the Spanish mainland. They rank with chestnuts in the market. If the price of chestnuts is high, the price of the acorn (ballota) is high. If the price of chestnuts is low, the price of the ballota is low.

We have found in America that oaks are not difficult to graft —if we want them grafted—but care must be observed as to hospitality of stock and cion. Information is needed there.

When I suggest that we deliberately set out to make a crop of the oak tree, I am sure that some typical American, accustomed to American speed, will say, "Too slow." At this point I wish to get the mind of the objector quickly shifted to the fact that *variations exist among individual oaks.*

A writer in the *Quarterly Journal of Forestry,* Vol. V, No. 2, pp. 119-123, says:

The *Quercus ilex,* growing at Holkham Station, England, is a tree of many types of habit, of infinite variation in the size of leaf and fruit. Speaking of bur oak in Sundance National Forest, Wyoming, Mr. Louis Knowles, Forest Supervisor, said in letter to the Bureau of Forestry, October 6, 1913:
Heavy seed crops occur about once every three years, while a considerable crop of seed seems to grow on individual trees *every year* [italics mine].

Dr. Robert T. Morris in *Proceedings,* Northern Nut Growers' Association, 1927, quotes Professor W. L. Jepson, University of California, to the effect that all varieties of acorn were used as food by California Indians, and further:

There is undoubtedly very great variation in the quality and yield of the various individual trees of one species, even in a given locality. Trees notable for their yield and especially for the quality of their acorns were the special property, in aboriginal days, of a particular family or small tribe. This fact of variation is true as well of the black oaks; for example, the coast live oak, *Quercus agrifolia,* varies so remarkably in the edibility of the acorns as borne by these trees that the yield of certain trees is esteemed by white men as a substitute for chestnuts. There are gardeners in the great Del Monte grounds at Monterey who gather the acorns of a certain coast live oak tree which stands in that area and eat them as they go about their work just as they might chestnuts.

Oaks have variations in precocity similar to those of chestnuts. Regarding the matter of slowness, I insist that it is unfair to judge as "slow" every individual oak of the whole genus of fifty American species, of which there are fifteen in California alone, and thirty in the eastern part of the country.

Some oaks grow rapidly.

Some oaks bear acorns when the trees are very young.

Some oaks bear heavily.

Some oaks bear regularly.

This is great material for the plant breeder.

Next I wish my objector to think also of the remarkable effects of hybridizing when one starts with trees of unusual excellence.

The poets who have written of the strength and the long life of the oak tree, pointing a moral of patience and achievement in the progress of mighty oak from the small beginning in an acorn, probably based their poetic utterances upon the few oaks that happened to grow in their own neighborhood and of which they had little accurate knowledge. Of other more speedy oak trees, poetic pens were silent. Trees are not used by poets to symbolize speed.

R. O. Lombard, Augusta, Georgia, had a water oak tree which he said he trimmed with a pocket knife when it was the size of a pitchfork handle. Nineteen years later I measured it— girth 67 inches, spread 51 feet.

A chestnut oak planted in a yard in the clay hill country of Georgia by Governor Lamartine Hardman made the following growth in thirty years: girth, 67 inches; reach 34 feet one way and 30 feet the other. So far as I know, there is nothing unusual about these trees except that their history seemed to be known with reasonable definiteness by persons who had lived with them.

Loudoun (*Arboretum et Fruticetum*, p. 1864) says of *Quercus palustris:* "Most beautiful of the American oaks. One hundred plants in London seven years from acorn and 15-20 feet in height. Most American kinds 10-12 years from acorn 20-30 feet in height." I add, apples rarely do better. Loudoun's figures come from cool Britain. Ness (Fig. 61) reports greater speed of growth in Texas.

As to precocity—note the facts reported by Hershey, this chapter. While walking through a Portuguese fig orchard, I came upon a little ilex tree (evergreen oak) shoulder high and full of bloom (Fig. 70). The ground beneath was littered with acorn cups. I asked the Portuguese laborer who was working nearby if this tree had borne nuts the previous year. He said it did, and pushed for the number, he said it might be two hundred. The figure may have been high, but the acorn cups under the lonely little bush in a neglected fig orchard are good evidence of some crop. I have seen bloom on hundreds and thousands of such small ilex oak trees in Portugal, Spain, Algeria, and Tunis, and was repeatedly told by residents that they bore fruit at an age which we in America think suitable for a young apple tree.

I came home from Africa and looked along my own Blue Ridge Mountain lane and found two chestnut oak suckers, each bearing fruit, and each in the seventh summer; one having grown from a stump the size of my finger and the other from a tree two feet across. (See Fig. 72.) I have no reason to think these were the best and most precocious trees of my Blue Ridge mountainside. They merely grew by my lane.

More remarkable in my opinion was the performance of dwarf oaks, locally called turkey oaks, on the top of the ridge. They grow in soil weathered from quartzite sandstone. This sandstone is cemented with quartz. It makes one of the poorest soils known. This particular tract had been further cursed by forest fires every few years for at least seventy years. The pines had long since succumbed and only huckleberries and turkey oaks remained. In April 1910 a fire killed everything to the ground, and in September 1913 the turkey oaks were full of fruit. (See Figs. 70, 71.)

CREATIVE WORK WITH OAKS—HYBRIDIZATION EASY, RESULTS PROBABLY STUPENDOUS

Like the hickories (see Chapter XIX) the oaks will hybridize themselves.

The idea of hybridization of oak trees is not new. In 1812 F. A. Michaux records, with description, one that is still called X *Quercus heterophylla.* In 1876 George Engelmann reported ten. In 1925 William Trelease named fifty-five, all of which are reported in an excellent monograph by Ernest J. Palmer, in the journal of the Arnold Arboretum, Volume 29, 1948. Anyone interested in this subject should scan these forty-seven pages.

After discussing the small chance of a hybrid acorn growing and making fruit, he gives references to hundreds of specimens that have already been found and says:

Hybrids among the oaks are probably much commoner than they were believed to be by most botanists even a few decades ago, and it seems probable that most, if not all, of the species within the sections of the annual-fruited (Leucobalani) and the biennial-fruited (Erythrobalani) are interfertile under the right conditions.

Several instances of experiments and results in the artificial production of oak hybrids have been referred to above in his monograph. But comparatively little has so far been done in this field, although it would seem to be a most promising one both from the standpoint of its practical value and its scientific interest.

Sargent reports (*Silva of North America,* Vol. VIII, p. 19) a number of hybrids of one species with several others. He further remarks that the hybrid offspring *grows more rapidly than the parent.* Professor H. Ness, Texas Experiment Station, who seems to have been a lone hybridizer of oaks (*Journal of*

Heredity, October 1918 and September 1927), summarizes his experience with hybrid oaks, at College Station, Texas, as follows:

> The indications observed on this soil are to the effect
> 1. That the hybrids descended from the live oak, as mother, are of a stronger growth in both the first and the second generation than either of the parent species.
> 2. That they are fertile at an early age in both generations, producing seed of normal viability.
> 3. That the first generation, while very variable in some combinations, is uniform and intermediate of the two parental forms in other combinations; but this uniformity is followed by the segregation of characters in the succeeding generations.
> 4. Because of the ease with which the hybridization of the live oak may be affected, the high fertility of its hybrids and other virtues already mentioned, to which very likely may be added improvement of the timber, there can be no doubt but that the breeding of new forms of oaks, as here indicated, has great economical and aesthetic possibilities.

Professor Ness found that during an unfavorable season these hybrid offspring made an average of three or more feet on every main limb, and nuts were borne in 1917 from an acorn planted in 1913. These are facts for pondering. Especially so is the fact that, in the second generation of breeding, some seeds planted in 1920 bore acorns in 1923 and bore a very large crop in 1925. Starting with their present amazing qualities, what may not hybridization produce among fifty American (and some foreign) species of oaks?

And after all this, the Texas Station has done nothing with oaks except to print a bulletin telling what wonderful shade trees the Ness hybrids have become. What on earth could wake up the Texas station? It is almost on a par with the Alabama Station that cut down the marvelous honey locusts.

As one of many promising foreign oaks, I submit the Valonia oak, *Quercus aegilops* (chêne velani f.). Trabut says (Bulletin

27, Gouvernement Général de l'Algérie, 1901, p. 56): "The acorns are sweet and edible and abundant when the 'valonie' is not harvested." The "valonie" (valonia) is the large thick cup of this acorn. It carries over thirty percent of tannin and is a regular article of commerce, gathered to the extent of several million dollars worth per year in the eastern Mediterranean countries. A tree producing such a double-barreled crop should be an interesting basis for breeding experiments.

California and New Mexico, please copy.

I submit that the conservation of our resources and the increase of our food supply alike demand that half a dozen men should be assigned immediately to the task of seeking the best oak trees and of breeding still better oak trees. There is excellent prospect of their being able to add millions to the wealth of the United States (also other countries) during the period of a normal lifetime.

But more important than the *cash* value of their work would be the conservation of the soil. Oak orchards could hold the hills that are now washing away as we plow them and attempt to grow cereals upon them.

Don't forget Dr. Hugh H. Bennett's report of five hundred thousand acres now being ruined every year and six million new acres of crop needed each year to give to each member of our increasing population the three acres per capita that we now use. In *three years* that equals the entire cropped acreage of Japan. We are indeed reckless stewards of God's heritage, if indeed we can be called "stewards." "Wreckers" is much more correct.

Thomas Q. Mitchell, 16 East 48th Street, New York 17, N. Y., reports that he has been breeding oaks with great enthusiasm for ten years. From voluminous manuscript reports of his work, I select the following quotations, which state some of his breeding experiments:

I think my most important discoveries are:

1. The "MASS-CROSSING" TECHNIQUE

2. That MANY hybrids have "INVERSE HETEROSIS" [See J. F. Jones's work with filberts, Chapter XI (Fig. 21) of this book.]

3. The value of "extreme variants"—especially those making TWO, THREE, or MORE successive growths each year.

4. That second growth can be induced—which is a valuable technique.

I can report that young oaks respond nicely to girdling, in order to force early bloom. I have girdled a good many, and never lost one. Cut away a ring of bark about ⅛ inch wide (before July) and keep the girdle open for 6 weeks. SOME oaks only 4 to 8 feet high (and probably 5 to 10 years old) have bloomed the following year. On the other hand No. 4 on my list, about 20 years old, did NOT bloom in the two years after girdling—and I have not seen it since.

From List of 47 New Hybrid Oaks Discovered or Raised by Thomas Mitchell:

1. (Bicolor x Macro) x LIAOTUNGENSIS. Young, pendulous, 6 ft. fruited 1944.

14. MONGO BI (Mongolica x Bi.) 9 trees, 4 fruited.

34. ACUTERRIC (CERRIS X ACUTISSIMA) 1 Boston, 1 Rochester, both fruited.

43. MACRANTHERA (IRAN) X DALECHAMPI 1 seedling, seed planted September 1946.

At 10 months old, No. 43 has made FIVE SUCCESSIVE growths. This is the best hybrid I have ever raised. Five other crosses from this same tree are very mediocre. . . .

I have made many thousand hand pollinations of oak blossoms and got nuts on about 1 percent of them—100 acorns. Some did not grow, but I have 20 hybrids now in hand. . . .

Using only 200 varieties to breed from, more than 20,000 first crosses, and more than 30 million "four ply" second crosses can be made. The inception of this program means the forthcoming end of world hunger, delayed only by the amount of time needed by the nurserymen to graft the millions of trees required. There will be no need to face alternatives of bread *or* meat. Erosion can be stopped. Floods will no longer be feared.

J. R. S. points out that the nurseryman can be excused from much of this chore. Let the farmer graft suckers and young trees on cut-over (not burnt) land.

Mr. Mitchell would like to correspond with other oak breed-ers. Since Mr. Mitchell is merely an individual, anyone telling him of discoveries should write in triplicate and send one copy to Morris Arboretum, Chestnut Hill, Philadelphia, Pennsyl-vania, to Morton Arboretum, Lisle, Illinois, or to St. Louis Bo-tanical Garden, St. Louis, Missouri.

CHAPTER XV

Nuts as Human Food

Farm animals eat much the greater part of the produce of the American acres. For that reason, we have thus far, in this book, been considering tree crops chiefly suited to feeding domestic animals; but now we come to the consideration of a series of crops which are grown, and should be, primarily for human food. I refer to the nuts. Walnuts, for example, like most other nuts, are a substitute for both meat and butter.

Tables of food analysis (Appendix) show that nuts have more food value than meat, grains, or fruits. Six leading flesh foods average 810 calories per pound. Half a dozen common nut kernels average 3,231 calories, about four times as much. Cereals at 1,650 calories are about half as nutritious as nuts. Fresh vegetables averaging 300 and fruits averaging about 275 calories per pound are less than a tenth as nutritious as nut meats.

HIGH QUALITY OF NUT FOODS

The quality of nut food is also of the very highest. Early food chemists called nut protein "vegetable casein," because of its close resemblance to the protein of milk. When the Chinese mother's milk fails, her babe is fed on milk made of boiled water and the paste of ground walnut, *Juglans regia*.

P. W .Wang, curator, Kinsman Arboretum, Chuking, Kiangsu Province, China, says (1922 *Proceedings* of the Northern Nut Growers' Association, p. 120):

In China there is no baby fed by cow's milk. When the mother lacks milk and the home is not rich enough to hire a milk nurse, walnut milk is substituted. The way of making walnut milk is rather crude here; they simply grind or knock the kernel into paste, then mix with boiled water.

Dr. J. H. Kellogg, of Battle Creek, militant nutivorous vegetarian and in his amazing person a substantial vindication of his theory, backs up the Chinese milk-substitute theory (practice) and further points out that while many vegetable proteins are hard to digest, those from nuts are very easy to digest. He further avers that nut fats (the other chief food elements of nuts) are "far more digestible than animal fats of any sort."

On pp. 83-92, 1920 *Proceedings* of the Northern Nut Growers' Association, Dr. Kellogg claims that animal feeding experiments show that twenty ounces of milk will furnish complete protein enough to supplement a vegetable diet otherwise deficient in complete protein, or that the same amount of protein

	Pints of Milk Containing as Much Protein as One Pound of Named Nut	Calories in Amt. of Milk Shown in Column One	Calories in One Pound of Named Nuts	Ounces of Named Nuts Needed to Replace 20 oz. of Milk
Acorn.........	2.4	780	2620	8.3
Almond........	6.4	2080	3030	3.2
Beechnut.......	6.6	2145	3075	3.0
Butternut.......	8.5	2762	3165	2.4
Chestnut.......	3.2	1040	1876	6.4
Chinquapin.....	3.3	1072	1800	6.4
Filbert or Hazelnut.....	5.0	1625	3290	4.0
Hickory nut.....	4.6	1495	3345	4.8
Pecan.........	3.6	1170	3455	5.6
Peanut........	9.2	2090	2600	2.2
Piñon.........	4.4	1430	3205	4.8
English walnut..	5.4	1555	3300	3.7
Black walnut....	8.5	2762	3105	2.4

can be furnished by the amount of nuts shown in column four in the table on page 203.

The housekeeper may probably do a little figuring on the relative cost of 20 ounces of milk and the nut meats listed in column 4.

The freedom of nuts from putrefactive germs and from ptomaine poisoning are points which we may esteem more highly as we increase our knowledge of what occurs in our digestive tracts.

The sufficiency of nuts as a substitute for meat in human diet seems well established alike by modern dietary experiment and by the experience of many primitive peoples. The sufficiency of a fruit-and-nut diet for humans is strongly hinted by its success with such physically similar animals as the orang-utan and the gorilla. Dr. Kellogg reports successful substitution of nuts for meat for a period of several months with a young wolf, a fish hawk, and many other carnivores. (*Proceedings,* Northern Nut Growers' Association, 1916, p. 112.)

I am well aware that vegetarianism is sometimes almost a religion, and although praising the nuts as food, I wish to confess my satisfaction in wielding the carving knife. Yet, I am not at all sure that I might not live longer on a vegetable-fruit-milk-nut diet. George H. Corsan, the "sassiest" man I know, and by far the spryest old man I know, is now past ninety. He classes milk as a poison, along with all flesh, and says that he lives on fruit, vegetables, and nuts. George Bernard Shaw, age 92, boasts of his vegetarianism, but is reported to take liver extract. As a matter of convenience when spending a day on the farm with my trees, I find the following lunch very satisfactory:

> 2 bananas
> 2 oranges
> a handful of nut meats
> a peanut-candy bar (for dessert)

In considering diet, it should always be remembered that WE LIKE WHAT WE EAT after we get used to it.

NUTS AND THE FOOD SUPPLY

Nuts offer a double opportunity for the improvement of our food situation. They can enable us to increase both the quantity and the quality of our food supply. During our frontier period of abundance, nuts were neglected in America both dietetically and agriculturally; but their use as food is increasing rapidly now, and their culture is receiving attention which promises a widespread industry in a short time.

The value of nuts as a means to increase the quantity of our food supply is forcibly suggested by the established practice of French farmers who expect a good English (Persian) walnut tree to yield one hundred fifty pounds of nuts per year on the average. (See p. 210.) These have food value greater than that of one hundred fifty pounds (live weight) of sheep, which is the total produce of a whole acre of good pasture for a year even in such good pasture countries as England or the United States. Good bluegrass pastures in the United States produce about one hundred fifty pounds of beef per acre, but by liberal use of fertilizer and legumes, the yield can be pushed above that.

Careful study of the ingenious table (p. 203), especially of the last column, indicates the high possibility of nuts as food producers, if the problem of food scarcity should ever present itself to us as even partly so acute as it has already presented itself to most of the world. The United States is still a frontier, a resource frontier. The nut trees appear to be veritable engines of food production waiting for us as the nation matures, and we steadily destroy our plow land (grain land).

CHAPTER XVI

A Meat-and-Butter Tree
for Man—The Persian Walnut

THE ORIGIN, DEVELOPMENT, AND SPREAD
OF THE PERSIAN WALNUT

The Persian walnut, *Juglans regia,* commonly called English walnut, is rival to the pecan as the nut most widely used in America at present. The peanut, also widely used, is not a nut in any agricultural or botanic sense, although it resembles a nut in its amazing food value. However, the English walnut is but a type. It might partly be replaced, in a few decades, by any one of half a dozen nut species now growing in the United States—black walnut, butternut, Japanese walnut (commonly called heartnut), pecan, shagbark, and shellbark. This assertion is supported by the following remarkable statement regarding the Persian walnut, which we all know to be a delicious and expensive article of food, as well as being a *large* nut with *thin* shell and easily accessible kernel:

The nut of the wild tree is small, with a thick, hard shell and a small kernel, and is scarcely edible, but centuries of cultivation and careful selection have produced a number of forms with variously shaped thin shells, which are propagated by grafted and budding.

At first it is hard to believe that statement, even though it is from the great authority Sargent, in his book *Silva,* a kind of horticultural bible.

How came such noble offspring in the garden from such ignoble parentage in the wood? The answer is that this prehistoric achievement *probably* resulted from artificial selection and chance crossbreeding. It *probably* happened in this way: In the beginning of Mediterranean or western Asian agriculture, some villager quite naturally brought from the woods the best nuts he (probably she) could find and planted them. The next generation took the best nuts from the village trees and planted them, the next generation did likewise, and so on. Thus we have tree generation after tree generation, each grown from the best-selected seed the people knew. This process of bringing the best trees together in the villages where trees were scarce gave a chance for both parent trees of the crossing to be of good stock.

This has been going on for an unknown period of time, certainly for many centuries, and has extended over a wide area—from Persia to Spain and from Persia to Japan.

As a result of this deliberate selection and extensive, though not deliberate, crossbreeding, there have been developed many excellent and varied types of Persian walnut.

The tree is thought to have been a native of Persia, whence it spread, going into the mountains of Syria, Asia Minor, and the borders of Palestine. Today, splendid trees of Persian walnut cast their lengthening shadows each morning and evening over the ruins of the temple of Baalbek and dot the adjacent gardens. They overhang the ruined walls of Constantinople, where Turks slew Greeks in 1453. They are scattered through Asia Minor, the Balkan States, through Greece, Italy, Spain, Portugal, France, Switzerland, and even in Scotland. Professor William Somerville of Oxford told me (in correspondence) of British trees seventeen feet in circumference and of some that ripened fruit at Gordon Castle in Scotland. Persian walnut trees are an important feature of the gardens of Persia. They seem to have traveled eastward as well as westward. I have seen them in Kashmir, in Eastern China, Korea, and Japan. Eventually, by coming to America with the European settlers, this tree has virtually girdled the world in the North Temperate Zone.

This process of planting the best nuts has been going on for two centuries in the eastern United States, where the trees are scattered from Massachusetts to Ontario, Michigan, and Georgia. In California (with its Mediterranean climate) progress has been more systematic and rapid, and an important orchard industry has been thoroughly established with a product of thirty thousand tons in 1925 valued at thirteen million dollars. This was about two-thirds as great as the French production. In 1938 (the last prewar year) our crop was fifty-five thousand tons, with farm value of twelve million dollars. In 1945 California's crop was sixty-four million tons and Oregon's seven thousand, with total farm value of thirty-five and six-tenths million dollars. Our import, over twenty thousand tons in 1925, had been stopped by 1938 and had been replaced by a slight export.

The California walnut industry is not of especial interest to the purpose of this book because it is not in need of aid. This book is written in the hope of starting something. The California walnut industry is well established. Furthermore, as now conducted it has little relation to the problem of soil conservation—the primary objective of this book.

The Old World walnut industry of the scattered trees has more significance than the intensively cultivated orchards of California with regard to future developments of importance to the human race.

THE HARVEST OF THE SCATTERED WALNUT TREES
OF EUROPE AND WEST ASIA

Having visited most of the European walnut districts, I consider it doubtful if anywhere in Europe one could find a neighborhood where there are half a dozen forty-acre orchards of walnut trees planted in rows and given systematic cultivation. In the province of Dordogne, one of the leading French walnut sections, orchards are almost unknown, but trees are exceedingly common along roadsides and field sides, in dooryards, and even scattered about the fields. The mature tree there is expected to yield one hundred and fifty pounds or more of marketable nuts on the average, and very large trees more than this. These trees stand alone, fine and shapely in the fields. Some farmers plant them about the fields at irregular intervals and then go on with their farming. If the walnut interferes a little with the growing wheat, oats, barley, potatoes, or hay, it pays for it in nuts, and the cultivation of the field crops helps the trees. The value of these trees is attested to by the rental practice of the locality. Landowners and farmers there rent one good walnut tree at the same rental as that received from an acre of plow land. Thus the fifty-acre farm, with fifty good walnut trees scattered about, rents for twice as much as a similar adjacent farm without trees.

When pushed for an explanation, the farm owner hesitated for a moment and then said, "You see, monsieur, it is zis way. It is income wizout labor."

I found an identical scattering of fine walnut trees along roadside, field edge, and farmyard in many parts of Switzerland and elsewhere in southern and southeastern Europe.

In the Grenoble district of France, near the city of that name, is the village of Tullin, the birthplace of one Mayette, a pioneer horticulturist, who lived about the time of George Washington. Mayette seems to have started the art of grafting walnut trees in that locality; he picked out the parent tree of the variety which bears his name. It is now widely scattered in France, California, and the eastern United States.

Mayette has a green and noble monument—his native village is embowered, almost buried, in the shade of Mayette walnut trees. They line the roadsides, the yards, the gardens, and in some cases they cover the surrounding hillsides, for here are some small orchards, often well cultivated. Most of the formal orchards are badly overcrowded, for it takes nerve to plant walnuts fifty to sixty-five, or even ninety feet apart, or to plant them closer and then take out trees when they attain a size that makes overcrowding a serious damage to them.

Professor Grand, Professor of Agriculture at Grenoble, the leading walnut district of France, told me that trees 20 meters apart would bear from 80 to 100 kilos (176-220 pounds) per tree and an average of 1500 to 1800 kilos per hectare (1300-1600 pounds per acre).

While journeying through the walnut districts of France, I was repeatedly told of trees that yielded 150 to 200 kilograms (330 to 440 pounds) of nuts.

It should be remembered that these trees usually *stand alone* with almost limitless root space and light. (See Van Duzee on pecan space, p. 257.)

I found one man who had planted out his whole farm with walnut trees ninety feet apart.

The wood is a substantial element in the value of the enterprise. In 1913 (note the date), I was told in Grenoble that a sixty- to seventy-year-old walnut had wood worth forty dollars, and a hundred-year-old tree was worth sixty dollars, while there was a local record of three trees one hundred and fifty years old having been recently sold for four hundred dollars.

Those trees, 90 feet apart, 5½ to the acre, will really be something a hundred years hence. They will be the kind that promoters use to catch suckers, multiplying that yield by 20 and calling it an acre, and selling you some land, or an orchard.

Graphic Summary of Farm Crops (Miscellaneous Publication 512, U. S. Department of Agriculture, 1940) gives a number of trees which, divided into the reported crop, gives twenty-seven pounds per tree. Of course that is *all* trees.

PERSIAN WALNUT IN OLD WORLD MOUNTAIN VALLEYS

The Persian walnut seems to appeal to owners of Old World mountain valleys—probably because of their air drainage. In the Old Testament, Solomon speaks of going "down into the garden of nuts to see the fruits of the valley." (*Song of Solomon*, Chapter 6, verse 11.) The Paris express from Milan climbs up to its Alpine tunnel through a valley where the walnut trees get thicker and thicker before the train finally dives into a tunnel to come out on the Swiss side of the mountain. This Swiss valley is also dotted with walnut trees. The upland valleys around Lake Geneva must have an average of two or three to the acre for a number of miles, so thickly are they scattered along roadside, field edge, village garden, and on pieces of land that are not easily tillable.

Onward through France to Paris and on to Le Havre they are a common sight. The German armies invaded France in 1914 under long avenues of walnut trees that lined the roadsides.

Asia Minor has interesting examples of mountain-valley orchards. The railroad from Tarsus, the birthplace of the Apostle Paul, to Istanbul (Constantinople), climbs up the Taurus Mountain wall through steep defiles and tunnels and then, near the top, comes out on a fine agricultural valley where hundreds of walnut trees are scattered about roadsides and fields. I have never seen finer specimens.

The interior of Asia Minor is too dry for the walnut except where irrigated, but this tree reappears on the other side of the plateau where the train comes down to Constantinople through a valley that opens out to the Mediterranean. This seaward-facing valley has more rain than the interior, and again the fine walnut trees appear scattered about as in France and Switzerland. I think I may say that for thirty miles there is scarcely an interval of two hundred feet without one of these magnificent trees. On a branch of the Morava River in Jugoslavia is another similar valley. There the walnut vies with cherry and other fruit trees for efficient use of corners of land.

Along the plain of the Danube and the Save in Jugoslavia, west of Belgrade, the walnut is a common shade tree for railroad-station yard and village street.

THE IRRIGATED PERSIAN WALNUTS

In the drier sections of Europe and West Asia the walnut goes into the irrigated vegetable garden, where it becomes a part of a two-story agriculture. For decades the United States received a substantial import of nuts from Naples. Most of them were grown on the slopes of Vesuvius and the nearby Sorrento peninsula, where it is a common practice to cover the vegetable garden with walnut trees. These trees stand up tall and spare like the common black locust (*Robinia pseudoacacia*) of the United States. Because they carry their heads high and because they leaf late in the season, the trees permit the Italian sun to reach the garden crops beneath, thus making a profit through two sources of income. The same type of gardening prevails in the gardens of Baalbek, in many other parts of Palestine and Syria, and throughout Persia, where one frequently sees the white branches and green foliage of the walnut standing above the wall that protects almost every garden of that hungry land. The California walnut industry is nearly all of the irrigation type, but because there is plenty of land in California, the two-crop system has not been highly developed. It is possibly a mistake that this has not been done. (See Chap. XXIII.)

THE ORIENTAL WALNUTS

The European practice of scattering trees about the farm and the village seems to be extensively worked out in many Oriental localities. One often sees a walnut tree or two near the mud-roofed houses and on the little farms of Kashmir. Nuts are one of the exports of this mountain valley, and carved-walnut work of great beauty is one of its most prized handicrafts and an important export.

The United States derives a substantial supply of walnuts from North China, where they are grown on seedling trees scattered about the farms in the hill country west and southwest of Peking. Here they grow in an interior continental climate closely resembling, but not duplicating, that of Iowa and eastern Nebraska, making this a very promising place to seek for parent trees for use in America. The trees from this area have met Zero a-plenty, but not so many warm waves followed by cold waves as those which, unfortunately, visit central, eastern, and southeastern United States.

THE PERSIAN WALNUT IN EASTERN NORTH AMERICA

In eastern North America we had the misfortune, though very naturally, to start the walnut industry with European strains in a climate that was strange to them. West Europe, namely, France and England, has an oceanic climate, which is characterized by cool summers lacking our hot humidity. It also lacks the sudden shifts from warm weather to cool weather in spring. In southern Europe, the Mediterranean climate is characterized by an open winter with rain and a hot dry summer without rain.

Now it so happens that the climate of eastern North America, being continental, has a spring and also an autumn with cold and warm spells alternating. Therefore, vegetation tends to start growth too early and plants like the apricot, peach, and Persian walnut sometimes get frosted. Furthermore, this eastern North America has a summer with heat, rain, and humidity to which the European plants are not accustomed, heat and humidity constituting the chief idea of heaven for fungi.

Consequently, many European plants come down with leaf blights when brought from England, France, or Italy to New York or Carolina or any other place east of the Rocky Mountains.

Spring frost or leaf blight is a detriment to most of the thousands of Persian walnut trees in the eastern United States. But here and there stands a tree so immune that an orchard of them would be very valuable. Nothing is more natural than to get a good walnut from such a tree and plant it, expecting to grow a tree producing fruit like the nut that was planted. Hundreds and thousands of people (myself included) have done this, not even knowing the source of the walnut, nor thinking that the resulting tree, being a seedling, is crossbred or a hybrid (Fig. 94) and therefore almost guaranteed not to produce fruit like the seed.

I cannot better illustrate this situation than by giving my own experience. With an interest in trees but no horticultural education, I had a vision of the old farm in Virginia waving green with English walnut trees and enriching me with their fruit. I knew of one tree in Washington, D. C., some fifty miles away, that bore barrels of nuts which were eagerly bought by local grocers at a good price. The seedlings from this tree perished in my yard the first winter, 1895. Then, in 1896, I sent to a New Jersey nurseryman, bought seedlings, and planted out three acres of three-foot trees. I thought that New Jersey stock would thrive in slightly warmer Virginia. The next year they were two-foot trees; the next year they were one-foot trees; and soon they died from the repeated winter killings.

I was no more stupid than many other people, but I did not know, and no one in the eastern United States seemed to know, that people had been grafting walnut trees in France and selling them for centuries. I knew of no place in America where grafted trees suitable for Virginia could be bought, or if they could be bought at all, and my walnut enthusiasm had to rest for several decades.

Meantime, many people here and there had succeeded with a seedling tree or two, and Mr. Daniel Pomeroy, near Niagara Falls, with an orchard protected from the warm spring days by the cool waters of lakes Erie and Ontario, was having such success with them that for a long time he sold his seedling trees far and wide. Several orchards of these trees were growing twenty years ago near the protecting lakes, but, being seedlings, they were not yielding heavily on the average, and we have long since ceased to hear of the Pomeroy trees.

Many of the Pomeroy trees have perished in the more changeable spring of localities farther south, as in Maryland and southern Pennsylvania, not so specially protected in spring by the tempering influences of water, and even farther south where they had a more premature start. Winter kill is usually spring kill, especially in the United States.

Here is a suggestive report on a Pomeroy orchard. The report is more than twenty years old, but it tells the seedling tree story.

80 South Lake Street
North East, Pennsylvania,
May 27, 1927.

The orchard was set in 1900; there are two hundred and sixty trees of which *two hundred are bearing*. They are set fifty feet apart, each way, with four rows of Concord grapes between two rows of trees; the grapes cover about half of the orchard, red raspberries are set on the other half. The trees do not affect small crops because the roots are very deep. Clean cultivation is recommended, and it is best to plow under a "cover" crop in the spring. The orchard bore 1800 pounds in 1924, 2400 pounds in 1925, and 4500 in 1926. Last year we gathered from five to six bushels off the largest trees.

Very truly yours,
E. A. Jones.

This report shows well the weakness of the seedling tree. At the end of twenty-seven seasons, with good culture, twenty-three percent of the trees were not bearing, while some trees were yielding two hundred and fifty to three hundred pounds each. Apparently ten percent of his trees, all as good as the best, would have yielded as much as, or more than, the whole lot. Because of similar performances, many seedling orchards in California have been top worked.

PERSIAN WALNUTS ON THE PACIFIC COAST

California, on the west side of the continent, in latitude 30° to 40° north, has the Mediterranean type of climate. There the whole Mediterranean flora is at home, including tender apricots, wine grapes, figs, and Persian walnuts; and, therefore, experimental plantings of Persian walnuts have thrived. The first commercial attempts succeeded, instead of being smashed out by frost as was my attempt of 1896. At that date the California walnut industry was well under way, and it has thrived ever since.

The California Walnut Growers' Association worked for

seven years to perfect a walnut-branding machine which puts their name on each nut. They offered a ten-thousand-dollar prize for the mechanical principle and then spent years in working it out into a machine that will brand two thousand nuts a minute, or a thirty thousand-pound carload in a day. In October, 1926, they had one hundred twenty-five of the machines running at a cost of only five cents for the four thousand nuts in a one-hundred-pound bag—one-thirtieth of the cost of small sealed cartons, on which they had previously been working.

The elimination of seedling orchards is increasing the California yield. The technique of the walnut-growing industry has developed, and now it has an association of growers carrying on national advertising and selling campaigns and marketing tens of millions of pounds of nuts each year.

E. F. Serr, Extension Specialist, wrote me December 20, 1947:

The average yield of bearing walnuts in California during the period 1930-1935 was about 6 to 800 pounds per acre. Between 1935 and 1940 this increased to about 1000 pounds. Between 1940 and 1946 there was additional increase until the average was about 1100 pounds per acre for the three years 1944, 1945, and 1946. Acreage 114,000 in California.

As the climate of Portland, Oregon, is a near duplicate of that of Paris, and Seattle's that of London, we see why western Oregon has promising Persian walnut orchards (over twelve thousand acres). However, occasional freezes, such as that of 1919 and later, more severe apparently than those that come to France, do great damage to the trees.

THE SEARCH FOR VARIETIES IN EASTERN AMERICA

While California successfully transplants another European industry, the Eastern states, having started with European strains and though deluded by them, are still experimenting and have spent years hunting for good parent trees among them. This long dependence on European strains is not unnatural. Professor F. N. Fagan, of Pennsylvania State College, made a partial survey of Pennsylvania and found walnuts growing in at least twenty-five counties. He estimated that the aggregate number in the State must be at least five thousand trees. Some of them are bearing regularly at elevations of fourteen or even eighteen hundred feet. Trees of local repute have been reported from southern Ontario, New York State, Massachusetts, and as far south as Georgia. Their total number runs into many thousands. Where are the best trees? Are they fit to become the basis of a commercial crop in the East? It is doubtful if anyone yet knows. Nearly all of the nuts have enough tannin in the pericarp, or brown skin of the kernel, to give the kernel a slightly bitter taste. Only a few of them are absolutely sweet. Most of the trees are subject to leaf blight; nearly all of them are winter killed in exceptionally severe winters. I may add, in passing, that I saw a surprising amount of winter killing on small branches on the commercial trees of France and Italy. Mostly, these European trees looked no more physically prosperous than some of the trees in the eastern United States.

It is probable that two or three dozen of these trees of the eastern United States are worthy of commercial propagation, but the location of the *best* parent trees requires a world search.

There has not been very much success with Persian walnuts west of the Appalachians. Mr. Riehl at Alton, Illinois (climate of St. Louis), reported failure with all he tried; but Mr. Otto Witte at Amherst, Ohio, thirty miles west of Cleveland and five miles from Lake Erie (lake climate), reports success with trees from German seed.

Upon the whole, the country west of the Appalachians is very debatable ground for the European strains with which we experimented prior to 1935—the date of arrival of the Crath strain.

I am not sure whether any of the Persian walnut trees I now have will prove commercially successful here, but some of them give fair hopes and may lead to something better. It is too early yet in my experiments to know what degree of success I may have in the near future. (Letter, N. F. Drake, Fayetteville, Arkansas, February 28, 1927.)

Mr. Hirschi, a private experimenter at Oklahoma City, reports (in 1947) general failure of all he had tried. There was too much alternating Louisiana climate and Nebraska climate.

As an example of the small botanic factors upon which commercial success may depend, I cite a variety called the Alexis. The parent tree stands about thirty miles from Baltimore, Maryland. The observant Mr. J. F. Jones, of Lancaster, Pennsylvania, owner of a very interesting test orchard, reported this variety as one of the most dependable known to him. This tree has borne regularly for Mr. Jones. This fact he explained as resulting from one habit of the tree, namely, that it makes a quick growth in the spring like the hickory, hardens its new wood, and makes no late summer growth. From varieties making a late-summer growth, the late-grown and therefore tender leaves are eaten off by a leaf chafer. This loss of late-summer leaf growth apparently weakens the twig, causing it to become easy prey to winter killing (actually spring killing).

The observations of Ford Wilkinson of Rockport, southern Indiana, give interesting confirmation of Mr. Jones's observation and of the widespread belief that the Persian walnut will not thrive in the Mississippi Valley. Mr. Wilkinson reported in 1926 that five of his forty English walnut trees had escaped fatal winter kill. One seven-year-old tree bore more than a half bushel of fine nuts in the fall of 1926. All of these five survivors were growing on the north side of large apple trees. They grew to a size and age where they produced from half to almost a bushel of nuts. "And then, one spring," said Mr. Wilkinson, "I cut the old apple trees down and plowed the ground, and that winter every one of the five trees died."

Mr. Wilkinson has investigated many Indiana trees of reputedly unusual hardiness and of actually unusual success, and he finds, in every case, lawn-grass competition or some other protection from late summer growth.

HOW TO GET BETTER PERSIAN WALNUT VARIETIES
FOR EASTERN NORTH AMERICA AND OTHER AREAS

We should have Persian walnut trees as hardy as our native black walnuts. They should be resistant to blight and of good bearing qualities. There is good reason to believe that such trees can be found, or else bred.

First, we need to search the world for the best trees that have already resulted from the chance labors of nature and man. It was a great agricultural and horticultural misfortune for America that we got our first plants and trees from Europe rather than from China and Japan. For reasons previously stated—greater cold, changes from cold to hot and to cold again; hot, humid summers, and fungi—hosts of European plants have failed in eastern North America, but the plants of East Asia (China, Manchuria, Korea, and Japan) have been subjected to great cold as well as to infernal heat and humidity.

That most astounding American, Thomas Jefferson, traveled much in Europe and was a keen observer and a diligent plant importer, yet he considered himself lucky if he got one out of one hundred European varieties safely across to America and fitted to the climate, but he kept on trying. There's horticultural heroism for you!

There are three great regions of the Persian walnut: (1) the Far East; (2) the Near East, including Persia, the Caucasus, Afghanistan, Himalayan slopes and vales, and Turkestan (yes, there are such trees in Turkestan), and (3) Europe. We draw from the worst of the three, Europe, and the worst part of Europe for breeding hardiness, namely, western and southern Europe, the region having a climate that permitted the survival of the tenderest trees.

We should institute a careful search of the Near East and Far East to find good and hardy parent trees. For example, Lorin Shepherd, M.D., who has served as missionary at Aintab, Turkey, reports having hunted through the mountain districts fifty miles north of that place in a locality which was depopulated by war a hundred and fifty years ago. In that time, the Persian walnut has run wild. Up near the snow line, at six thousand feet elevation, it associates with the beech. It grows in thickets, and the trees are only four or five feet in height, because they are kept nearly prostrate by the great burden of snow that lies on them each winter. Yet Dr. Shepherd reports that they bear good nuts.

Caravans have been carrying walnuts back and forth across Asia for unknown centuries. The splendid walnut trees and nuts in the Vale of Kashmir and its adjacent hills are well known, but there are many other valleys opening out of the Himalaya Mountains and west of them in Afghanistan, and east of them in southwestern China.

It is interesting to speculate on the crop-tree resources of this old region, where each valley is a plant world to itself, and where valleys are separated from valleys by snow-covered ranges or bleak plateaus. For many centuries, skillful farmers have been making their living in these remote mountain valleys and in the remote provinces of China. The Chinese province of Yunnan, with a great variety of elevations, seems to be a veritable tree laboratory with interesting walnut tree developments, as reported by Joseph Rock of the U. S. Department of Agriculture. (See National Geographic Magazine, 1925.) We should have a half dozen men exploring this region, but none, I am told in Washington, has been there since 1929.

We might begin on walnuts with deliberate science where the Asiatics left off their chance improvements.

It should be a comparatively simple matter for us to get the best walnut trees from the many thousands whose crops have been exported to us from North China. The process of importation is simple.

Korea and Japan also offer promising fields for search. The Japanese walnuts are perhaps the most promising of all, because they have been subjected to great heat and humidity and therefore, like most Japanese trees, should have leaves very resistant to blight.

I have seen very fine trees bearing very good-looking nuts (which I did not taste), and this (at Kamisuwa) in the part of Japan which has skating for a winter sport and rice (indicative of humidity) for its chief crop. These trees were in Buddhist temple grounds at Kamisuwa and in several villages a few miles to the north. The whole locality merits careful exploration.

The Yokohama Nursery Company has already demonstrated what can be done in this line. For years they have propagated trees for export. They propagated for me nine out of ten different strains of Korean and Chinese persimmons from cions which I secured in person or by friends in Northwest China and sent from Shanghai to Yokohama by mail. The Yokohama Nursery Company grafted the cions upon native stocks and shipped them to America under the care of the U. S. Department of Agriculture.

The Horticultural Department at the University of Nanking may be at almost any time again ready to propagate trees, and its students from nearly all parts of China, and its faculty, have better opportunity than has the traveling explorer to observe and secure desirable parent trees. With a small expenditure of American money, some commercial arrangement could probably be made; at least it could have been made before China got into war. Many wide-awake American missionaries, provided they can continue in China, would also be glad to cooperate in this work of local observations and in sending specimens of nuts and cions.

Now that I have outlined the general plan of finding the good Persian walnuts in territory that has been but little explored, I wish to submit a couple of choice nuggets of fact.

A dozen years ago, a Protestant missionary by the name of

Paul Crath traveled through Poland and western Russia, whence he brought back reports of huge and productive Persian walnut trees, centuries old, bearing large crops of good nuts in climates where temperatures of –30°F., or even –40°F. were not uncommon. Crath also brought a lot of seed from which seedling trees have since been widely scattered for testing, and there is every prospect that among the hundreds of seedlings of this origin, something very valuable will be found. I have seen some of them that have survived –34°F. without flinching and are said to bear good nuts. (If interested, see *Proceedings,* Northern Nut Growers' Association, J. C. McDaniel, Secretary, Department of Agriculture, Nashville, Tennessee.)

Another hardy strain of Persian walnut is reported by Lynn Tuttle of Clarkston, Washington, who tells of some people, with walnuts in their pockets, who came from a place in northeastern Rumania where the ice on a lake sometimes froze five feet thick. Mr. Tuttle reports that one tree from this source survived, in perfect order, a recent cold winter which destroyed practically all the Persian walnuts in Washington State east of the Cascade Mountains. This tree is patented under the name "Shafer." But I fear its early budding habit.

Between the Shafer strain and the Crath strain (mentioned above), or by crossing them, we should get something that will be far hardier than any of the old varieties that we obtained from France, a country with oceanic climate, mild winters, and cool, not very humid summers.

At present writing, 1950, the Broadview Persian walnut, originally from near Odessa, Russia, has shown remarkable vigor, hardiness, and productivity at a number of places in the northeastern United States. It has survived winters that injured all others in certain frosty localities. It is also bearing well, along with some Craths, for Mr. Royal Oakes, at Bluffs, Illinois. Opinions differ as to quality of nut, but people come back to Mr. Riley Paden, of Enon Valley, Pennsylvania, for second orders after buying the first one.

BREEDING PERSIAN WALNUTS

With a world collection of walnuts as material to work upon, it seems plain that much improvement would result from scientific breeding. A tract of land and a staff of men should become busy on this work at once.

The Persian walnut is especially alluring to plant breeders because of its great variation within the species—variation as to blight resistance, frost resistance, speed of growth, size, shape, quantity, and flavor of fruit, thickness of shell, and in other ways. One of its chief troubles is early spring growing and consequent frost injury. Yet there are strains here and there that remain dormant to an unusually late period in the spring.

For example, I happened to be walking through some orchards near Grenoble, France, on June 10, 1913, and inquired what had killed a tree that stood leafless in the orchard. The owner replied, "It is not dead. It has not come out in leaf yet." This incredible fact was evidently true. A perfectly healthy tree it was, just beginning to show the first sign of growth. Across the road, cherries were ripe, farmers were making hay, and the wheat was in head. This late-blooming tree was not of the best, but its nuts, though scanty, were of quality good enough to cause the tree to be kept.

This type of variation is not rare. As I rode from Milan to Paris on May 18, 1926, I saw from the car window, shortly after entering Switzerland, a number of trees that were much less advanced in foliage than their fellows nearby. Trees with similar habits have been found in America.

We have one tree among our hybrids that continues dormant until about the first of June, about four weeks later than the normal, but after it puts forth its leaves it makes three or four times as much growth as the other trees of the same age. (J. W. Killen, Felton, Delaware, February 8, 1916.)

Our English walnut tree is thirty years old and for the last fifteen years it has borne annually. I think that it has averaged about one

and three-quarters bushels for the last five years. It has over two
bushels this year. It blooms so late it looks like a tree in midwinter
up to the first week in June and then the leaves grow very rapidly.
I am not positive that it bore in 1899 (a year of terrible winter), but
I think it did. I am positive that it bore a good crop last year and
that the mercury was 20 degrees below the previous winter. (Asa
M. Stabler, Spencerville, Maryland.)

I have one of these trees growing beside my lane.

Using these late-blooming strains (and there are doubtless
many others) as a base for breeding, gives reasonable certainty
of getting an almost frost-proof Persian walnut. Working with
East Asiatic strains should give us blight-proof walnuts. It appears reasonable to think that a hardy Persian walnut might
eventually be found or bred to grow and produce well almost
anywhere that our native black walnuts now grow. It may be
that the Crath and the Shafer strains, or their progeny, will
do it.

So much for this one species, but it hybridizes with other
species and does it with great ease. For example: A farmer
near Camden, New Jersey, having one tree which he liked, had
a nurseryman grow a hundred seedlings from its nuts. The little
trees did not look like their mother, and when they began to
bear they indicated that they had a butternut father. The
farmer dug up ninety-nine as worthless, and the hundredth was
scarcely worth keeping. Episodes like this have happened again
and again, but this method might produce great results if
skillfully used.

There is a locally famous Persian walnut tree in Berks
County, Pennsylvania. It is supposed to be two hundred years
old. The owner has been offered five hundred dollars for the
tree as lumber, but it bears so many good nuts that he holds
it. Innumerable seedlings have been grown from this tree by
the thrifty Pennsylvania Germans. Almost invariably, they are
worthless. Apparently, they are the hybrid progeny of a butternut father which stands a quarter of a mile to the northwest.

A Persian walnut, hybridized with butternut, produced a

nut which looked like a Persian walnut. Planted at Catham, Ontario, in the fall of 1910, by December 1924 it was 41 feet high, 19 inches in diameter, with 52-foot spread; it grew in good alluvial soil, with plenty of room. It produced a few nuts, large, thick shelled, with a small kernel which came out almost entire. The tree stood the severe winter of 1917-1918 without the slightest injury. (Information from Professor James Neilson, Port Hope, Ontario.)

These facts indicate both the troubles of growing seedlings and the possibilities of breeding better trees.

Professor Ralph E. Smith, of the University of California, says (Bulletin No. 231, pp. 154-155, *The Economics of the English Walnut*) that in almost every case the crosses between a black walnut and a Persian walnut "show a rapid development and within the first four or five years they assume a size and rapidity of growth several times as great as that of other seedlings [*i.e.,* either parent]. The rapid growth of some of these trees is truly astonishing. Professor Smith points out that these hybrids stand excess of water and drought better than does the Persian parent. Then he tells one more exceedingly suggestive thing. While trees of this cross are almost invariably worthless for fruit, he reports *one* that "seems to produce every year a very large crop of nuts." It only takes *one tree* to found a variety. All the Baldwin apples and all the navel orange trees in the world started from one tree.

The application of well-known methods to the development of the various species and hybrids of walnuts is an interesting task with promising possibilities. It needs imagination and patience backed with money.

CHAPTER XVII

Another Meat-and-Butter Tree—
The Eastern Black Walnut

A NEGLECTED GIFT OF NATURE

The American black walnut, *Juglans nigra,* also called "Eastern" black walnut, may become a greater future asset to the human race than its now more appreciated rival, the Persian walnut. It will surely become a greater asset than the Persian walnut if it should be as much improved. While the Persian walnut started as an almost worthless product of a wild tree, nature has, at the beginning, produced in the American black walnut a product of substantial merit and of some commercial value.

Nothing has been done to improve the black walnut except to hunt for the best wild trees that chance has produced. We have even mercilessly destroyed many fine nut-bearing trees in the quest for its valuable timber. Yet, it has helped to fatten countless American frontier herds of swine. The American Indian made use of the walnut as food. It has been a food of some importance to the colonial American.

Hogs do exceedingly well on walnuts. Stock hogs will winter nicely on walnuts exclusively, but small hogs cannot break the walnuts. Brood sows, for example, will do well on walnuts, needing corn only while suckling actively. Tons of the very richest poultry and hog food can be produced on one acre of land. Two or three mature walnut trees will supply food enough for from one to two dozen hens the three winter months in Kentucky. (Dr. P. W. Bushong, Edmonton, Metcalf County, Kentucky, August 12, 1913.)

Governor Hardman of Georgia told me that he sometimes heard the following sequence of sounds: first, falling walnuts; second, hogs in motion toward the tree; third, the popping of walnuts in the porcine jaws.

Several years ago one fine fall day I was over in Kentucky scouting for a pecan tree I had heard of and went to the cabin of an old Negro whom I knew and found him hulling a very large pile of black walnuts he had gathered. "Uncle Abe, what are you going to do with all of those walnuts?" I asked. "Cap'n," he replied, "I'se gwine to eat these this winter when I don't have any meat." (Letter, J. Ford Wilkinson, Rockport, Indiana.)

For generations the gathering of black walnuts has been a joyful autumn labor of the American country boy, and much rural sociability has gathered around apples and walnuts beside the autumn and winter hearth fire.

Mr. Albert Chandler, lawyer of St. Louis, writes: "Walnuts keep Ozark hens laying in winter. You see a heap like a coalpile back of a good farmhouse (mountaineers in cabins are too shiftless). Nuts are shoveled out on a flat rock and broken with a wooden mallet. The hens pick the meats out clean." (This was easily verified at Linn Creek, Camden County.)

THE INFLUENCE OF FOOD FACTORIES ON THE WALNUT

Now the wild black walnut is participating in the new food era—the era of machine-prepared foods.

Since the commercial manufactures of candy and ice cream have become an established American industry, there has sprung up a surprisingly large trade in wild-walnut kernels.

For two reasons eastern Tennessee and adjacent States are an important region in the production of walnut kernels. One

reason is that it is a good place for walnut trees, and the other
is that it is a country of limited available resources and rather
overcrowded population. Many of the families are large, and
many boys, girls, and women have few opportunities for employment. Picking out walnut kernels offered profitable additional
employment. It was like the cottage loom of Revolutionary
days.

Persons who have never gathered walnuts fail to appreciate
the great productivity of these trees in localities where they
grow abundantly. The *Madison Survey,* a Tennessee paper published by a vegetarian disciple of Dr. J. H. Kellogg, who runs
a school for mountain boys and girls not far from Nashville,
reports that the school went to the autumn woods with a picnic
dinner one day in October, and brought back in trucks and
wagons over two hundred bushels of black walnuts in the hull.

Picking out walnut kernels for sale is a home industry of
many years standing in the Southern Appalachians. One dealer
at Sneedsville, twenty-seven miles from a railroad in northeast
Tennessee, claimed to have sold two hundred fifty thousand
pounds of kernels in 1934.

Small growers can make faster home hulling of black walnuts
by using an old corn sheller—sixty bushels to the hour—or they
can hoist the hind wheel of an automobile, put under it the shoe
of a truck tire, and let the walnuts pass between the tire and the
revolving wheel. It throws nuts and hulls fifty feet. They need
to be separated by hand or by simple mechanical devices, not
difficult to make.

Government inspection to enforce sanitary conditions has
driven most of this business into the factory.

MACHINERY PROMISES TO HANDLE MORE AND MORE
OF THE CROP

Tom Mullins, Mt. Vernon, Kentucky, put one million three
hundred thousand pounds through his cracker in thirty days in
1947. His machinery cost twenty-two thousand dollars. He has
a real establishment, typical of modern food-factory industry.

Mr. Lefkovits, of the Lefkovits and Fleishman Company, Nashville, said in 1947 that his company and another Nashville firm used annually seven to eight million pounds of black walnuts in the shell, bought principally in middle Tennessee and Kentucky.

The Smalley Manufacturing Company, of Knoxville, Tennessee, was paying farmers four dollars fifty cents a bushel in 1947, cracking them by machinery and wholesaling the kernels at seventy-two cents a pound.

Mr. Spencer Chase, of the Tennessee Valley Authority, states that twenty-five million pounds of black walnuts in the shell are produced on the average in the territory served by the T.V.A. This is thought by others to be a conservative figure. Usually the demand is ahead of the supply, now that factory-made food and ice cream are so greatly on the increase.

THE BLACK WALNUT ORCHARD

All of the above-mentioned commercial facts have depended upon wild nuts—the chance product of nature. Few of the readers of this book have seen any black walnuts except the wild variety. An industry is now starting on the basis of commercial propagation of a few varieties of black walnuts—the best wild trees that have been found. The parent trees of these varieties have been selected from millions of wild trees. The search for varieties was made by the Northern Nut Growers' Association, working in conjunction with the Department of Agriculture and a few members of State staffs. One year, prizes were offered for the best nuts. Over sixteen hundred lots had to be examined, a sizable job. That was only one of many prize contests.

This new industry depends upon four facts:

1. The technique of budding and grafting nut trees, which has been recently developed in America. By skillful use of the new technique we may multiply any tree we choose and make of it a variety with an indefinite number of specimens.

2. Several parent trees of superior merit are now available for propagation.

3. An increasing demand for black walnut kernels.

4. The new industry has possibilities of heavy production because of the wide range of territory suited to the black walnut.

This technique has been worked out to the point where it is safe to say that trees of any desired variety can be had in any quantity in a comparatively short space of time. Many private experimenters scattered over the country are successful in grafting walnuts—black, Persian, Japanese—and butternuts, and also many varieties of pecan and hickory. Success in grafting nut trees is by no means as sure as with apples, and the degree of success seems to vary greatly from year to year, probably due to the fact that we do not yet know nor observe all of the controlling factors.

This technique seems to have been first explained for the layman in terms easy to follow in a book called *Nut Growing* by Dr. R. T. Morris (Macmillan, 1921). Anyone interested in nuts should read the book because of its valuable information —and its sense of humor. The essentials of Dr. Morris's methods of grafting are explained in various bulletins. See those of the U. S. Department of Agriculture. I myself have taught the art to at least half a dozen farm hands. I would not hesitate to take any dozen illiterate mountaineers, good whittlers or fiddlers (fiddlers preferred), and if they tried, I could make eight or ten good (but slow) grafters out of the dozen in two hours' time.

One school teacher wrote that she had set ten black walnut grafts (as directed in Dr. Morris's book) and got ten successes. That is better than I usually do. I salute the lady!

Incidentally, some propagators use ring budding exclusively.

As a result of extended search, several varieties of black walnut are now considered worthy of commercial propagation. Several others are believed to be of great merit, but since they have been only recently discovered, there has not been time to test them fully. So many varieties are now being tested in so many localities that the findings of any one year are likely to be reversed. However, I believe that, twenty years from now, we shall know that each of half a dozen varieties is best suited to *some* area. If you want to keep up with this, join the Northern Nut Growers' Association!

The Thomas variety has had the most testing, and it has done well from Canada to Carolina and Texas.

The Thomas variety has fruited at Fairhaven, Vermont, where the winter temperature was $-30°$ F.

The first orchard of black walnut trees to make a commercial income was that of E. A. Riehl of Alton, Illinois, who planted some gulch banks and bluff sides overlooking the Mississippi River to walnuts and chestnuts.

As early as 1915 Mr. Riehl reported, "We have found by actual test that the Thomas gives over ten pounds of meats to the bushel, and with care, ninety percent are unbroken quarters."

The flavor of these nuts is of unusual excellence. The tree is a fast grower, though somewhat subject to loss of leaves in late summer from fungus. I have had seven-foot trees of this variety produce nuts in the nursery row.

A crop of black walnuts to occupy winter hours of farm labor appears to be a very effective item in farm economy, especially on the family farm.

A number of plantations, mostly of the Thomas variety, are now in bearing. I had to pay eight dollars a bushel for some Thomas variety seed in the autumn of 1947.

Concerning an orchard on the Eastern Shore of Maryland, Mr. H. Gleason Mattoon reports as follows:

This year (1947) the Andelot orchard produced 557 bushels of Thomas, 73 Stabler, 53 Ten Eyck and 162 Ohio. It is the poorest yield for Ohio since 1939 and the best ever for Stabler.

The original planting in 1932 was 180 Ohio, 180 Thomas, 180 Stabler and 125 Ten Eyck. The losses which have not been replaced amount to about 50 trees. They were planted on 50-foot centers with peach interplants. I do not have accurate records of the orchard, but the 165 Ohio trees have borne between 300 and 500 bushels every year since 1940, with the exception of the present one. Stabler has always borne poorly. Thomas biennially bears a nice crop.

Call it 330 bushels of Ohio on the average from 1940 to 1947 (and that is low). That is two bushels per tree for the 165 trees at 16 trees to the acre (there is space for 17), 32 bushels, 320 pounds of walnut kernels. At 3,105 calories per pound, that is the food equivalent of more than 1,000 pounds of round steak (950 per pound) and this from trees interplanted with three peach trees to one walnut tree.

This seems to show that the walnuts are very much greater food producers than almost any other source of protein crops except the bean and the peanut.

In my opinion the real future of the black walnut is in permanent pastures, about six to ten trees to the acre.

There is no reason to think that the best varieties of black walnut have yet been found, and it is highly probable that trees better than any now living have been destroyed in the slaughter of trees which has marked the whole era of the white man in America. Some of these tree tragedies have probably destroyed parent nut trees that would have been worth millions if propagated in sufficient numbers. The following episode is a good illustration of this point.

Mr. Harry R. Weber, a lawyer of Cincinnati, for whom nut trees are an avocation, found in two successive years the shells of a black walnut resting on a wing dam in the Ohio River near Cincinnati. The shape of this shell both inside and outside bore such a resemblance to that of a Persian walnut that its kernel must have been very easy indeed to remove. Where had this nut tree grown—in all the wide reaches of the Ohio Valley above this dam upon which it had floated? Wide search, correspondence, and newspaper publicity (I wrote to dozens of country newspapers in New River Basin) all seemed finally to fix the place in Floyd County, Virginia, near the headwaters of the New River. Mr. John W. Hershey, of Pennsylvania, made a five-hundred-mile journey to investigate the hillsides of Floyd County. When he arrived, the farmers showed him a stump in a bare pasture field. The lumbermen had cut this probably matchless tree, and not a sprout remained.

THE DEMAND FOR BLACK WALNUT MEATS

The black walnut is unique among commercial nuts in retaining its flavor when cooked. Cooking makes many other nuts lose flavor, but the black walnut comes through as tasty and attractive as ever. This is a great advantage in this age of factory-made food—ice cream and candy, nut bread and nut cake.

With the movement on in America for good health and physical efficiency, and with the present high cost of meat, there is an increasing emphasis upon the meatless diet. A large increase in population will force us in that direction through scarcity of meat. Under such conditions, tasty black-walnut bread made of whole-wheat flour is not only good, nutritious, and wholesome, but is almost a complete substitute for bread, butter, and meat.

Ice-cream manufacturers have been trying to buy walnut meats in twenty-thousand-pound lots. Apparently, the future demand for the black walnut might far outrank the demand for the Persian walnut, which must be eaten uncooked and is not the equal of the black walnut for candy, cake, or ice cream. Therefore, the Persian walnut has less potential value than the black walnut in American and European diet.

THE BLACK WALNUT AREA

The territory for the black walnut industry in the United States is wide. In this respect it exceeds corn, our most widely scattered important crop. One single species, *Juglans nigra,* thrives in northern New York and in southern Georgia, in north-central Wisconsin and in south-central Texas, and from central Massachusetts to western Kansas, Nebraska, and Oklahoma. with a substantial slice of South Dakota and Minnesota included in its range. G. E. Condra, who knows Nebraska from end to

end and corner to corner, reports walnut trees in every county
in that State. Roughly, this walnut belt covers most of the Corn
Belt, most of the Cotton Belt, and tens of thousands of square
miles of Appalachian and other eastern hill country on which
no type of agriculture can survive but grass, trees, or terraces.

BETTER VARIETIES OF BLACK WALNUT

We should never lose sight of the fact that at the present
moment the black walnut industry depends on chance wild
nuts and that we may find better specimens any day. Certainly
we should expect to breed better nuts, much better, perhaps
rivaling the Persian walnut in physical form or at least in avail-
ability of kernels. This can be brought about by deliberately
breeding the best black walnuts we can find and perhaps hybridizing them with other species of walnuts. As a step in this
direction, I now have rows of black walnut trees from Thomas
nuts pollinated by Ohio and other good nuts. That is breeding
of a kind, and the Thomas seems to be somewhat prepotent,
that is, seedlings resemble parent.

America should have two or three persons employed on this
task of testing out several thousands of these hybridized seedlings of promising ancestry. Much time could be saved by grafting these young seedlings onto mature trees and thus bringing
them to fruit sooner than by waiting for them to grow large
enough to produce nuts on their own tops.

THE BLACK WALNUT AS A FOOD PRODUCER

The food values of the black walnut can be seen by studying
the food table (Appendix) and comparing the food yields of
some other crops, especially pasture, because pasture produces
meat, the rival of nuts in the production of protein as food. The
good pasture of England or Illinois gives about one hundred
and fifty pounds live weight per year of mutton or beef. Of this,
nearly half is waste in slaughter, and there is considerable waste
in the meat. Therefore, twenty pounds per year of nut meats

from a black walnut tree come close in actual pounds of edible food to the product of an acre of blue grass.

There is a noticeable resemblance here between the French equivalence in the rental of an acre of land and a walnut tree. (See p. 209.)

The nutrition value of the nut meats from an acre of black walnuts reported above by Mr. Mattoon puts an acre of grass to shame as a unit of the nation's food production.

Grass grows well beneath black walnut because of its deep root system and its thin open foliage, which casts only a light shade. (United States Department of Agriculture, Farmers' Bulletin No. 1392, p. 8.)

James Dixon, landowner and bank president of Easton, Maryland, says that wheat beneath walnut trees seems to be actually better, and Mr. Ford Wilkinson, of Rockport, Indiana, says, "A catch of red clover can be gotten under a black walnut tree almost any season whether ground has been limed or not."

The cost of extracting the kernels of the walnuts from an acre of land is probably more than that of slaughtering and dressing the meat from an acre of land. However, I am not certain of this. Machinery is being developed for the black walnut, as it has been developed for the California walnut industry. Perhaps some kind of a mechanical crusher might open a market for inferior black walnuts for poultry food on the farm, the chickens picking out the meats.

Dried walnut hulls are easy to carry in an automobile and have a remarkable capacity for making a machine stick to an icy road.

CHAPTER XVIII

A Group of Meat-and-Butter
Trees—The Other Walnuts

In the preceding chapters I have given much space to presenting facts about two species of walnuts and the philosophy of the subject in general. In brief, this is: (1) Find the best existing strains—and we apparently have the basis of a good industry now. (2) Breed better strains by selecting, crossing, and hybridizing, and we have the basis of a better industry.

This philosophy is applicable in varying degrees to each of the other species of walnuts, which are as follows:

1. Butternut, *Juglans cinerea*
2. California walnut, *Juglans californica*
3. Texas walnut, *Juglans rupestris*
4. Arizona walnut, *Juglans major*
5. Chinese walnut, *Juglans regia*, var. *sinensis*
6. Manchu (Siberian) walnut, *Juglans mandshurica*
7. Japanese walnut, *Juglans sieboldiana*, var. *cordiformis*, sometimes called "heartnut."

For a fuller discussion of various species, see *Nut Growing*, by Robert T. Morris (Macmillan). There is always, of course, Sargent's *Silva*, for an exhaustive presentation of questions of species and Standard Plant Names for the latest bulletin from the battlefront of the name callers. I've named a tree or two myself without any license whatever. It's terrible, anyone can call a tree any name he pleases and there is no one to protect the poor thing.

The space limitations of this book prohibit full discussion of these several species, but I wish to emphasize a few points of especial significance.

The butternut (No. 1, above) grows in colder climate than the black walnut. In the 1922 *Proceedings* of the Northern Nut Growers' Association, p. 72, J. A. Neilson, Professor of Horticulture, Port Hope, Ontario, says:

The butternut is much hardier than the black walnut and has a much wider distribution in Canada. It occurs throughout New Brunswick, in Quebec, along the St. Lawrence basin, and in Ontario from the shore of Lakes Erie and Ontario to the Georgian Bay and Ottawa River. It has been planted in Manitoba and does fairly well there when protected from cold winds. West of Portage la Prairie the writer observed a grove of seventy-seven trees. Some of these were about thirty-five feet tall, with a trunk diameter of ten inches, and had borne several crops of good nuts.

The kernels of the better specimens come out of the shell more easily than do those of black walnuts. The butternut offers interesting crop possibilities for the northern section of the United States. Some people prefer its nuts to the black walnut. A selected grafted variety, the Deming, was reported by J. F. Jones, Lancaster, Pennsylvania, to bear when it is two feet high.

The Chinese walnut (No. 5, above) was long classified as a separate species, but botanists have now become convinced that it is merely a variety (suggestive fact) of the Persian walnut. Therefore, it has been discussed, either directly or by implication, in Chapter XVI.

The Japanese walnut is a species of exceeding promise. In its native home it grows throughout the climatic range of Japan, embracing climates as dissimilar as those of Nova Scotia and Georgia and all between, and accentuated by the reeking humidity of the Japanese summer with its strong fungus tendencies.

The tree also thrives in a great range of soil from sand to clay. Apparently it is a veritable goat in its feeding habits. This makes it a very rapid grower, and in rich soils a single leaf is sometimes a yard long.

It is precocious, some seedlings producing fruits at four or five years'of age. It bears its fruit in long clusters and is very prolific. It should be noted, however, that the tree is by no means long lived like the black walnut. Professor Neilson further reported significant facts about the Japanese walnut, heartnuts.

Seed planted in the spring of 1924 at Winnipeg. July 20, 1927, tallest tree was twelve feet high, one and one-half inches in diameter. They are also growing nicely at St. Anne's in Quebec, near the mouth of the Ottawa River.

In a not especially favorable location in Sharp's backyard at Riverton, New Jersey, is a fifteen-year-old Japanese walnut, which receives no especial encouragement but produces annually four bushels of nuts. (Letter, Joseph H. Willits, Professor of Industry, University of Pennsylvania, October 10, 1912.)

We found that the Japanese walnut was happy from the start, and three years after planting produced an abundance of nuts combining the good qualities of both the American butternut and the black walnut, with meat much thicker than the butternut and not nearly so oily, an improvement on the black walnut and butternut as well, and a vast improvement on these trees in respect to leafing, as the Jap is one of the earliest trees to put out its leaves in the spring, far ahead of the black walnut of equal size and at a much more tender age. It is one of the most interesting trees we have because of its bloom at the end of the branches and its marvelously long, plump catkins scattered along the trunk, and the nuts, instead of being formed singly, are produced on long stems like an elongated bunch of grapes, having as many as twenty-two nuts on a stalk. (Long Island Agronomist, Vol. VI, No. 6, January 1, 1913.)

Mrs. R. S. Purdy, 218 South Willard Avenue, Phoebus, Virginia, has the j. seiboldiana sample I sent you. The year that I was at Lancaster (1912) I visited this tree; it was then eighteen years old and bore that year sixteen bushels of shelled or rather hulled nuts. . . .

They grow in clusters of twenty-four nuts. The tree was planted by a little girl twenty years ago from a nut she got from a sailor at Old Point Comfort. The tree is about a foot in diameter. It is very powerfully rooted. (Letter, G. H. Corsan, University of Toronto, December 23, 1914.)

The wood, unfortunately, is soft and of little value, but we can scarcely expect one tree to have all the virtues until after breeding work has been done. It has recently shown a tendency to be attacked by an unidentified virus.

The Japanese walnut merits much attention at the hands of plant breeders. It is at present in rather bad repute because ignorant or unprincipled nurserymen have scattered its seedlings widely over the United States, calling it the English walnut; but the specimens were only seedlings of no particular merit. The result of this deception has been to dampen the ardor of many planters. Other nurserymen, in good faith, sold seedling trees produced from nuts borne on Japanese walnut trees in this country. These trees turned out to be hybrids that could scarcely be distinguished from butternuts of average quality or less. They had resulted from the very active hybridizing susceptibilities of Japanese trees growing within the wind-blown pollen range of butternut trees. It seems almost as if the Japanese walnut chooses butternut pollen rather than its own if it has the chance. Therefore, deliberate hybridizing of selected trees becomes very easy.

The heartnut, of which grafted trees are now available in several nurseries, is merely a variety of the Japanese walnut. Some heartnut trees produce nuts whose kernel comes out in one piece, a fact of great commercial significance.

Recently, some butternut x Japanese hybrids have been commercially propagated under the name Buartnut. They are pretty trees for the yard and give people in the northern tier of States and in Canada a chance to grow nuts.

Consider the crossing of good specimens and the hybridization of the various species of walnut with each other. Consider this in connection with all the qualities above mentioned, as well as the fact that Japanese, Manchurian, and Chinese specimens come from sections of Asia having cold winters and hot, humid summers, thereby showing climatic relationship to the cold northwestern sections of the Mississippi Valley.

In view of these considerations, it seems clear that we have the opportunity to make a great number of walnut varieties suited to a great number of climates and conditions. It might easily be within the range of possibility that we can produce some kind of walnut to be grown especially for pig and poultry food. The ability of walnuts to retain feeding value while lying on the ground for some weeks enhances their suitability for pig feed.

You will be interested to know that a letter from Dr. Edwin D. Weed, Duluth College, Duluth, Minnesota, states that Chinese walnut trees I sent him went through the winter 1925-1926 well and are growing vigorously. These (the seed) were from ten thousand feet elevation in North China. (Letter, J. F. Jones, 1927.)

CHAPTER XIX

The Pecan—King of Hickories—
A Type Study in Tree Crops

The hickories are a great family of food producers. They will be even greater in the future if scientific agriculture prevails.

The pecan is, at present, the king of the hickory family and affords an excellent and nearly completed case to illustrate the idea that wild trees can become the basis of new crops.

The pecan has passed rapidly through a number of interesting stages in its utilization by man.

1. It started as a wild tree, covering a large area and producing large quantities of nuts, valued as mast, mostly unused otherwise by man.

2. Trees of superior producing quality were selected out of the mass of mediocrity, and the attempt was made to propagate them by planting seed.

3. Seedlings from superior trees produced many variations (Fig. 94), and almost without exception these seedlings were inferior to the mother tree. The result was paralysis of human enthusiasm and general neglect of the species.

4. The technique of propagation was worked out. Then in a manner exactly comparable to the development of varieties in apples, selected trees were propagated, giving rise to named varieties such as Stuart, San Saba, Schley, Busseron, Indiana, and others.

5. Grafted and budded pecan trees were planted by the hundreds of thousands. Orchards were developed. An industry was achieved.

The new industry gave proof of its reality with a product worth millions of dollars; a national association of growers; widespread attempts to control diseases and pests and to solve the cultural problems; establishment of a national pecan experiment station; a beginning by various State experiment stations in study of the industry, and finally, a flock of bulletins from many States and from the United States Department of Agriculture.

6. A final stage of the industry has been reached with laboratories for research and experimentation as to the use of the product, and their natural accompaniment of factories for the manufacture of pecan foods for distribution by bottle, carton, and can. This puts it in the rank of established American food industries.

The pecan has arrived. It is not merely prospective or possible, as with so many of the things discussed in this book.

The industrial record includes one more phase so typical of new and alluring American industries, namely, promotion, speculation, and swindling enterprises. The pecan has been a shining example of this. Yet more, bringing the pecan forward has developed a substantial mythology which still has its faithful believers, especially as to where the pecan grows and where it will grow. For unknown centuries, the tree has braved the blizzards of Missouri, Illinois, and Indiana; yet millions of our people think it is only a neighbor of the orange and a denizen of the Deep South.

THE NATURAL RANGE OF THE PECAN

Before the white man began to spread the pecan, the tree was native to a corner of Iowa and to a large part of the Mississippi basin south of Iowa. The pecan was found on the Ohio River from southern Ohio westward; up the Mississippi to southeastern Iowa, and thence southward almost to the mouth of the Father of Waters. In the Missouri River Valley it reached the extreme northwestern corner of the State of Missouri. Thence southwestward across eastern Kansas, extending into this State about one hundred and twenty-five miles along the southern boundary. Pecans lined the streams in the greater part of Oklahoma, almost all of the streams of eastern Texas, and on into Mexico. Eastward of the Mississippi the pecan was found in great abundance in the Ohio Valley up to and including a corner of Ohio; also through central Kentucky and Tennessee and in a few parts of Alabama. Altogether, the area within these boundaries includes something over a half million square miles, including parts of thirteen States (Ohio, Indiana, Illinois, Iowa, Missouri, Kansas, Oklahoma, Texas, Kentucky, Tennessee, Alabama, Louisiana, and Mississippi).

Over this area there were many millions of wild trees. Professor of Horticulture E. J. Kyle, in Texas, claims seventy-five millions of wild trees in that State. For an unknown time, these and previous millions of pecan trees have been producing hundreds of millions of pounds of nutritious crops. They went to decay, or for the food of wild animals, but to some extent to feed the American Indian. During the regime of the tribal leaders in the old Seminole Nation in Seminole County, Oklahoma, they had a law that fined a person five dollars or more for mutilating a pecan tree. (*American Nut Journal*, August 1927, p. 29.) Yet, some people call the Indian a savage. Whoever calls the Indian a savage should go look at the gullies we white men have made in Oklahoma where the Indian made *none!*

THE PECAN CLIMATE AND EXTENSION
OF THE PECAN AREA

Pecan trees of great size bearing excellent nuts grow wild in the Ohio valley, but by chance the pecan received earlier and more attention in the South than in the North. Accordingly it spread more rapidly eastward through the South than through the North. Hence the origin of the still persisting myth that the pecan belongs only in the Cotton Belt.

This belief in Cotton Belt exclusiveness is an example of the ease with which patent error survives, for the first settlers of Illinois, Indiana, and Missouri found on their lowlands thousands of pecan trees from two to three feet in diameter and one hundred feet high. It was (and is) a common practice to leave them when clearing. Many stand in cornfields today. Stately pecan trees planted by George Washington in 1775 ornament the lawn of his home at Mount Vernon. Large, beautiful, healthy pecan trees are scattered through northern Virginia, northern Maryland, and southern Pennsylvania, in a climate typified by that of Philadelphia.

According to the *Proceedings* of the Northern Nut Growers' Association, 1925, p. 98, Thomas Jefferson presented George Washington with some pecan nuts which he planted with his own hand around Mount Vernon, March 25, 1775. According to the late C. S. Sargent, Director of the famous Arnold Arboretum, these trees, respectively 86, 97, and 98 feet high in 1925, "probably have not lived out half their lives."

DeCourset, a Frenchman who served with Washington, left a record that "the celebrated gentleman always had his pockets full of these nuts, and he was constantly eating them." It is amazing and also suggestive to know that Washington's fruitful diary speaks of them as paccane or Illinois nuts.

I know a pecan tree near Hughesville, in Loudoun County, Virginia, forty-five miles northwest of Washington, at an elevation of five hundred feet, in a climate almost identical with that of Philadelphia, except that it has greater extremes of cold ($-30°$ F.). That tree is about eight feet in circumference, eighty feet high, with a spread of seventy feet, bearing fruit, and according to the oldest inhabitant of a generation now gone, it is about one hundred twenty-five years old. A few miles away at the county seat of Leesburg there is another old pecan tree with a girth of eight feet four inches and a spread of ninety feet.

Harry R. Weber, of Cincinnati, Ohio, reports a southern pecan tree about one hundred years old at Lebanon, thirty miles northeast of Cincinnati, girth twelve feet eleven inches, spread ninety-three feet, height eighty feet.

The Illinois origin of a perfectly healthy specimen at Mont Alto, Pennsylvania, altitude 1,100 feet, raises the interesting question of the origin of many of these northern pecan trees of great size.

In the village of Mont Alto, Franklin County, Pennsylvania, a tree is growing with the following data:

Diameter breast high (D.B.H.) 22 inches; total height, 55 feet; clear length of trunk, 15 feet; height of crown, 40 feet; width of crown, 40 feet; age 47 years.

This tree bears fruit every year. The quantity is, however, small, considering the size of the tree. The owner said the yield was about seven quarts above the amount that his own and his neighbors' children ate. I must admit that I would not wish to estimate the annual consumption by the children. I was told by a reliable person that this tree grew from a small tree that was brought from Illinois by a son of the then owner of the property. The son was later elected mayor of Quincy, Illinois, but as to whether he got the tree at or near Quincy, Illinois, I am not able to say. (J. L. Illick, State Forester, Mont Alto, Franklin County, Pennsylvania, Pennsylvania Department of Forestry, State Forest Academy, June 5, 1915.)

A tree at Colemansville, Lancaster County, Pennsylvania, is nine feet, ten inches in circumference at two feet from the ground, stands on a rocky hillside, and is reported to be bearing well. (J. F. Jones, Lancaster, Pennsylvania.)

C. A. Reed, U. S. Department of Agriculture, says that Easton, Maryland, contains the largest planted pecan tree known: girth (1920) 15 feet breast high; reach 129 x 138. In 1927 it measured 16 feet, 1 inch girth at 4 feet, 6 inches.

In a park in Hartford, Connecticut, there was a pecan tree ten feet in circumference, perfectly hardy, standing alone. It was planted as a nut in 1858 by Frederick Law Olmstead. It ripened at least one nut in the season of 1923. It ripened that nut because Dr. W. C. Deming fertilized the blossom by hand with bitternut pollen, *a very significant fact.*

This information comes from Dr. W. C. Deming, long-time Secretary of the Northern Nut Growers' Association. The tree would probably produce abundant crops, except for a habit very common among pecans, namely, that its pistillate blossoms do not mature at the same time as do its staminate blossoms. This tree has since become a hurricane victim.

On Spesutia Island in the Chesapeake Bay, latitude 39° 15′, is

. . . a giant one hundred and six feet tall. It has a spread of one hundred and ten feet. It has two limbs, respectively fifty-seven and sixty feet long, and is thirteen feet in circumference, three feet from the ground. It is an annual bearer of thin-shelled nuts that, though rather small now, are mighty good to eat. . . .

A seedling from this tree is eighty feet tall with an equal spread and is a particularly beautiful tree—when I saw it there were two or three nuts on nearly every twig end. They are fair size, too, very thin-shelled, and very pleasant-tasted. (Extract from newspaper article by Wilmer Hoopes, Forest Hill, Maryland; information from Robert H. Smith, Spesutia Island, Perryman, Maryland, January 23, 1915.)

The pecan is native to North America. Therefore, it is accustomed to spring frosts by hundreds of thousands of years' experience. Therefore, it sleeps late in the spring. Therefore, it can survive winters in places where the summer will let it ripen its fruit rarely or possibly not at all. Hence such surprising facts as these:

1. Thrifty trees at Michigan Agricultural College, East Lansing, grown from Iowa seed planted by Liberty H. Bailey.

2. Trees grown from western Texas seed, latitude 35° 30′, longitude 100°W., enduring −10° in latitude 39 in northern Virginia, but not ripening seed. Shortly before the year 1900, Mr. Thomas Hughes, a schoolmate of mine, sent some pecan

nuts from his home in Sweetwater, Texas, latitude 35° 20′ north, longitude 100° 20′ west, to Mr. A. B. Davis, nurseryman of Purcellville, Loudoun County, Virginia—Philadelphia climate. Trees from this seed seem to be perfectly hardy, but seldom if ever have time enough in which to ripen their fruit. When topworked to good Indiana varieties, they bear in four or five years and ripen the nuts nicely.

3. Trees from Iowa seed ripening nuts at Lincoln, Nebraska.

4. Pecan trees thriving and ripening seed (rarely) fifteen miles north of Toronto, Canada; also trees from Georgia seed thriving in southern Ontario. James Neilson, Professor of Horticulture, Port Hope, Ontario, told me of a thirty-foot pecan tree from Georgia seed growing on the farm of Theron Wolverton, Grimsby, Ontario; of one thirty-five feet in height at Simcoe, Ontario, on the farm of Lloyd Vanderburg, who brought the seed in person from Missouri; of a group of five trees at Richmond Hill, Ontario, thirty-five feet high, fifty years old, standing in sod with no attention, seed from southern Indiana. Young grafted northern pecan trees are doing well and enduring ten degrees below zero on the farm of John Morgan at Niagara-on-the-Lake, Ontario. Professor Neilson also reports (Northern Nut Growers' Association Annual Report, 1923, p. 25) pecan trees fifty years old, thirty-five feet tall, perfectly hardy on the farm of C. R. Jones, at Richmond Hill, fifteen miles north of Toronto, latitude 43-45° north. They rarely ripen, but in the year 1919 did so. I have seen these trees. They are fine.

5. Very surprising is a communication from J. U. Gellatly, Westbank, British Columbia (*American Nut Journal*, April 1928, p. 65), reporting successful fruiting of good pecans five years after planting in the orchard. This is at an elevation of fifteen hundred feet in the Okanogan Valley.

6. Most remarkable of all, perhaps, is a thrifty pecan tree at Fairhaven, Vermont, latitude 43° north, altitude 530 feet. This tree is the lone survivor of many attempts by Mr. Zenas Ellis,

an enthusiastic private experimenter. It blooms, but being alone it does not set fruit. It is very suggestive breeding material.

All these facts go to show that the pecan has great possibilities as a shade tree in a large area where it cannot be a commercial dependence, but may produce an occasional crop. It is a beautiful and majestic shade tree, with alluring possibilities through hybridization.

Another piece of pecan mythology is to the effect that the pecan is limited not only to the Cotton Belt but to alluvial soil. Most people east of the Mississippi believed this in 1910. Perhaps this piece of mythology spread eastward from the West. It is true in the southwestern pecan country, because in many parts of Kansas, Oklahoma, and Texas, natural tree growth is limited to the valleys, beautifully pecan-bowered valleys reaching back with their long ribbons of green through the upland pastures of the hills yellow and brown with drought.

In a large part of the Texas area, good tree growth of any species is limited to the river valleys, in an area where the upland is often too dry for agriculture or good forest. Indeed, the river valleys of a great area in central and western Texas are forest islands nursed by the waters of the adjacent streams and nearly or quite surrounded by slightly arid land. In places one can stand on a plateau with a rocky, shallow soil a hundred feet or more above the stream. On this height the rainfall of twenty or twenty-five inches will support only a scrubby growth of drought-resisting scrub trees, but from this point one can look down into the valley which a stream has carved from the plateau. Because of the moisture from the stream its banks and flood plain are covered with magnificent trees, among them tens of thousands of pecans. These pecan islands reach far back, almost to the headwaters of the streams that drain the Edwards Plateau in west-central Texas.

William A. Taylor, Chief of the Bureau of Plant Industry, U. S. Department of Agriculture, Washington, D. C., wrote me January 14, 1920:

A cousin of my father, who located in what was then Tom Greene, a county of approximately the size of Massachusetts, told me thirty years ago that pecans were being wagoned to San Angelo from points 100 to 120 miles further up the north fork of the Concho and its tributaries (latitude 32° north, longitude 102° west).

Seventy-five miles to the southeast of Sweetwater at Coleman . . . and at Brownwood, a little farther east, it, the pecan, is or was altogether the dominant river and creek-bottom tree twenty years ago, and I presume it is still a conspicuous feature of the Texas landscape.

Shortly after the publication of the first edition of this book, I received a letter from one of its readers, a forester in the employ of the Soviet government. He wrote me from some place in Turkestan. He sent me pictures of scenes identical with the dry Texas upland, from which one looked down into the narrow valley full of pecan trees, in this case walnuts—Persian walnuts, I think.

My correspondent invited me to come to Russia and ride with him and see the trees, but my attempts to reach him by mail failed. I hope they did not cause the poor fellow's liquidation.

There was small reason for the people *east* of the Mississippi to believe the alluvium myth.

Testimony as to the upland growth of the pecan east of the Mississippi River can be piled up almost indefinitely.

A great many of the large trees of the North and East, including most of those previously mentioned, are on upland.

Mr. B. T. Bethune, Georgia, wrote in the *Rural New Yorker*, February 3, 1912:

In Middle Georgia in the "Red old hills of Georgia" underlaid with granite, we have pecan trees more than three feet in diameter three feet above ground, whose branches reach a height of seventy feet, with a spread of sixty feet, and which have born successive crops without a single failure for three-quarters of a century. . . .

Elsewhere in his article Mr. Bethune speaks of a tree from which over nine bushels of nuts had been sold: "That tree grew on a poor ridge in the pine woods a few miles south of this city." Alas, I cannot learn what city.

Soil is one thing, but frost is another. Late frosts are as deadly to pecans as to most other blooming plants. Hence the following advice from one of the pioneer pecan experimenters, Mr. T. P. Littlepage, who had an orchard between Washington, D. C., and Baltimore:

Under no circumstances should northern nut trees be set on low land. The northern pecan on my farm is just as subject to frost as peaches, or more so. As a result I do not think I will have a peck of pecans this year on thirty acres. But I am equally sure that were they on nice high peach land, the average successful crops would be equal to those of peaches or apples.

THE IDEAL PECAN SOIL

The facts about the pecan seem to be that it was native to alluvium; therefore, having had opportunity to get food easily, it has not developed ability to fight for food in less favorable locations. The pecan, therefore, needs deep friable and moderately moist soil such as would naturally produce a forest of white oak, hickory, and walnut trees.

W. C. Reed and Son, of Indiana, pioneer experimenters with grafted northern pecans, reported as follows on their 1926 crop:

Crop varied from twenty to fifty pounds per tree; think two trees bore seventy-five pounds each.

Trees were planted twelve years ago on high clay land.

They have been cultivated regularly.

Were not fertilized, but were on good, strong land.

Trees are from thirty to thirty-five feet tall.

Mr. Reed sold these nuts to nearby grocers at thirty cents a pound.

J. Ford Wilkinson, Rockport, Indiana, another pioneer, reports in a letter of January 26, 1928:

My budded pecan trees growing on high-land clay soil are bearing
remarkably well; in fall of 1926 transplanted trees from 10 to 13
years old produced from 25 to 85 pounds of nuts each, younger trees
bore accordingly, some 5- to 6-year-old trees not transplanted produced from 5 to 10 pounds each.
These trees are growing on good land, are fairly well cultivated,
but have never been fertilized.

With soil, as with probably everything else, continued experience will probably disclose new problems.

I am sending you the record of one row of eighteen trees that illustrates the utter impossibility of conveying precious facts briefly.

Year	10	11	12	13	14
Average of first six trees	13	28	13	22	17
Average of last six trees	½	3	2	3	5

Here is the interesting fact that, because of the difference in the
soil at one end of this row from that at the other, there is the difference of a profit or a loss. This difference was not to be discovered
by the average man, for the land appeared to be much the same, and
only years of careful observation have made clear the importance
of selecting orchard soils with infinite care. (Letter, Mr. C. A. Van
Duzee, Cairo, Georgia, May 11, 1927.)

A HARDY TREE

After thirty years of experimenting with it and submitting
it to many rough tests, I find that the transplanted pecan tree
needs distinctly more petting in its early stages than the apple
tree. In contrast to this the seedlings are very tough.

The pecan can scarcely be called a tender tree. Once it is
established, its great root system makes it hard to kill. Mr. Ford
Wilkinson, of Rockport, Indiana, reports three-foot seedlings
with roots nine feet long.

An orchard of fine-looking trees has this history:

This grove is located at New Harmony, Indiana, and was the first
pecan grove planted in this State, and it has had a varied history.
The seed was saved from a very fine pecan by John B. Elliotte of
New Harmony and planted in the fall of 1876. Trees were grown
in Elliotte's Nursery for two years and then planted in the grove by
Jacob Dransfield. The first winter after setting, the rabbits cut them
all back to the ground. They came up nicely the next spring, and
Mr. Dransfield, to keep the rabbits off, set a four-inch drain tile
over each one, and as they grew and the wind switched them around,
it cut every one of them off; so that the damage was the same as that
done by the rabbits. Mr. Dransfield then gave it up. After the Ohio
River had overflowed several times, he paid no attention to the trees
for several years but cultivated the land in corn. The pecans, however, were not so easily gotten rid of and kept coming up each season
until finally they let them grow and the grove is the result. (Letter,
W. C. Reed, Vincennes, Indiana, June 19, 1916, who said information was from Mr. Elliotte's son and Mrs. Dransfield, who were still
living.)

After digging up a pecan nursery, I have trouble with the
shoots that come up from the roots.

Unfortunately, the blossoms are not so hardy as the trees. I
have not tabulated the record, but I have noticed that in some
seasons a combination of weather factors will kill pecans, walnuts, and hickories, while apples and peaches come through
with fair crops.

NATIVE PRODUCTIVITY OF THE PECAN

In its wild condition, the pecan is a tempting tree, but the
industry had to await the recent coming of the art of propagation. The wild pecan was an important asset to the early settlers
of the central Mississippi Basin.

In 1916 I planted bunches of pecan seed and black walnuts
together in the blue grass (clay soil, good for about thirty-five
bushels of corn per acre) along my upland lane in Loudoun

County, Virginia, altitude seven hundred fifty feet. They were not fertilized, cultivated, or in any way protected. It was a test. Both species were able to fight it out with the grass and make a high percentage of survival. In the drier places it took the seedlings ten years to get five feet high, but they were very stocky and by the end of the decade they had begun to grow more rapidly. There is little doubt that most of the pecans will eventually become large trees if let alone. Where clumps of blackberry bushes invaded the grass, the trees are larger than the others. Grass is a deadly enemy to small trees, and in many (or most) cases it may smother them fatally in the infant stage.

In the valley of the lower Ohio, as in the vicinity around Evansville, there are almost solid forests of pecan. I have seen one tree there six feet in diameter; I have seen them towering twenty or perhaps thirty feet above the top of the white oak forest. This locality is one of many in the middle and lower Mississippi Valley.

Up to four or five years ago wild pecan trees were very abundant along all the streams in certain sections in southwestern Missouri, particularly Bates County. They were so abundant that it was the practice of many rural residents to harvest the pecans in the fall by cutting down the trees. . . . In many a woodchopper's cabin these wild pecans filled an important place in the dietary of the family. The same was true with the early settlers along the bottom lands of the Missouri and the Mississippi rivers in the state of Missouri. Within my own recollection I have known cases where families looked upon their winter supply of nuts, including the wild hazelnut, black walnuts, and pecans, as necessities rather than luxuries. Of course, this order of things is entirely changed now except in remote regions. (Letter, W. L. Howard, Assistant Professor of Pomology at the University of California, January 16, 1917.)

Professor C. J. Posey, University of Kansas, tells me that between 1881 and 1886, when he was a boy, on the Kaskaskia bottoms fifty miles east of St. Louis, the land law was that each man had to fence his own crops against roving stock. The bottoms were open, chiefly wooded, and it was customary to let the hogs run. The farmers would gather up the sows and pigs in the spring before the young had left their mothers. Each owner marked his own with his particular brand, usually nicking their ears. He would let them run, giving them a little feed so that they would stay within reach. In the autumn the young were nearly as wild as deer and were sometimes ready to be slaughtered without feeding but were fed a little at the edge of the clearing to keep them within reach.

"With a little corn for bait, the farmer would go to the rail fence at the edge of the clearing and holler. With merry grunts up gallops your year's meat supply!"

No wonder the early settlers of Illinois settled in the timbered lands along the streams and thought the prairie worthless. (See *North America*, J. Russell Smith, p. 297.) By 1918 all this had changed. Each man had to fence in his own stock, and the erstwhile waste became property.

As late as 1910 some persons known to Professor Posey were making ten or fifteen dollars a day gathering pecans, then the nuts became so valuable that the owners began to keep the people away from their trees.

As boys, Professor Posey and his brother gathered ten to fifteen bushels of hickory nuts in a day.

Occasional trees are very productive and yield nuts of fine flavor. Mr. J. F. Wilkinson, of Rockport, Indiana, an intelligent and careful observer, says:

I have gathered the crop from a particular tree four years as follows: 1906—eight bushels; 1908—six bushels; 1910—twelve bushels; 1912—nine bushels; making thirty-five bushels in all. The tree is an every-other-year bearer, but has borne lighter crops of from one to three bushels in its off years; and as far back as I have known the tree, it has borne a good crop every other year and a light one between. After 1912 the land changed hands and the owner has gathered this tree, but kept no definite record except the crop of 1922, which was 600 pounds.

This tree was 90 feet high, 100 feet in spread, trunk four feet in diameter. An acre could hold only four of them, with chance for sunshine on all branch ends.

Such trees are not very common, but there have probably been thousands like it; and there are now probably hundreds of them alive and bearing at this moment. Their number would have been vastly greater but for the forest habit of crowding.

Native trees here have a habit of producing a full crop about once in two years. Many native trees have a record of over five hundred pounds' production in one year. (F. R. Brison, County Agent, Cooperative Extension Work in Agricultural and Home Economics, San Saba, Texas, letter, February 14, 1925.)

M. Hull, Assistant Horticulturist, Louisiana State University, reports (*American Nut Journal*, October, 1926, p. 57) that a pecan tree twenty feet in circumference at waist height, is one hundred and fifty years old, has a spread of one hundred and thirty-two feet, and has borne approximately sixteen hundred pounds of nuts in one season.

It grows on the farm of G. B. Reuss at Hohan Solms, Louisiana, about thirty miles south of Baton Rouge. The local postmaster said, with apparent sincerity, that it had borne twenty-seven hundred pounds. (Information from C. A. Reed, U. S. Department of Agriculture.)

In 1925 Felix Hermann went before a notary at Bexar, Texas, and swore that he had gathered twenty-two hundred pounds of pecans from a tree with a spread of two hundred and twenty feet. Commenting on this, F. W. Mally, County Agent at San Antonio, who had not seen the tree, said:

However, I may say that while this is more or less an exceptional tree, there are a great number of very large pecan trees along the banks of the rivers in this territory. Whether they are as large as this one is not known, because they have not been measured. It is not unusual for many of these large trees to produce from twelve to fifteen hundred pounds of nuts in a season, and they would probably reach a ton if they were all gathered and weighed.

It is not surprising that man should have a desire to start an industry when nature has given such object lessons all the way from Indiana to Texas and from Texas to Georgia.

THE PECAN INDUSTRY STARTS

The pecan industry started with wild produce, and wild nuts are still marketed in large quantity, larger quantity than the produce of the grafted orchards.

Texas, with her millions of pecan trees, is usually the leader at the present time, although Oklahoma claims the honor for 1947. Albany, Georgia, with over a million grafted trees of bearing age within fifty miles, is the greatest single center and by far the leading center for the growing of named varieties.

Before the last quarter of the nineteenth century, all attempts at propagation were limited to planting seed, a practice that notoriously results in fruit unlike the planted seed (Figs. 80 and 94).

Mr. H. Fillmore Lankford, Princess Anne, Maryland, said, "The nuts from my large tree are delicious, as I have said, but nuts grown from seedlings from this tree in some cases are as bitter as quinine, and in other cases are as sweet as the nuts from the parent tree." Then came the conquest of the technique of grafting and budding. This was acquired for pecans before it was used for other nuts. Promptly thereafter, the pecan industry started almost like a conflagration. Run-down cotton plantations were cheap, and with the aid of commercial fertilizer and crops of cowpeas, velvet beans, and peanuts their soil was quickly restored. Some of the early pecan plantations were carried at almost no cost by hogging down crops of legumes and corn and oats.

Grafted and budded pecans were planted by the ten thousands during the first fifteen of twenty years of this century. Some of them were planted by near swindlers, who worked something like this:

1. Obtained the record performances of individual trees such as I have quoted.

2. Took a lead pencil and figured on a basis of the biggest yield that ever happened on the best tree on record.

3. Let the prospectus show a similar yield for each of twenty trees to be planted on an acre of ground.

4. Made it happen every year, beginning very early.

5. Sold the future nuts at a pleasing imaginary price.

The figures are indeed impressive if one does not see the following fallacies. The trees do not bear as early in orchards as on paper, or as often, or as much, and it is quite impossible for twenty full-grown pecan trees to be accommodated on an acre of land. The pecan tree reaches proportions so gigantic that an orchard of trees as large as the largest reported would require more than an acre of ground per tree; yet the first planters put in twenty and often more.

As long ago as 1914 some *bona fide* growers began planting four trees to the acre, realizing that four big pecan trees would require that much land. Dr. C. A. Van Duzee wrote in 1927 (letter to the author):

My opinion is that at twenty years, pecan trees will require at least one-eighth of an acre, and more would be better; at thirty years, one-fourth of an acre, and soon after that one or two trees to the acre would be quite enough.

As factual basis for this conclusion, Dr. Van Duzee, who kept an amazing lot of actual tree records (*facts*), gave the following:

The outside row (23 trees) has the adjoining field to extend its roots into and is exposed to sun and sky on that side; its roots are

out in the field one hundred feet; it gave us 3,744 pounds of nuts
during the last five years under my care.

The second row, which divides the fifty-foot space between it and
the first row and a similar space between it and the third row for its
root pasture, gave us 1,745 pounds of nuts during the same period
of time.

These trees are of the same variety, were planted at the same time,
and received practically the same treatment: each tree in the first
row has occupied a root pasture of 6,250 square feet. The inside
trees had but 2,500!

Mr. Fullilove, Shreveport, Louisiana, a large grower, considers three or four trees per acre the proper number.

I think it quite possible that the pecan orchard may eventually have all costs but harvest and fertilizer carried by the
pig or other animal that pastures beneath and beside the pecans—one or two trees to the acre, with legumes growing
beneath them all the time.

THE PECAN SWINDLER

A national magazine carried in June 1910, as part of a full-
page advertisement, the following:

Surest Pecan Land. A pecan grove of five acres nets $2,500 yearly.
No work—no worry—no loss of crop and little cost of upkeep. . . .
The paper-shell pecan tree begins bearing at two years, produces
fifty to two hundred pounds of nuts at seven years, and two hundred
to two hundred and fifty pounds at ten years, increases yearly thereafter, and lives to the age of one hundred years in North Florida.
Five acres will keep the average family in comfort the year round.

And the really curious part of it is that people bought their
five acres with sweet and innocent faith.

Another Chicago vendor of distant lands wrote me in 1912:

From a careful investigation the following estimate seems to be
a conservative one:

Age	Per Tree	Per Acre	Per 5 Acres	5 Acres at 25¢ lb.
Fourth year........	1 lb.	20 lbs.	100 lbs.	$25.00
Eighth year........	45 lbs.	900 lbs.	4,500 lbs.	$1,125.00
Fifteenth year......	220 lbs.	4,400 lbs.	22,000 lbs.	$5,500.00
Twentieth year.....	350 lbs.	7,000 lbs.	35,000 lbs.	$8,750.00

It is not surprising, of course, that the yarns of the near swindlers' prospectus should have resulted in little but disappointment, with each of their acres with its twenty trees, a fabulous number, bearing a fabulous crop with fabulous regularity.

Indeed, figures are especially deceptive when one gets to multiplying yields per tree by a number of trees per acre. It is so easy to plant trees too close together both on paper and on land. Individual trees perform wonders occasionally, but somehow, when they are set out in rows and given a term of years, they fail to perform every year on the average as the rare genius tree does once in a while. George Washingtons really are scarce among men and among trees.

Colonel C. A. Van Duzee, Cairo, Georgia, reported at the 1912 meeting of the Northern Nut Growers' Association that the best Frotcher pecan out of four thousand trees in a two-hundred-and-thirty-five-acre plantation yielded eleven pounds in its seventh summer. I saw his orchard the next year growing great crops of cowpeas, peanuts, and corn and hogs, and the trees were in a thrifty condition.

One of the early show trees of the South belonged to Mr. J. B. Wight of Cairo, Georgia. It stands in a garden. It has all the room it can use. It has been fertilized without stint.

Colonel Van Duzee, one of the keenest and best-informed students of the pecan said of it (Letter, July 24, 1927):

Mr. Wight's big Frotcher is the most fruitful tree I am familiar with. It was about eighty feet in height and breadth when I measured it a few days ago, and it is said to have given Mr. Wight a net income of over a hundred dollars on the average for each of the last ten years. It is estimated that the tree has a root pasture of two-thirds of an acre of fertile soil, and the clear space about it would average nearer forty feet than twenty.

At 21 years of age, the spread of this tree was 84 feet x 71 feet, and in its 21st year it bore 306 pounds.

Meanwhile, the experts at the Georgia State Experiment Station had been sticking to it that one thousand pounds of nuts per acre would be a good average, and this figure has not materially changed. I have spent considerable time checking on it. It is my opinion that, a few decades hence, the yield of good pecan orchards will be substantially increased over the present. It is surprising how long it takes to shake an industry down and then build it up to its optimum. For example, experiments in eastern North Carolina, a hundred miles apart, show marked differences in the performance of trees of the same variety as to regularity of bearing, amount of bearing, time of blooming, and other very important characteristics.

Then there is the question of varieties, as well as care! Edward Bland, manager of a large orchard at Albany, Georgia, wrote me in February 1948 as follows:

We produced in the Simpson orchard in 1945, 219,000 pounds of poor-quality nuts on about 1,700 trees, and did not produce any in 1946 on the same orchard. In fact we did not even go into the orchard to harvest. It came back with 140,000 pounds in 1947, and no doubt the crop will be even larger in 1948. With us, the producing orchard seems to run in about three- or four-year cycles, the third or fourth year being a failure. We have orchards that are producing practically nothing, and we have others that will probably produce 1,000 pounds per acre per year.

This is a good showing at food production compared with the results of pasture (Appendix) or any other meat production or even grain production (Appendix). It is a rich agriculture, and we may expect more productive varieties in the future.

It is not yet time to say whether a well-placed northern pecan orchard can do as well as the southern. Perhaps it can.

Many foolish investors were swindled in the small-unit absentee-ownership enterprises of the southern pecan boom, as they were with apples in the Pacific Northwest. That did not stop an honest and legitimate development of a large industry in the South, prosecuted by bona-fide farmers who are looking after their crops as a farmer should.

This industry is of two kinds. East of the Mississippi, orchards are planted. In Texas and Oklahoma, bushy meadow pastures full of wild pecans are being thinned out to give the big trees a chance at sunlight, and the young trees are being grafted and budded to good varieties. The pecan industry of the cotton country has reached the point where it promises crops of such large size that the great problem will be to find a market. An energetic national organization is struggling with this problem. In New Mexico, the dry air reduces fungus attacks and simplifies the spraying problem, which is a major expense in the more humid Southeast.

D. F. Stahmann, Las Cruces, New Mexico, writes (letter, February 23, 1948):

> We began planting pecans in 1936. I have individual trees that are 12 years old that produced 150 pounds of pecans last year. We have one 900-acre block (in the irrigated Rio Grande Valley) that we planted in 1937, 1938, and 1939, about one-third each year, that produced 54,000 pounds of nuts in 1944, 110,000 pounds in 1945, 227,884 pounds in 1946, and 231,346 pounds in 1947. The year 1947 was a poor pecan year.
>
> We plant these trees 30 feet x 60 feet in intercrop with cotton, cantaloupes, or alfalfa. We find that alfalfa and Bermuda grass greatly reduce the yield, as they are not compatible with the pecan trees. We have some varieties in our own orchard that will not yield five percent the yield of our best varieties.

That last sentence gives eloquent exposition of the immaturity of the industry. The area of irrigable land in New Mexico is not large, and there will be fierce crop competition for the acreage—*milk* vs. *nuts!*

A NORTHERN PECAN INDUSTRY

The pecan industry north of the Cotton Belt is still in the pioneer stage. It had its start in the enthusiastic work of a group of tree lovers who lived in southern Indiana, where the native pecans tower above the oak trees and produce nuts by the barrelful. These men (especially Thomas Littlepage and J. Ford Wilkinson) spent time and money scouring the river bottoms in search of the best tree among thousands. They have brought out half a dozen good varieties.

Another factor in the promotion of the industry has been the formation and work of the Northern Nut Growers' Association. It was started in 1910 in New York City by people with an idea rather than an industry—T. P. Littlepage, corporation lawyer, Washington, D. C.; Robert T. Morris, M.D., surgeon, New York City; W. C. Deming, M.D., Connecticut; John Craig, Professor of Horticulture, Cornell University. Dr. Deming, age 88, is the sole survivor.

This Association has been, and still is, the great repository and clearing house for information concerning nut growing in the North, and any person thinking of planting more than one tree should join the Association. Its *Proceedings* record the advances of knowledge, and it has a certified list of nurseries selling nut trees. Write to J. Colvin McDaniel, Secretary, Department of Agriculture, Nashville, Tennessee.

After the Indiana pioneers had searched out parent trees of unusual merit along the Ohio and its branches, experimental grafting began, and the trees of northern origin have been widely disseminated throughout the belt of marginal pecan territory from Connecticut through southern New York, Pennsylvania, Ohio, Indiana, Illinois, and southern Iowa and thence southward.

In almost every case these plantings have been a combination of ornament and experiment. They have proved that these particular varieties of pecan trees are hardy over most of the country where there is a good corn climate.

As to the bearing records, the trees of some varieties, especially the Busseron, show bearing habits that may be likened to the apple, so far as precocity is concerned. Given reasonable cultivation and good soil, grafted nursery pecan trees come into bearing in from seven to ten years. Topworked trees, where the cions are put on vigorous pecan stock, usually begin to bear in four or five years (Fig. 91). Such has been my experience with several of the Ohio Valley varieties in the Philadelphia climate of northern Virginia.

The large numbers of these trees that are scattered over the middle North will tell us, as the years go by, about the soil and climate necessary for the pecan tree. As trees, the Ohio Valley varieties, Butterick, Busseron, Indiana, Greenriver, and perhaps some others, have proved perfectly hardy at Cornell University, Ithaca, N. Y., but they can rarely ripen their fruit.

Unfortunately, the preliminary conclusions from many of these experimental plantings may be necessarily harsh for the pecan because of its blossoming habits.

It seems true that some of the varieties, and possibly all of them, are incapable of *self*-fertilization. A tree may be wrongly condemned for sterility when it only needs a mate. In my own observation I have found that the Busseron and Butterick varieties are capable of fertilizing each other. A Busseron tree bloomed helplessly until a Butterick bloomed near it, after which both bore nuts. This seemed to indicate that experimenters should always have two varieties; and unless satisfactory reciprocal pollinating habits are known, it is better to have three. Don't get excited and buy Butterick. The parent tree has had a wonderfully productive record for ninety-four years, and most of the grafted offspring have been disappointingly nonproductive. According to my best knowledge today:

Busseron is the earliest to bear.

Busseron and Indiana are earliest to ripen.

Greenriver ripens later but bears more if it gets ripe.

Major is considered most productive of all, although the nuts are smallest of the lot.

J. Ford Wilkinson, probably the most consistent student of the pecan, says that Indiana pollinates Busseron and Major; Greenriver is self-fertile in southern Indiana, and the Indiana variety usually blooms at the time to pollinate Greenriver. He adds that the varieties in blooming do not always maintain the same calendar relations to each other. Why? Don't ask!

Any of these varieties may be surpassed and replaced at any time by an invader from the experimenter's test blocks.

In the light of experience, people here and there are beginning to plant small orchards of northern pecans. The quality is quite as good as that of the southern, and most of the varieties can be cracked by pressing them against each other in the hand. The nuts are not so large as those in the South, a factor of great importance for the commercial grower, who sells whole nuts, but not so important for the vendor of kernels, and kernels are likely to be the chief outlet.

BREEDING PECANS AND THE FUTURE OF THE PECAN INDUSTRY

In the named varieties of the pecan which we already have, there are a number of great crop trees. Nearly all of these are of chance origin. But, like the walnut, the pecan is capable of improvement by breeding. E. E. Risien of San Saba, Texas, has crossed the San Saba pecan with other good pecan trees. Nuts produced from the resulting seed have shown 69 percent weight of meat after drying for six months.

The fact that the pecan hybridizes even in nature with several of the hickories is extremely suggestive. If unaided nature has produced such a surprise as the McCallister hiccan (Fig. 18), what may scientific breeding be expected to do? There is reason to believe that breeding experiments can give us a much better log of pecan and pecan hybrid crop trees for the different sections of the country.

Especially promising should be more crosses between the northern pecans and neighboring selected hickories. Thus far, these hybrids have shown great vigor, rapid growth, and low yielding habits, but we have not tested many. None of these crosses can be dismissed until we have tried a few thousand and found what might pollinate them. The McAllister hybrid, a beautiful rapid growing tree, produces the largest known nut of the hickory genus, but it usually produces only an empty shell. However, in one or two places where there are many hickories, it fills. What pollinates it? You can best find out by visiting and studying the collection of Fayette Etter, Lemaster, Pennsylvania, each day for a week or two at blooming time. Iowa Agricultural College produced some twenty thousand hybrid apples to get five meritorious trees. Plant breeders need patience and stick-to-it-iveness.

According to Ford Wilkinson, nature has produced one known good hiccan (pecan x shagbark) hybrid with a good nut and a good bearing record—the Burton variety.

PECAN REGIONS

We have four, if not five, distinct climatic areas, each of which should have its own group of pecan, or pecan hybrids, or both. First, and at present the most important because of the large commercial orchards, is the southeast, or humid part of the Cotton Belt. This region extends from eastern Texas to North Carolina. The dominant characteristics of the climate here are moisture, humidity, and a long growing season. These are favorable to great development of fungi. Therefore, the trees for this area need to be particularly resistant to this enemy, for if the trees must be sprayed as apple trees are sprayed to keep the foliage effective, great cost is added to production. It is time consuming, difficult, and costly to spray a tree eighty or one hundred feet high and with an eighty- or one-hundred-foot reach.

The southwest pecan area, including most of Texas and Oklahoma, has less rain, less humidity, less fungus activity, but even more summer heat. Varieties may thrive there which could not thrive in the more humid East. Pecans can be grown successfully by irrigation in California and in Arizona, but there is small reason to expect commercial competition here with the vast areas of naturally watered land in the eastern United States, unless freedom from spraying in the arid air might turn the trick. The pecan is not sufficiently productive per acre to compete for this almost microscopic area of irrigable land—microscopic in comparison with the vast rain land of the East.

The northwest pecan section (eastern Kansas, southeastern Nebraska, southern Iowa, and most of Missouri) has very severe winters, but it has pecan summers, and pecans are scattered over most of it. Doubtless, breeding from the best wild strains of that locality would produce good producing trees for 100,000 square miles of the western part of our Corn Belt down the Cotton Belt.

The Ohio Valley region, which has given rise to all the varieties of pure-bred northern pecan now under cultivation, may yet have better pecans in the woods than have thus far been found, and doubtless the proper combination of qualities, resultant from breeding, would give us many better trees than we now possess. For example, the Busseron seems to be the most precocious. Messrs. J. F. Jones of Lancaster, Pennsylvania, J. Ford Wilkinson of Rockport, Indiana, and W. C. Reed of Vincennes, Indiana, have repeatedly had them bearing in the nursery row the second or third year after grafting. I have had the same experience myself on the slope of the Blue Ridge Mountains in Loudoun County, Virginia. This precocity, combined with some tree that may be more productive or more regular in bearing or a more rapid grower or earlier in ripening its nuts, offers interesting possibilities.

Such improved trees have a possible area of usefulness covering a quarter of a million square miles in the large stretch of territory between the Mississippi River and the Atlantic Ocean, and between the Cotton Belt, the Great Lakes, and New York City, and also below one thousand or fifteen hundred feet elevation.

Possibly a set of early ripening varieties of pecan and hybrids can be produced in the Northeast, the strip from central New Jersey and central Iowa to Massachusetts, southern Vermont, and southern Minnesota. The present varieties are hardy in most of this area. There is Mr. Zenas Ellis's one pecan tree at Fair Haven, Vermont, waiting for the hand of the hybridizer.

Who is going to breed these pecans and make these new engines for the production of fat and protein?

CHAPTER XX

More Meat-and-Butter Trees—
The Other Hickories

If anyone knows of better nuts than some of the shagbark hickories of the northeastern United States, I beg him to write me at once and send me a few, for I wish to know this superior thing.

The hickories were food producers of importance to the American Indian and are of great economic promise for scientific agriculture in the future, if we should ever settle down to use and save this continent rather than to skin it and ruin it as at present. There are several reasons for this promise.

In the first place, there are many varieties, and many more possible varieties. There are doubtless from one hundred to five hundred varieties (perhaps more) of hickory worthy of commercial propagation now growing in the United States. I mean varieties in the horticultural sense, like varieties of peaches or cherries—a tree so good that an orchard like it would be valuable. I cannot prove this fact, but the following statements make it appear reasonable. Sargent, in "Notes on North American Trees" (*Botanical Gazette*, September 1918), lists fifteen species of hickory, twenty-two varieties from a botanical standpoint, and seventeen hybrids.

The truth is that the great Sargent scarcely got started on the hybrids. Willard G. Bixby, of Baldwin, New York, had more than that many hybrids in his nursery, and the J. F. Jones Nursery of Lancaster, Pennsylvania, had rows of them which Mr. Jones made by crossing the two best trees known to him.

When one considers that there have been seven thousand named varieties of the one species of apple, my estimate of one to five hundred varieties (perhaps more) of hickory may be too low, for we start with fifteen species scattered over a million square miles and multiplied by an almost indefinite number of natural hybrids.

Willard Bixby of Baldwin, New York, perhaps the greatest authority on the horticulture of the hickory tree, reported as much as one in thirty hickories in certain Indiana forests to be hybrids. It was this locality that produced the McCallister (Fig. 18).

Every prize offered for better varieties of hickories brings forth a number of specimens which will deliver their kernels in whole halves. These are mostly shagbarks, but sometimes there are other species, and the total number of such trees in the United States even now must, I am sure, be several hundred.

As an elemental part in this variation, the fruits of different trees vary in flavor almost as apples do. Even the squirrels race through tops of hickory trees to get to the particular one they fancy.

The nuts vary in thickness of shell, shape of kernels, and in almost every conceivable way. Some of these wild nuts yield kernels in unbroken halves. I have no doubt that there are several thousands of such trees growing in America today. In 1926 the Philadelphia Society for the Promotion of Agriculture offered prizes for the best hickorynuts and received at least a dozen shagbarks that yielded kernels in unbroken halves. Some hickories bear fruit late, or sparingly, or rarely; others bear heavily and with considerable regularity. For example, the Pennsylvania hybrid tree called "Weiker" has a well-established record of twelve bushels at a single crop. An Illinois correspondent writes:

This tree (a seedling of course) is the finest tree in our section of Illinois, and we have lots of bearing hickories here. This tree bears every year from three to ten bushels of splendid nuts. (Letter, C. H. Walter, Canton, Illinois, January 31, 1916.)

The different varieties within the various species vary much in speed of growth, and furthermore the different species vary much in speed of growth. For example, under given soil conditions a shagbark may make from three to nine inches of growth in a season, while the bitternut will riot away with two or three feet, and the pecan makes nearly as much as the bitternut.

The habit of growth among most shagbark hickories is a rush in spring. This ends with the formation of a terminal bud in about six weeks. After this, no amount of coaxing or fertilization or cultivation will make it grow again until the next year. On the other hand, the pecan species grows all summer; therefore, a horticultural disappointment results when the experimenter grafts pecans on hickories. They grow and make a nice foliage, but the pecan top is used to being fed all summer, while the hickory root has the habit of doing most of its work in the spring. Therefore, the pecan top is underfed and comes through with little or no fruit, or with fruit of diminutive size. As a consequence, the graft of pecans on hickories is of value only as a curiosity and as a means of making a quick climatic test of the variety. In some places it can be brought to fruit on an established hickory more quickly than on the roots furnished it by a nurseryman. I have benefited by this fact in my own experimental work.

The hickory is a food tree of great economic promise because we now have learned how to graft it and can make a hickory orchard just as we can make orchards of other trees.

The hickory is a food tree of great promise, also, because of the great geographic range of the genus. A single species, the delicious and beloved shagbark, grows in southern Maine, the whole of New York State, a substantial strip of Ontario, thence to southeastern Minnesota, southeastern Nebraska, eastern Texas, and eastward to the western edge of the Atlantic Coast Plain. According to Forest Service Bulletin 80, *The Commercial Hickory* (pp. 23, 24), it covers all or parts of thirty-three States.

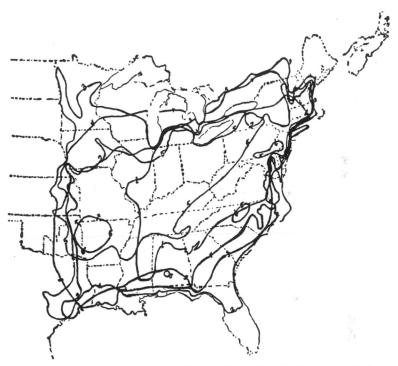

Courtesy J. S. Betts, U. S. Dept. of Agri.

NATURAL RANGE OF FIVE LEADING SPECIES
OF HICKORY

Photostat copy of map showing various species of hickory

1. Pignut, *Carya glabra*
2. Mockernut, *C. tomentosa*, syn. *alba*
3. Bitternut, *C. cordiformis*
4. Shagbark, *C. ovata*
5. Shellbark; bignut shagbark, *C. laciniosa*

Note the extreme range of the bitternut, and that the shellbark seems to avoid mountains and has outlying areas in Pennsylvania and in Carolina, probably due to migration in an Indian pocket, if the Indians had pockets in those days.

The mockernut and pignut, other species of hickory, cover almost the same area (p. 271).

The bitternut, northernmost of all the hickories, has a range almost identical with that of the shagbark except that it reaches northward to Georgian Bay in Ontario, and to the tip of Lake Superior at Duluth in Minnesota. Sargent points out that it is common southwest of Montreal, and one of the commonest trees in the forests of western Ontario.

The other thirteen species of hickory grow in different parts of this territory. These species do not cover their ranges solidly. I have never heard of a wild shagbark in my section—Appalachia, Potomac River, and Blue Ridge Mountain—but it is found both northeast and southwest of me.

The "bignut shagbark" (No. 5 on the map, p. 271), often called "Kingnut" and "shellbark," seems to be perfectly at home far to the eastward of its native land. As a result, almost any county may have from seven to ten or even more species at home, or capable of being at home, within its boundaries.

Another reason why the hickory tree is a food plant of great promise is because of the great range of soil in which the different species can make themselves at home. Many of the species live in and must have rich soil. But the mockernut will grow in poor sand and gravel, and the bitternut, unwelcome because of its flavor, grows well on a myriad of high hills. Thus, these two species offer great possibilities as stocks on which choice varieties which naturally have poor roots may be grown on sand, gravel, and high hills.

This process of getting roots to fit a particular ground condition has been tried many times. For example, the British gave up growing cinchona in Ceylon because the good variety did not thrive. The Dutch are growing it in Java with the same good variety grafted on stocks of a rank-growing but almost worthless variety of cinchona. The story of the Phylloxera and the European grape industry on American roots is another example. (See *Industrial and Commercial Geography*, J. Russell Smith and M. O. Phillips, pp. 513, 514.)

Again, the hickory tree is a food plant of great promise because of the easy hybridization within the genus.

Hybridizing opens interesting and unpredictable possibilities of breeding varieties suitable for the particular needs of particular localities. Indeed, the hickories hybridize themselves with great freedom.

The hybrid is a law unto itself, occasionally outdoing either parent. For example, the Weiker tree, a cross between *Carya ovata* and *C. laciniosa*, has yielded twelve bushels, although a crop of three bushels is exceptional for trees of either of its parent species.

By this same independence and undependability of hybrids, the tree grown from the seed of a hybrid tree is almost certain to be unlike the parent tree. The offspring tends to revert to combinations of the qualities of the two or more types that have been blended to make it: plant hybrid seed and you get the parents again. Therefore, each natural hybrid is unique and, like those made by man, it can only be reproduced by grafting, budding, or other vegetative reproduction.

The McCallister hiccan (Fig. 88), a natural Indiana hybrid with a nut often more than two inches long, raises speculation as to the wonders that we may yet find in the forest or may produce by hybridizing, and we regret the many wonderful trees that have grown and perished and are perishing even in our day with no one to save them.

Now that grafting of nut trees has come to be a simple thing, the hybrid is at our service to become the foundation tree of a particular variety, just as all the Baldwin apple trees have come from one original foundation tree. A number of these hybrids are growing on the grounds of experimenters, and some are for sale in commercial nurseries.

This phenomenon of hybridization opens a field of utility for the bitternut. Save for one disadvantage, this bitternut is magnificent. I well remember the first one I saw. The ground beneath a majestic pasture tree was littered with large, white, plump nuts. The shells were so thin that I could crack one by pressing it between others in my hand. The rounded lobes of fat meat looked like brains in a skull. I carefully weighed some of these nuts and meats and their shells and found that as they came green from the tree they were 52.4 percent meat. This is nearly as high as the southern pecan, which ranges from 50 to 60 percent. Mr. Bixby reports the Vest shagbark tests 49 percent, the Kentucky 47 percent, the Triplett 46 percent, while pignuts bought in the market by Edward B. Rawson, Lincoln, Virginia, yielded 25 percent meat.

To the taste, my pretty white nuts were as bitter as dead sea water; but nature, alone and unaided, hybridizes bitternuts and shagbarks, and some of them are sweet and good to eat.

Of one such tree, the Fairbanks, S. W. Snyder of Iowa, ex-President of the Northern Nut Growers' Association, says:

Regarding the Fairbanks hybrid hickory, will say it is one of the best bearers to be found among the hickories, that is, as it grows here. The parent tree has had a wonderful record in the production of nuts. Mr. Fairbanks informed me at one time that the old tree had not failed in a full crop in a period of twenty years; and what is considered a full crop for it is something remarkable for a hickory tree, as it bears considerably heavier than the average variety of hickory. (Letter, Center Point, Iowa, January 19, 1927.)

My own experience of twenty years with this variety confirms Mr. Snyder.

The Fairbanks differs from its shagbark parent by growing much more rapidly than shagbarks usually grow and by bearing much earlier than they usually bear. We may expect to wait eight to ten years for fruit from grafted transplanted shagbark.

Harvey Loose, Upper Red Hook, New York, says in a letter:

I have had quite a lot of experience in grafting hickories. . . . Shagbark upon shagbark, I found, usually took about eight or ten years to come into bearing.

I have found that by spading in some well-rotted manure and later sprinkling a few handfuls of bone meal with a little bit of nitrate of soda and extensive cultivation, one can push the shagbark along tremendously after it has once become established.

My own experience confirms this. Feed your shagbarks! Feed them unreasonably, and they will grow almost with the maples. I have had the Hagen shagbark make three to four feet in the nursery row. You can't have nuts until after you get a tree.

I have had Fairbanks trees fruiting in the third season, when grafted on wild bitternut, growing in an uncultivated field on top of the Blue Ridge Mountain in Northern Virginia. I do not want to stampede people into growing the Fairbanks. It is good, but not quite so good as the pure shagbarks, although many could not notice the difference. I merely mention it to illustrate the point about breeding hybrids rather than waiting for nature to produce them. Dr. W. C. Deming reports that several pure shagbarks have borne in three, four, five, and six years when topworked on vigorous stocks.

The subject need not be expanded to one who has read this book. Searching out the best wild hickory trees now growing in America would give the basis for a profitable industry. I am sure that my two-hundred-acre farm would furnish me a competence if it had on it a full stand of mature hickories as good as the best now growing wild. With two hundred acres of such trees I could live at ease.

By hybridizing we can get much better trees. Therefore, the need is to search to get the best wild trees and then employ a staff of men to test and breed.

THE FUTURE SERVICE OF THE HICKORY TREES

The past services of the hickories suggest their future. For one thing, they have helped to feed the American Indian and also to produce the mast that has fattened the hog of the American frontiersman.

Hickory nuts have very hard shells but excellent kernels with which, in a plentiful year, the old hogs that can crack them fatten themselves and make excellent pork. These nuts are gotten in great quantities by the savages and laid up for stores of which they make several dishes and banquets. (Lawson, *History of Carolina,* p. 98, quoted in Sargent's *Silva,* Vol. VII, p. 133.)

These hickories might again render this service, but their excellence as human food probably puts the forage crop out of the running for the future. It is not likely that the hickory can compete with the oak in weight of pork per acre. We should not forget that the acorn, like the chestnut, is solid meat.

The past use of the hickory as food by the American Indians and the Colonists is suggestive of its future use in the age of science. Hickory cream formerly made by the Indians might easily be made again in this age of machinery. In *Silva* (Vol. VII, p. 133), Sargent thus describes it by quoting William Bartram's *Travels in North America:*

The fruit [of the hickory] is in great estimation by the present generation of Indians, particularly *Juglans exaltate,* commonly called shellbarked hickory. The Creeks store up the last in their towns. I have seen above an hundred bushels of these nuts belonging to one family. They pound them to pieces and then cast them into boiling water, which after passing through fine strainers preserves the most oily part of the liquid which they call by a name which signifies hickory milk. It is as sweet and rich as fresh cream and is an ingredient in most of their cookery, especially hominy and corn cakes.

The Indians sometimes crushed roasted sweet potatoes in this hickory milk as a kind of gravy, and hickorynut oil from the shagbark seems to have been a staple article of diet in the Virginia colony.

Robert T. Morris's book, *Nut Growing,* gives further testimony:

The wild walnut, or hiquery tree, gives the Indians by boyling its kernel, a wholesome oyl, from which the English frequently supply themselves for its kitchen uses. Whilst new it has a pleasant taste, but after six months, it decays and grows acid. I believe it might make a good oyl, and of as general an use as that of the olive if it were better purified and rectified. (Thomas Ash, *Carolina, or a Description of the Present State of that Country,* p. 12, quoted by Sargent, *Silva,* Vol. VII, p. 133.)

Partly evaporated hickory milk keeps a long time in jars, a quality of great importance for food.

We think of the hickorynut kernel as being troublesome to pick from the shell. For most wild varieties this is true, but in this day of machinery it appears to be a simple problem in chemical or mechanical engineering to work out the technique whereby we may imitate the cider mill, shovel the nuts into a machine, crush them, extract the oil by mechanical processes or boiling, and bring the hickory oil or hickory cream or hickory butter into competition with cow's butter, olive oil, coconut oil, corn oil, lard, goose grease, tallow, margarine, *et cetera,* as food fats of the future, fresh or canned.

Meanwhile, hickorynuts are still delicious for eating by the fireside, for dessert, and for candies. But they cannot rival the black walnut as the standard for general cooking purposes. For agricultural uses see Chapter XXIII.

Some live organizations (of which there are too few) might render a service as follows: Work out a pilot-plant process of making, canning, and cooking hickorynut cream. This might shift the emphasis of search from the tree that gives nuts that crack out as complete halves, to the tree that produces the greatest *quantity* of nut meats, regardless of shape of kernel or thickness of shell. *Quantity!*

Tennessee Valley Authority, please take notice! Also the Rural School and Sanitarium, Madison, Tennessee, or any vegetarian stronghold, mountaineer school, or other group interested in nutrition, food, health, or conservation, or life in the hill country. Please take notice!

CHAPTER XXI

Some Suggested Lines of Work—The Unexplored Realm

This book makes no attempt at being the last word on any topic. Its avowed purpose is to establish an idea, to convince persons of creative intellect that here is a great field meriting scientific, experimental exploration. The preceding chapters have attempted to establish with some degree of thoroughness an idea of the importance of certain fields for work. A tree crop may receive considerable space in this book, but that does not necessarily mean that it is of more ultimate importance than are others which have received briefer treatment or have even been omitted altogether. The limitations of time, space, and knowledge have prevented covering the whole field with uniform thoroughness. Therefore, some of those which appear to have been slighted may, after a few decades of experimental work, prove to be of greater importance than those which have received more space here.

This chapter, therefore, makes brief mention of a number of possibilities—mere suggestions.

THE BEECHNUT

The beeches, *Fagus* species, are northern members of the mast family, as the chestnut is a southern member of the same family. In Europe, the chestnut has taken the Mediterranean shores and the beech those of the Baltic. In America, the chestnut has taken the Appalachian upland and the beech those of northern New England and southern Canada.

The beechtree, like spruce and pine, runs down the Appalachians with the cold climate of its high elevation.

The people fence their woodland with woven wire, and when the beech crop hits they are ready to buy hogs and turn them in about the middle of November. In this way there are thousands of pounds of meat made in this country. (Letter, G. T. Shannon, Willow Shade, Kentucky, December 19, 1915.)

The abundant nuts of the beechtree have helped to make the hog what he is. Beechnuts have long been important in Europe as food for the wild boar of the forest and for the semiwild hog of Europe. The oil from beechnuts is used as a substitute for butter. The residue makes a good food element for animals.

As encouragement for those who might consider making a beechtree capable of producing a modern commercial crop, I cite the following:

The search of Dr. Robert T. Morris and of Willard G. Bixby for beechnuts large enough to merit experimental propagation has resulted in the discovery of specimens (varieties) of the American beech varying much in size. This fact is promising for the plant breeder.

There are a dozen species or more of the beech. This also is a promising fact for the plant breeder. Along with these two facts, we should keep in mind what has happened to the Persian walnut (p. 207) as a result of being propagated by man.

The beechtree can be grafted. This fact has given us the ornamental purple beeches. In the future it may give us wide expanses of grafted beechtrees stretching over the unplowable stone lands of New England and furnishing forage for beechnut bacon, dessert nuts for the table, and oil material for the kitchen of the future housewife. Pots of beechnut butter nearly two thousand years old have recently been unearthed in Poland.

Thus far, scientific agriculture seems to have utterly neglected the beechnut.

THE PISTACHE

This choice dessert nut is gathered wild in southern Turkey and in other widely scattered locations, and for many years it has been an article of commerce. For a long period of time unusual specimens have been propagated by grafting and budding. The great superiority of these few unusual varieties of unknown origin is suggestive of further improvement of the species.

The pistache is a tree having wide climatic adaptation. Commercial varieties have been introduced in the United States from Italy, Sicily, Algeria, Greece, and Afghanistan. Other varieties have been introduced from Palestine, Syria, Libya, and France, and also from China, whose climate differs greatly from that of the other countries just mentioned. The fact that the pistache comes from Afghanistan and from China, as well as from Mediterranean countries, indicates a wide area of possible adaptation in the United States. It is reported thriving at Nucla, Colorado, 5,800 feet altitude, on the grounds of U. H. Walker. (*American Nut Journal*, June 1926.) Also forty miles west of Wichita, Kansas. (Letter, Merrill W. Isley.)

Merrill W. Isley, an American missionary in southern Turkey, told me that the tree grows wild in an area one hundred miles by one hundred miles near the northeastern corner of the Mediterranean Sea, just north of the present boundary of Syria. There, on the mountains, the pistache tree grows wild as the scrub oak grows wild. It has survived and made profit on what perhaps is the roughest land used for agriculture anywhere in the world. (See Fig. 106.) The tree seems to thrive on thin soils, especially on that underlaid by limestone. It can cling in the crevices of rocks where there is almost no soil to be seen. Mr. Isley tells me of land in Turkey where it is so steep that only goats and men can climb, yet the diligent Armenians had there grafted the pistache trees and were establishing valuable orchards before the desolations of World War I.

The pistache tree has two habits that are rather discouraging to the commercial planter. Mr. Isley reports that his orchard of seedling trees grafted in place required fifteen years of cultivation before producing commercial crops. This was on land good for grapes and wheat but subject to the very dry summer of the Mediterranean climate. The pistache also has very irregular bearing habits. Mr. Isley states that in his Turkish pistache country the trees will be loaded with fruit once in five, six, or seven years. In other years the trees commonly set full crops, but the nuts drop off. In view of the fact that the trees are male and female, it is perhaps possible that the dropping could be overcome if the planters were to provide pollen-bearing insects to fertilize the trees.

My own observation and my interviews with pistache growers on the rocky lava slopes of Mt. Etna in Sicily confirm Mr. Isley's Turkish data in every respect—wild trees in rough places; grafting to improve strains; spasmodic bearing with rarely a heavy crop—but high prices. This rare fruiting should not be accepted as final.

Fortunately the nuts keep well—so well, in fact, that in Turkey before World War I the Ottoman Bank would lend heavily on bags of pistache nuts.

THE NUT-BEARING PINES

Dr. Robert T. Morris, in his book on nut growing, makes some statements that give one a shock. For example, he says that research has led him to place the nuts in the pine cones near the head of the list of nut food for human use. He thinks that even now, pine nuts are second only to the coconut.

This statement is open to question, because it is difficult to estimate the amount of crops that are little used in trade and that are chiefly consumed where produced, while no statistical record exists for them. To strengthen his conclusion, Dr. Morris states:

1. That he thinks there are thirty species of nut-bearing pines to be tried out between Quebec and Florida.

2. One species, the bunya bunya pine, produces nuts the size of the average English walnut. Others grade on down to a nut the size of a grain buckwheat.

3. The pine trees, with their wide array of species, produce nuts with a wide variety of edible qualities. [See table, Appendix.] Some of them are sugary and sweet. The sugar pine, *Pinus lambertiana*, that lordly timber tree of the Pacific Coast, bears a nut rich in sugar and oil. Its sap evaporates, leaving a solid sugar. Many of the other pine nuts are rich in both protein and fat.

4. Some of them have the quality, less common among the nuts, of furnishing starch. For example, the *Araucarias* of the southern hemispheres long furnished a starch food to the natives of South America, South Africa, and Australia.

5. Many of the pine nuts are good when eaten either raw or cooked. Starchy varieties must be cooked. Some of the oily kinds may be covered with water and then pressed, thereby yielding a thick, milky substance which can be kept for a long time and which is essentially a substitute for meat.

A few facts about one of these pines may serve to show that this genus merits the attention of the botanical creators.

The piñon, *Pinus edulis* (also called pinyon), of southwestern America can be seen displaying its virtues to good advantage in the vicinity of Sante Fe, New Mexico, in a climate that is hopeless for agriculture without irrigation and where pasture is of low yielding power. There the piñons raise their beautiful heads and stretch by the thousands across the landscape as far as one can see. The land may be hopeless for agriculture because of the steepness of the slope, but the piñon is perfectly at home on the hill, dry and rocky though it be. For ages the nuts have been a mainstay for the native, furnishing both fat and protein food. They are gathered chiefly by the Indians and Mexicans. When the railroad came, the Indian found a market; and now the pine nuts are shipped by the ton, by the carload, almost by the trainload. The annual value of the harvest amounts to tens of thousands of dollars.

Though the piñon is a wild product of the open forest, and of the otherwise treeless range, no one knows what might result if the best piñon tree of all the millions that nature has already produced were chosen as parent stock. This one species of the many edible pines is scattered over a half million square miles in our own Southwest and northern Mexico. Next comes the question—what might the plant breeder do with this genus and its many species?

A private experimenter who starts a collection of the nut-bearing pine trees will add beauty to interest, for what is more beautiful than a group or collection of pines?

It is probable that a million-dollar endowment could be profitably employed for the next century on nut-pine investigations alone.

THE ALMOND

The almond is a cousin of the peach. The two trees are so similar in appearance as to be indistinguishable to the inexperienced. Some people claim that the peach is merely a specialized almond. Even when the fruit is nearly ripe, the almond looks like a small undeveloped peach. Upon closer examination, it appears to be a peach with a thick skin, practically no pulp, and a seed with a large kernel that lacks the bitter flavor of the peach-seed kernel. The shell surrounding the kernel varies greatly in thickness and hardness.

The world's commercial almond crop, exclusively a product of the Mediterranean climate, is worth many millions of dollars a year. It is a well-established industry in California, Spain, Italy, Sicily, and in other locations of the Mediterranean climate, namely, Chile, South Africa, and South Australia. In this almost world-wide distribution, its behavior is standard for the list of crops from Mediterranean countries—Valencia orange, lemon, raisin, and the rest.

The fruit of the almond has already entered into the machine- and factory-production stage which characterizes our present and prospective food supply. You can go to the store and buy candied almonds, salted almonds, almond meal (or paste) used to make macaroons and other cakes either in the homes or in the factories.

There is reason to think that the almond area might be greatly extended by applying known methods of plant improvement.

The tendency of the almond to early blooming is its chief limiting factor, but there are some strains, chiefly those with hard shells, that seem to survive and thrive in a climate like that of Connecticut, southern Vermont, and many other parts of the eastern United States. A specimen of my own bore abundantly under the same conditions that produce a major part of the Virginia apple crop, namely, a good soil plus fertilization without cultivation. One species of hardshell almond is reported living on the grounds of Zenas Ellis, Fair Haven, Vermont. It produced fruit on the grounds of G. H. Corsan, Islington, Ontario (near Toronto).

It seems reasonable to expect that breeding and selection would produce much more desirable strains suitable to the climate of the eastern United States. Apparently, it could become a crop for untilled hill land, especially if planted on horizontal terraces to preserve rainfall. (See p. 324.) I see no reason why a small amount of work over a period of years might not start breeding with these hardy hardshells, and by the use of cracking machinery produce almonds at less cost than in California. But that requires some originality, and our society has not yet learned how to use its few (oh, so few!) original minds.

The wild almonds of the desert hold out an interesting dare to the constructive botanist, a kind of special promise as a crop for arid lands. Indeed, most of the crop is now grown on unirrigated soils in the Mediterranean lands of rainless summer. Its structure, namely, a peach without juice, suggests aridity, and adaptability thereto.

Some species have elaborately developed root systems, hair on leaf and fruit, and other drought-resisting characteristics. Silas Mason, Botanist, U. S. Department of Agriculture, discovered several species that grow wild in the deserts of Nevada. In the *Journal of Agricultural Research* (Vol. I, No. 2, November 10, 1913) he describes a group of species which he calls "wild almond," *Prunus andersonii;* the "wild peach," *P. texana;* the "wild apricot," *P. eriogyna;* the "desert almond," *P. minutiflora,* and the Mexican "wild almond," *P. microphylla.*

They range from northern Mexico to southern Oregon. Some of them are surviving desert conditions where the rainfall in some seasons is not more than one inch a year. Mr. Mason cites an interesting adjustment. Under desert conditions they must, among other things, sprout quickly when water comes. In a week's time a greenhouse specimen had a plumule one centimeter long and a root nine centimeters long. The eleven natural hybrids listed by Mr. Mason are very suggestive for the plant breeder who would use this desert specialist genus to make crops for desert lands.

These species await development into usable crops. There is a fascinating life work, or avocation, for someone who likes it, and it might easily give a crop worth millions a year on land now almost unused. A professor of botany (with imagination) and a garden and a job in some arid college, could support himself with teaching and quite likely make a world reputation by starting an industry. A question that puzzles me is this: what is there about plants that makes botanists' minds so numb to economic values? Many botanists will race to the end of the world and fight like cats to decide whether it is one species or two, and then drop them both even though they are bristling with economic possibilities that any farmer could see.

THE APRICOT NUT

The kernel of the apricot seed is much like that of the peach and the almond. At canneries it appears that the seeds of apricots have twenty to twenty-five percent kernel. In California, before World War I, these seeds were sometimes put through mechanical crackers. When free from the shells the kernels were roasted and sold for almonds to some confectioners. The Pure

Food Law apparently stopped the fraudulent but useful and probably in no wise injurious practice. Also in California during World War I, oil was extracted from apricot kernels. It was very much like cottonseed oil and was used for cooking and as oil for canning sardines.

Under postwar conditions, the cottonseed oil is cheaper, so apricot kernels have found more sophisticated uses. Professor W. V. Cruess, of the California Agricultural Experiment Station, tells me (1948) that the kernels are now crushed, sweet oil for milady's face lotions expressed, and the press cake used as fertilizer, because the prussic acid it contains prevents its use as stock food. We should attempt to breed an apricot with a bigger and better seed, letting the flesh atrophy, thus specializing the apricot for particular purposes as we do with cattle to make breeds, milk breeds, and ox breeds. They might become better nuts than almonds.

For an interesting discussion of peach, prune, almond, and apricot structure, origin, and relationship, see Bulletin 133, Bureau Plant Industry, U. S. Department of Agriculture, "Peach, Apricot, and Prune Kernels."

THE CHERRY-TREE NUT

One of the clear-cut memories of my youth on a Virginia farm is the joyful noise of pigs as they cracked the seeds of fallen cherries beneath the trees in the orchard where the pigs pastured. This suggests developing another tree forage crop for midsummer pig pasture. The pigs do the work, don't forget that aspect! Cherries come after mulberries and before persimmons.

The structure of the cherry is essentially analagous to that of the peach, a juicy pulp surrounding a kernel which contains in the center a germ carrying the food for the young seed, which also happens to be nutritious to animals. Bulletin No. 350, Bureau Plant Industry, professional paper, *The Utilization of Cherry Byproducts*, 1916, estimated that the seeds then thrown away at canneries in the United States contained 300,000 pounds of oil worth $60,000. It is true that the Mazzard cherry has a tremendous variation in size and flavor of pulp, in size of the seed, both shell and kernel. This tree is very fruitful and grows absolutely wild in the eastern forest margins and along fence rows from southern Ontario to the Cotton Belt. I have seen cherry trees in northern Virginia whose trunks were two feet in diameter and from fifty to sixty feet in height.

I recall one tree from which I saw fifty gallons picked in one season. Then, curious as to the amount remaining, I walked around the tree estimating its burden. Upon looking up at it I could not tell where the fifty gallons had come from. I made a rough estimate of it, branch by branch, and concluded it must have at least two hundred gallons of fruit left as they began to rot and fall off on July 5. Estimating one-sixth for seed and one-third of the seeds for meat, we have eleven gallons of meats —call it eighty pounds—probably as nutritious as almond meats.

I see no reason why breeding should not produce Mazzard cherry trees with seeds that carry enough meat to make of the fruit a profitable crop when picked up by pigs. This is a possibility worth investigation. A cherry has been crossed with the apricot and the peach. (*Journal of Heredity*, July 1916.)

FILBERTS AND HAZELNUTS

Two species of bushes, the hazels, producing nuts that are not bad, grow wild and fairly run riot in many American fields. In some places they are considered a troublesome weed, because foraging animals do not touch the leaves and the nuts are not marketable and the bushes spread. I have seen acres of them in Iowa pastures.

The growing habit of these species is baffling to the horticulturist who would graft a good specimen. The plant becomes a clump through spreading underground stems; the clump grows on the outsides and dies in the middle. It resembles the raspberry in this respect. Therefore, the grafted specimen soon dies. Dr. R. T. Morris tried grafting one of these to the greedy European filbert. The filbert made such a demand upon the stem of the hazelnut that it kept on feeding the filbert instead of making more underground stems. (See his book, *Nut Growing.*)

At the present moment some worthy varieties of hazels and many worthy varieties of filberts are available in commercial nurseries and are ready for growth in a wide area in the cooler parts of this country.

Here is a sample of the complex relationships between climates and trees which shows the necessity for widespread experimentation: Winter kills Italian red filbert at Lancaster, Pennsylvania, for J. F. Jones, while Mr. S. W. Snyder, an equally careful observer, reports it to be the hardiest he has at Center Point, Iowa, near the northern edge of the Corn Belt, but this merely illustrates the point that winter killing is commoner in the Cotton Belt than in the Corn Belt.

One hazelnut, the Winkler, is self-pollinating and, on good rich soil, very productive of large, well-flavored nuts.

The filbert, a cousin which hybridizes with the hazels, is an important commercial crop in the Mediterranean countries and the Caucasus, whence we still import considerable quantities of nuts.

The late Professor J. A. Neilson of Port Hope, Ontario, reported filbert trees fifty years old, one foot in diameter, twenty-five feet tall, bearing well at Ancaster, near Hamilton, Ontario, and also a plantation of European filberts eighty-five years old on Wolf Island at the mouth of the St. Lawrence River, latitude 44° 10′ north.

Some European filberts blight when grown in the eastern part of the United States, but they thrive in western Oregon, Washington, and British Columbia. Several thousand acres of commercial plantings have been made, and they are now reported to cover twelve thousand acres in Oregon with all the paraphernalia of a commercial crop. In the decade before World War II the local supply cut our imports in half, and many small plantings are thriving in the eastern half of the United States.

In making some hybridization experiments (hazel x filbert), J. F. Jones, Lancaster, Pennsylvania, found that the same crossing produced nuts from which grew bushes six inches in height and ten feet in height, with all the intervening heights represented (Fig. 21). Mr. Jones produced hybrid nuts larger than those of either parent, and some of his hybrids have produced at the rate of a ton of nuts to the acre.

This work of Mr. Jones is highly promising in the plants already produced, and still more highly promising if more systematic work is done.

In speculating on possible results to be obtained by breeding in this group of species, the following facts should be kept in mind:

1. Tree hazels of European or Asiatic origin (often called Turkish Tree hazels) grow, bear, and are prefectly hardy at Rochester, New York.

2. These trees have been grafted for centuries in Turkey, and they are now being used for stocks in commercial filbert orchards in Oregon. I find that they are reasonably easy to graft.

3. The wild hazels are now growing in Newfoundland, Labrador, the shores of Hudson Bay, and the Peace River country of Northern Alberta.

I have seen the hazelnut growing as far north as Hudson Bay, and it is very hard to distinguish it from the elm. The hazelnuts grow to a height of from twenty to twenty-five feet, and the elm comes down to about that height. (Robert T. Morris, 1922 *Proceedings, Northern Nut Growers' Association.*)

The wild hazels thus seem to grow and be adapted to the whole stretch along the southern edge of the sub-Arctic forest, and they reach across the continent north of the present limits of agriculture.

It is interesting to speculate upon what a selection, crossing, and hybridizing from these hardy wild stocks and their good Eurasian cousins might produce in a decade or two. A private experimenter, J. U. Gellatly, of West Bank, B. C., reports that he has hazels from the Peace River Valley, Alberta, district where they have to stand 60° below zero, and some from Manitoba that take it at 35° to 40° below. These he has crossed with his large filberts, and they are starting to bear very promising nuts.

Professor George L. Slate, Geneva Experiment Station, Geneva, New York, has done notable work, making about two thousand known crosses and watching and fruiting each of the two thousand seedlings. This is a perfect example of the process for the increase of which this book is written. Salute Mr. Slate!

SOAPNUT TREE

This tree produces a fine cabinet wood and a nut whose *hull* contains an excellent saponaceous principle said to make a perfect lather. Some authorities claim that it has cleansing qualities superior to manufactured soaps. The *kernel* of this nut has an oil claimed to be the rival of olive oil. It will grow in large sections of the United States. (See *American Forestry*, November 1917.)

THE HOLLY TREE

A holly tree which grows wild near Vancouver Sound has leaves reported to possess a nutrient value that rivals cereals. According to George W. Cavanaugh of Cornell University, the analyses are as follows:

	Proteins Percent	Fats Percent	Moisture Percent	Ash Percent
Holly leaf............	14.56	13.56	5.24	4.
Oats................	11.8	5.		
Barley..............	12.4	1.8		
Wheat..............	12.0	2.0		
Corn................	10.5	5.		
Rye................	10.5	1.7		

It is claimed that simple processes can extract these fats in form suited to human consumption.

Holly grows over a wide area, and breeding may be expected to double the fat content of the leaf.

THE GINGKO TREE

The straight-growing gingko tree bears heavily of nuts. They have an offensive smell as the pulp decays, but the chemical qualities are suggestive of possible industrial uses. In some places the nuts are roasted, and when eaten they taste like roasted corn. The tree comes to us from Japan and thrives in much of the eastern United States. Why not—? ? ?

THE PAPAW, *Asimina triloba*

This is a native American tree of great beauty if given space in which to expand. Such a beautiful tree should be planted on our lawns. Its fruit is nutritious, having 435 calories per pound and having big seeds high in protein. It is liked exceedingly by some persons. disliked by others. To a very few it is somewhat poisonous.

The fruit may be found to a small extent in some markets, but it ripens in the autumn glut of foods. It has large seeds much like lima beans in size and shape. They may (almost must) carry most of the nutriment. The tree has the disadvantage of being hard to transplant, but it has the great advantage of having foliage that seems to be abhorred by all pasturing animals. Sheep, goats, cows, and horses apparently will scarcely touch it with their feet. It can spring up and thrive in the most persecuted pasture, and it grafts readily. The late Dr. G. A. Zimmerman, of Piketown, Pennsylvania, had scores of varieties of papaw under test. You can read more about it in the *Journal of Heredity,* July 1916 and January 1917. Also a mimeographed bulletin by H. P. Gould, U. S. Department of Agriculture.

THE HORSE CHESTNUT

The horse chestnut tree, so highly prized as a beautiful shade tree, bears abundant crops of a nut which analyzes high in nutrients:

Starch and starchy matters. 42   percent
Albuminous matter   5   "
Oil. . . . . 2.5   "
Saccharin matter.   9   "
Mineral matter. . . . . 1.5   "
Water. 40   "

This nut has long been used in Europe as food for deer, for the zoo elephant, and for game, and is regularly gathered for that purpose. During World War I, horse chestnuts were used to some extent for human food, but especially for forage. The bitter element was removed by boiling the crushed nuts, a method apparently similar to that used by the American Indians to remove the bitter elements from acorns. I'd like to have ten thousand tons a year of nice shiny horse chestnuts. My cow-feed factory would hum.

THE OSAGE ORANGE, *Maclura pomifera*

This tree grows in the country once occupied by the Osage Indians, who used its wood for bows. The French explorers called it *bois d'arc,* wood of the bow, now corrupted to "beaudarc" in Texas. The tree produces timber of fine quality and also bears heavy burdens of large fruit (one to one and one-half pounds) that may possibly have commercial value if processed in a chemical works of special design.

Professor W. R. Ballard, Maryland Agricultural Experiment Station, wrote me, June 14, 1916:

Professor Norton tells me that, while at Shaw's Gardens, St. Louis, he discovered that the fruit had almost as much starch in its composition as the potato. The abundant resin has no doubt prevented its utilization for food.

Careful analysis shows the fruit of osage orange to be rich in resins, protein, fat and starch, and apparently well worthy of much more experimental work than has been bestowed upon it. (For details, see J. S. McHargue, "Some Important Constituents in the Fruit of the Osage orange," *Journal of Industrial and Engineering Chemistry,* 1915, Vol. 7, p. 612.)

The fact that the osage orange tree has a magnificent wood should always be kept in mind and the resin might be extracted and used.

It is possible that more palatable varieties could be selected as well as varieties which would be more productive. I have seen trees in large hedges which were loaded with fruit, while many others were without fruit, and others bearing very sparingly close to them. (Letter, J. S. B. Norton, Maryland Agricultural Experiment Station, December 1, 1927.)

THE SUGAR MAPLE

The sugar maple is native or adapted to a large area of northeastern United States and southeastern Canada, and is the basis of a considerable industry. So far as I know, neither selection, breeding, nor grafting has ever been attempted as a means of increasing the efficiency of this tree. If the sugar maple is capable of being improved as the beet was improved, the results might be revolutionary for the hills from Nova Scotia and Quebec to Minnesota, with a detour through Appalachia.

THE PRIVET

Upon reading the first edition of this book, W. H. Mills, Professor of Rural Sociology at Clemson College, South Carolina, called my attention to the privet. He said (in his letter):

It is commonly used with us as a hedge plant, but after birds scatter the seeds it is found growing wild. For some years I have been feeding my chickens its seeds in increasing quantities, as I have found no ill effects from its continued use. The chickens begin to eat the seed as soon as they ripen. Such is the abundance of the crop that the seed usually hang on the trees until the middle of February or March when the cedar waxwing birds in the migration north stop long enough to clean the trees. I believe that we can have at practically no cost a supply of poultry feed for two or three months during

the winter from this source. I am enclosing an analysis which I got
the chemist of the South Carolina Experiment Station to make some
two or three years ago.

	Dried Berry	*Original Berry*
Moisture...........	8.93 Percent	69.04 Percent
Ash................	3.46 "	1.17 "
Fat................	18.39 "	6.25 "
Protein............	5.94 "	2.02 "
C. Fiber...........	16.28 "	5.54 "
N. Free Ext.........	47.00 "	15.98 "
	100.00 "	100.00 "

J. H. Mitchell, Chemist
S. C. Expt. Station

QUEENSLAND NUT

The Queensland nut, *Macadamia ternifolia,* produces a nut
with an epicurean flavor but an almost impossibly hard shell.
Some of these trees are growing in southern California. An Aus-
tralian experimenter has produced one tree producing nuts
with a soft shell, but what is a hard shell to a stone crusher? (For
details, see U. S. Department of Agriculture, Bureau of Plant
Industry, Bulletin 176.)

ARGANIA

Walter C. Lowdermilk reports that in Morocco the Argania
tree is of great service to the natives. French foresters have
worked out a system that seems to be satisfactory both to the
foresters and the natives.

The leaves of the tree are relished by the goats as food. The
fruit also is relished by goats and camels, and it makes an oil
which the natives prefer to olive oil.

A multiple-use system has been worked out as follows: Many
of the trees are cut off low and make many suckers. At intervals
the goats are turned in to harvest the leaf crop. Meanwhile some
of the trees are left to grow tall and produce oil—thriving on
rainfall of ten inches, Mediterranean climate. California and

northern Mexico please take notice, also Australia and South Africa.

Walter C. Lowdermilk also reports that in the Kybelian section of Algeria the population is as dense as that of Belgium and trees are so valuable and important that a person sometimes owns one branch of an olive tree. The property problem is simplified by the whole village gathering its olives on the same day.

This fractional ownership of a tree has resemblance to a similar practice among the oasis dwellers of Tozeur, Tunisia, where a man will sometimes own one-fifteenth of a palm tree. At harvest time, the owners are all present to divide.

THE WATTLE

This tree, whose bark contains much tannin, is cultivated to the extent of two hundred thousand acres in Natal. Since Natal has the climate of southeastern United States, it would seem possible to introduce the industry. Australia also has promising species of wattle.

THE TUNG OIL TREE

The tung oil tree, *Aleurites fordii*, of central China yields fruit from whose seeds is extracted one of the most valuable drying oils known in commerce. This oil is imported into the United States to the extent of forty or fifty million pounds a year and is worth several million dollars. The United States has many climatic areas similar to that in which this tree grows, and here we have a case in which we are working at crop introduction more diligently than with the Chinese chestnut (with which we have succeeded). The trade stoppages resulting from the recent wars in China have put up the price and also, of course, have stimulated the chemists to search for a substitute.

The U. S. Department of Agriculture and various experiment stations in the Southern States can tell you about tung. Of course, it comprises nothing but the moving of an industry, like bringing oranges from Spain to California, but that usually turns out to be a real job, especially as our Cotton Belt climate differs from that of China to a greater degree than that of California differs from the climate of Spain.

WILD PLUMS, CHOKECHERRIES, AND SAND CHERRIES

This is a group of remarkably hardy natives of the northern Great Plains region. Prof. S. S. Visher says of them, "As a boy in South Dakota, I was much impressed with the value of these small trees not only for human food, but for hog food and for food of prairie chickens, bobwhites, etc."

These trees start with a great hardiness, and the previous discussion of cherries and apricots shows that they might be specialized into two groups of crops, fruit food and nut food for man or beast, especially beast.

Upon reading the first edition of this book, Mr. James Knox, writing from Riverside, California, said, "The Colorado desert *Prunus* have ten times more variation than the prairie species have."

999 ? ? ?

No botanist, only God, knows how many more trees might become crop trees if man did his best with them.

Food for man, feed for beasts, tanning materials, dye materials, fibers, rubbers, gums, medicines? ? ? ?

There is no reason to think the figure 999 is fictitious or exaggerated. We have scarcely nibbled at the corners of our possibilities with the trees. If anyone should wish to check on that statement, I suggest the perusal of *Food Plants of the North American Indians*, U. S. Department of Agriculture Miscellaneous Publication No. 237, issued July 1936. Its 64 pages list and give brief statements of food use of 1,112 species in 444 genera and 120 families (as botanists count families). Who is ahead, the Indians—or the white man with his experiment stations, Departments of Agriculture, arboreta, botanic gardens, and professors of botany? Plainly, the Indian has them all licked.

I salute the memory of the late Boyce Thompson, multimillionaire copper magnate, who asked himself, "What will men need most, a hundred years from now?" And he answered his question by saying, "Plants, knowledge of plants." And then he endowed the Boyce Thompson Institute, Yonkers, New York, and they are doing a good job, although a lonely one.

Meanwhile, Dr. Hugh H. Bennett, Chief of the Soil Conservation Service (in the U. S. Department of Agriculture) says that we are still destroying five hundred thousand good, arable acres each year.

Ownership of land in the United States is still a license to kill it and therefore to kill a part of your country. Now, just what is treason, anyhow?

CHAPTER XXII

A Peep at the Tropics

THE PRODUCTIVE TREES THAT ARE WAITING

The temperate zones are rich in crop-yielding trees. But many of these trees are largely untested, almost unexplored from the standpoint of economic botany. If I have succeeded in establishing that fact, what can I say about the tropics?

The vegetation of that large and little-known realm is almost wanton in its productivity; and in the total of economic possibilities the trees seem to have more than their share of fruits and other useful products.

"A History of Fiji," by Alfred Goldsborough Mayer, *Popular Science Monthly*, June 1915, p. 527, declares:

An extraordinary number of the forest trees of the Fijis furnish food for man. Such are the bread-fruit, which grows to be fifty feet high with deeply incised glossy leaves, sometimes almost two feet long. The Malay apple, or kavika (Eugenia), grows to a great height and bears a delicious fruit, which when ripe is white streaked with delicate pink and most refreshing and rose-like to the taste. The cocoanut palm clusters in dense groves along the beaches, the long leaves murmuring to the sea breeze as they wave to an fro, casting their grateful shade upon the native village. Of all trees none is more useful to tropical man than the cocoanut. . . .

Bananas and the wild plantain (fei) grow luxuriantly in the forest, as do also oranges, lemons, limes, shaddocks, guavas, alligator

pears, the papaw, mango, and many other smaller shrubs and vegetables. . . . Famine is indeed all but impossible in the high islands
of the tropical Pacific.

To advance from gathering wild produce in the forests to the
systematic planting and cultivation of the tree as a crop is an
easy and natural step. The process, though old, may perhaps
be only at the beginning of a series of new developments. It may
be seen today in the active stages of its development. Consider
rubber cultivation.

A RAW MATERIAL—RUBBER—AN EXAMPLE

Rubber is the perfect example of the wild trees whose uses
were quickly demonstrated, after which the technical processes
were worked out with great rapidity, and manufacturing and
agricultural industries created with almost magic speed.

For two centuries the small need for rubber was supplied by
spasmodic tapping of the wild trees of the forests. Then the
bicycle and automobile made a sudden increase in demand
for rubber. A few small experiments in cultivation were producing much talk by the year 1900, and by 1910 the highly profitable result of a few small plantations had started the full-fledged
rubber boom with capitalistic organization, scientific prosecution, technically trained European supervisors, and thousands
of Oriental coolies. By 1918 the success of the growers had been
so great that overproduction and glutted markets with rubber at
eighteen cents threw consternation into the camps of the men
who had begun their cultivation on an extensive scale and had
planted their trees when rubber was two dollars per pound.
That is almost a norm for new industries.

At first, any wild rubber tree was used. When cutthroat
competition had forced intensive and scientific development,
orchards were planted with selected seed rather than chance
seedlings. This gave a substantial percentage increase of produce, but a severalfold increase of output is had from the orchard

of *grafted trees*. Better rubber orchards now have a leguminous cover crop to feed them with nitrogen. All this development, and even more, took place in a quarter of a century.

SILT PITS IN THE RUBBER PLANTATION

We are indebted to the rubber growers for the economic large-scale development of a device which in Malaya they call "silt pits." This old but essentially unknown device promises to be of great value for the tree-crop agriculture of many climes. Holes are dug in the orchard near the rubber trees. The rows of holes are connected by a ditch on contour lines. Thus the fine soil carried by running water is caught in the pits. The pits also hold surplus rain water and increase the yield of the trees. (See Figs. 23, 110, 130, 131, 132, 135, 136.)

A STAPLE FOOD—THE DATE—AN EXAMPLE

The productivity of tropical crop trees is well established by the performance of the date. Considered as human food, this is possibly the king crop of world agriculture. Certain oases such as Tozeur in Central Tunis were described in the first century by Roman travelers. The cases still correspond to that description and apparently have been yielding dates every year in the intervening eighteen centuries. Crops of wheat and corn are not expected in Illinois year after year on the same bit of land. Yet these dates yield more food per acre by far than wheat or corn and have done so year after year for centuries. They are probably supported in this seeming miracle by windblown dust.

The foliage of the date, being feathery at the top, permits sunlight to come through and fall upon an underorchard of olives, apricots, and figs, and beneath these, beans and other leguminous crops will grow—literally a three-story type of agriculture so rich in yield that only a portion of the date oases need to be worked so diligently. If it were all worked diligently, there would be a glut of vegetables.

THE PEJIBAYE—A WET-LAND TROPIC RIVAL
OF THE DATE

The pejibaye is similar in food value to the well-known date, but it is almost unknown. I doubt if one percent of the people of the United States or England have ever heard of it, but I mention it to prove the productivity of tropic trees and our ignorance concerning them. According to an article in the *Journal of Heredity,* April 1921, this tree, which is a palm, is a rival of the date in productivity. (See Fig. 109.) The fruit is more productive than the banana, but it differs from the date in having starch instead of sugar for its main constituent. For months at a time the pejibaye is the chief food supply of the native peoples of southern Costa Rica and the lowlands of Colombia, Venezuela, and Ecuador. It was mentioned by the early Spanish travelers, has been in use ever since, and yet few indeed have even heard of it.

A RAW MATERIAL—KUKUI OIL—AN EXAMPLE

In the attempt to establish the idea of the fecundity of the tropical tree I have already mentioned a basic raw material, rubber; a staple food of the Old World, the date, and one of the New World, the pejibaye. Now I cite another industrial commodity, a paint oil, which we might make from the kukui, or candlenut tree, whose oleaginous seeds make a brilliant flame that lighted Polynesia for an unknown period of time, before the Standard Oil can brought a cheaper illuminant.

This tree grows, or will grow, in many wet tropic lands.

Kukui, *Aleurites moluccana* (synonym *triloba*), is generally distributed throughout Polynesia, Malaysia, Philippines, Society Islands, India, Java, Australia, Ceylon, Bengal, Assam, China, Tahiti, Hawaii. It has been introduced into the West Indies, Brazil, Florida, and elsewhere. The tree has wide-spreading branches, attains a height of forty to sixty feet, and is characterized by large, irregularly lobed leaves of a pale green color and nuts about two inches in diameter, containing one or two seeds. In Hawaii, kukui is common on all the islands, being the dominant native tree of the lower mountain zone and easily recognizable at a distance by the pale color of its leaves.

Everyone knows that the ground under kukui trees is literally covered with nuts, of which few are used for any purpose at present.

At two hundred pounds of nuts per tree and eighty trees per acre there would be a yield of eight tons of nuts per acre. It has been found that algaroba yields from two to fourteen tons of beans per acre. A good stand of kukui will give a larger product per acre, and a conservative estimate would be five tons of nuts. On 15,000 acres the annual crop of nuts would thus be 75,000 tons.

We may probably assume 15,000 acres as a safe estimate of the kukui in Hawaii.

From our experiments it appears easy for a man, woman, or child to pick up five hundred pounds of nuts per day. The nuts are, of course, to be gathered free from the soft outside husk. Only an extremely small percentage of the nuts spoil or turn rancid even after lying two years on the ground. The spoiled nuts float in water and may thus be easily separated from the sound ones. At thirty cents per one hundred pounds, the laborer would receive one dollar and fifty cents for five hundred pounds, a day's work. The average oil content of the meat or kernel is sixty-five percent. The kernel equals thirty percent of the weight of the nut. About 19.5 percent of the nut is therefore oil. In the Sunda Isles, where kukui oil is an important article of export, experiments have shown that ninety percent of the oil is obtained by commercial methods through the use of presses. The oil recoverable by commercial methods would thus amount to 17.5 percent of the weight of the nuts. From one hundred pounds of nuts 17.5 pounds of oil would be obtained, or a value of one dollar and seventy-five cents at ten cents per pound.

Kukui oil has been shipped from various islands of the Pacific to the United States for the past seventy-five years for use in making soap, paint, varnish, and artists' oil. The market price is the same as or slightly higher than that of linseed oil and varies with the price of the latter. (Hawaii Agricultural Experiment Station, Honolulu, Hawaii, Bulletin, No. 39, issued February 8, 1913, "The Extraction and Use of Kukui Oil" by E. V. Wilcox and Alice R. Thompson.)

No one is making large commercial use of it at present, but it seems to be a remarkable waiting resource.

PALM OIL—BOTH FOOD AND RAW MATERIAL—
A SUGGESTIVE METEOR OF COMMERCE

The palm is said to be second only to the grass family in the number of its species. When one considers its present productivity as resulting chiefly from unscientific chance, it opens interesting speculation as to what it might produce after a few decades of selection and breeding.

The meteoric rise of the African palm-oil trade is one of the miracles of recent commerce. The oil palm, widely scattered in the forests of western and central Africa, produces a big fruit with an oily pulp surrounding a hard nut containing an oily kernel. For ages the African native has been boiling this fruit and skimming off the oil for butterfat.

About the beginning of this century some of this palm oil reached Europe. It attracted the attention of manufacturers, and Lord Leverhulme, English soap magnate, father of Sunlight Soap, sent scientific explorers to investigate the commercial possibilities. Their reports sounded almost too good to be true, so this lord of soap checked up their accounts with personal journeys. As a result, he established large plantations of oil palms in western Africa. This was very easy to do, for the trees stand many to the acre over large areas of wild forest. The trade shot up, and Europe now imports hundreds of thousands of tons of palm kernels and palm oil. This industry has sprung up almost as quickly as the rubber industry. Between 1911 and 1926 the European import leaped from 25,000 to 500,000 tons and became an established part of European food and African economy. The Dutch successfully introduced the palm into the East Indies.

Oil palm gives three distinct kinds of oils:

1. The edible oil boiled from the fruit.
2. The inedible oil pressed from the fruit.
3. The oil of the kernel. The kernel has long been extracted

from the nut by the African women, working with two stones. A cracking machine now does the work.

Meanwhile, this tree is running wild in Brazil, having been brought there by the early importations of Negro slaves. However, Brazil is not solely dependent upon the African oil palm, for it has one of its own, babassu, growing wild over a large area about the latitude 5° south and now in the process of being introduced into commerce.

WHAT HAPPENED TO THE COCONUT

The ancient and well-known coconut, with its rich oil and myriad uses, need not be here expanded upon, except to point out one fact that can revolutionize the economics of almost any kind of vegetable oil. Chemical researches have recently added something to the oil which turns the strong liquid coconut oil into a sweet-flavored tallowlike substance which now graces millions of European and some American tables in place of the more expensive butter.

Save for sugar, most of the other standard Tropic exports are tree crops—cacao, coffee, tea, cinchona, spices, and Brazil nuts. The Brazil nut, *Bertholletia excelsa,* of our market, grows in a wild tree that towers above the Amazon forest as the pecan towers over the oak tree of southern Indiana. These nuts go to waste by the millions of bushels. Only a small fraction of the crop is gathered and sent to the market. There are many other trees in these forests producing edible or oil-giving nuts.

I must mention two other important tropical tree foods. One is the papaya, a fruit resembling a cantaloupe. It grows in huge clusters from the upper part of the trunk, which grows almost as rapidly as corn and bears astonishing loads of fruit in a few months. (See Fig. 108.) A few are grown in Florida, for use as a table melon, and the jelly is sold as a delicacy.

The other is the avocado, or alligator pear, and this tropic delicacy offers an interesting dare to the commercial genius of man. The fruit is as large as a Bartlett pear, sometimes even twice or three times as large, and has a thick buttery meat which analyzes from fifteen to thirty percent fat. Puerto Ricans sometimes chop it into little cubes and mix it with their rice as we might mix butter with our rice. This makes the rice (dietetically) into bread and butter. As a constituent for salad, avocado furnishes the oil.

The avocado tree thrives from sea level to six thousand feet in Guatemala. A few are growing in California and Florida. But the real problem is to extract the oil so that it will keep—dry it or can it—or find some means of handling it fresh. I do not know what a deep freeze would do to it.

In giving brief mention to this long list, I make no attempt to be complete. I wish merely to try to suggest the riches of the Tropics in tree-crop possibilities.

Messrs. Dorsett, Shamel, and Popenoe, in Bulletin 445, Bureau of Plant Industry, U. S. Department of Agriculture, list twenty fruits in Brazil which they describe as little known. They have little doubt but that there are several hundred trees native to the tropical world and producing an important fruit not now of commercial utilization, nor capable of it, yet capable of improvement through selection and plant breeding.

SOIL DESTRUCTION, OR TREE CROPS IN THE TROPICAL FORESTS

The enormous growth and density of the tropical forest in the rain belt near the Equator stands in the minds of millions as the highest type of rank luxuriance, but it is erroneous to assume that this luxuriance signifies great fertility. The tropical forest grows with great speed, reaches great height, produces a vast quantity of vegetation, but often the forest rests upon a base of laterite—namely, soils high in aluminum or iron and very low in fertility. In some places the soil has so much of one of these metals that we shovel it up as ore. The elements of fertility are soluble and have been leached out by rain and the continuous heat, but on top of this sterility stands the forest. There is a certain resemblance here to a glass case (Wardian case) which the botanist puts in a window to make a little torrid zone where he can keep plants. He puts some earth and some water in the case of glass—adds a few plants, and seals the case. There are records of these little forest worlds having lived for fifty years in this case. The plants drink their little supply of moisture, exhale it, drink it again, as they take up fertility, make leaves, and drop them. The leaves decay and feed the plants; thus fertility and moisture run round and round the cycle of life and death. The process bears a resemblance to the water in the boilers of an ocean steamer, the same fresh water being used for the boilers over and over again—boiling and condensing; boiling and condensing, hundreds of times as the boat crosses the ocean.

The plants of the tropic forest produce leaves, fruits, stems, and trunk. The trees drop leaves, twigs, fruit, and finally they die, rot, become plant food, and are picked up again by the plants and thus rebuild the forest once more. The circuit is complete. There stands the magnificent forest like a giant Wardian case, using a small quantity of endlessly circulating fertility, with some continual reenforcement of carbon oxygen and nitrogen from the air.

Man comes with axe and fire. Man cuts the trees of the forest, burns the wood, sets out a plantation, cultivates the soil. The hot sun destroys the humus, and the rain carries fertility away. The plantation fails, the forest is gone now, it cannot replace itself. Coarse grass takes its place. In all continents the equatorial forest is shrinking by this means.

It is estimated that by this process the equatorial forests of Africa are perishing at the rate of half a million acres a year (see *Road to Survival*, William Vogt.) The native system is to deaden the trees, plant a patch of garden, use it for a season or two, then move on before the possibility of forest regrowth has disappeared. The forest and the Negro and other tropic denizens have lived together thus for ages. The white man, with his system of forest destruction, is the greatest destroyer that ever trod the earth.

If much of the tropic forest is to be preserved, we must make use of tree crops. Tree crops will safeguard fertility while producing food for man. In most cases there can be an undergrowth of leguminous nurse crops of small tree and bush to catch nitrogen, hold the soil, make humus and feed the crop trees—nuts, oils, fruits, gums, fibers, even choice woods.

There is nothing revolutionary about this idea. Note the chief exports of the tropic forest areas today—rubber, cacao, palm kernels, palm oil, coconuts, cinchona, mango, Brazil nuts, ivory nuts (tagua, Ecuador), babassu nuts. There might be a host of other nuts and fruits for export or home consumption if they were to receive half the effort that has been given to the apple or orange.

In some cases, such as the coffee and cacao plantations of today, the nurse tree crop towers above the harvest tree crop, furnishing nitrogen and partial shade. Perhaps a fruitful legume might be found for this purpose.

The need is for a realization that the choice is—*tree crops or nothing but coarse grass of almost no value*—*Cognales*, they call such wastes of the tough cogon grass in the Philippines.

H. L. Shantz, after much research in Africa and elsewhere, thinks that much of the Equatorial forest of Africa has already gone to grass.

TROPICAL NEEDS AND THE EROSION PROBLEM

Certain parts of the tropical world are in need of having tree crops developed much more extensively than at present. In a large part of the tropical world the main food supply of native peoples is a cereal. The system of production is to burn the forests, plant a crop of two of corn, and after harvest let the thicket grow again while another piece of land is cleared and burned to make a fresh field. This method destroys the forests. It devastates the soil. If a permanent field is made, the destruction is often even more sure and final. The torrential character of most Tropic rainfall lifts field erosion to the plane of an economic terror.

See (1) Bulletin of Agricultural Research Institute, Pusa, No. 53, Calcutta, 1916, "Soil Erosion and Surface Drainage," Albert Howard; (2) "Afforestation of Ravine Lands in Etawah District, United Provinces," E. A. Smythies, Indian Forest Records, Calcutta, 1920; (3) *Proceedings* of the Second Pan-American Scientific Congress, Section 3, Conservation of Natural Resources, O. F. Cook, p. 573, "Possibilities of Intensive Agriculture in Tropical America."

The soil destruction in India and Central America described by the papers just referred to shows that the Tropic denizens are destroying their lands as rapidly with cereals as we are destroying ours in the southern part of the United States with corn, cotton, tobacco, and gullies, and like ourselves they are in need of development of tree-crop agriculture if the lands are to continue to serve the race. Mr. Vogt, Specialist in Conservation for the Pan-American Union, tells pitiful tales. For example: The people of El Salvador, Central American republic, have a diet of 1,500 calories. They are tilling steeper and steeper land, and it fairly slides away, often in great landslips. In laying out an Andean airfield in Venezuela, the engineer is told to provide for ten inches of rainfall in thirty minutes. A dark and hungry future faces the Andean corn grower, and he is so hungry that he hasn't time to wait for tree crops, even had he the stomach (figuratively) to do so. (Be sure to ponder Fig. 61 in this book.) Starvation, in your time, looms over the Andean mountain tops.

TREE CROPS AND TROPIC FAMINE

On both sides of the equator, in latitudes from six to ten or twelve degrees, the rainfall is concentrated into one season, and as a result the forest gives way to open parklike country called savannah, where trees are scattered over the grasslands. As distance from the equator increases, the rainfall, trees, and grass diminish until finally the desert prevails. (See map, Fig. 138.) This grassland zone of the rainy season and the dry season is a latitude of famine, for the reason that the Tropic rainfall is the most unreliable in the world. An examination of the distribution of rain by months (Fig. 128) shows how difficult is the problem of growing a cereal, which is at present the chief staple of the people. When should they plant it? The rainfall is so unreliable that they may plant two or three times and fail. But an established tree can wait for rain and use it when it does come, and it therefore has a better chance of harvest than any annual crop like sorghum, millet, and corn. Especially would this be the case with an extensive use of water pockets such as are used in connection with the rubber in Malaysia. (See Fig. 110 and Chap. XXIII.)

FORAGE CROPS—BEANS (GRAIN SUBSTITUTE)—SOME EXAMPLES

There are many tropical trees producing beans whose forage value and use are much like the honey locust and keawe. For example, babul is the most widely distributed tree in India. I saw babul trees in my first moments in India as I landed on the coasts of Coromandel from Ceylon. They were growing in white sand. I saw them at the foot of the Himalayas clinging to rocky slopes, and again at the extreme west as I neared the port of Karachi.

Everywhere, the goat herder leads his flocks to these trees. Often he cannot wait for the beans to fall, but with a long hook he cuts down branches so that his wards may eat the beans and also the leaves of the tree. In a year of bad drought these trees will be beheaded thus by the million. Often they stand in land too dry for dependable agriculture. In good years they store up branches for lean-year nibblings.

The gigantic saman tree of India yields sugar beans. They are greedily eaten and are said to improve the quality of milk.

In Cuba the guasima is left when pastures are being cleared, exactly as persimmons are left in the pasture fields of Georgia. The bean is eaten hungrily by all farm stock, but I can get no measure of its actual productivity per unit of area.

In my section of the country in eastern Cuba, specially in Camaguey province, the cattle country of Cuba, the guasima tree and the algaroba are considered as valuable trees on account of the fruits that they bear in the dry season when the pastures are exhausted; in winter both fruits constitute a valuable food for the stock. Horses, cattle, mules, and hogs eat them. A great value is given to the fruit of the guasima because the native farmer considers that fruit is specially adapted to feed the horse. Our cattle man, our native *guajiro*, collects the fruit of the guasima and feeds his best pony on it, because his animal will grow on a fine coat of hair. (Letter, Dr. Emilio L. Luaces, Santiago de las Vega, August 25, 1916.)

The guasima tree is a native of Cuba but also of all tropical America, and the species *Guazuma tomentosa* is a native of Java too. Two species are known in Cuba as guasima. *Guazuma ulmifolia*, am., and *Guazuma tomentosa*, H. B. K. of the natural order *Sterculiaceae*. Both are equally common and produce pods eaten by cattle.

The guasima is perhaps the most vastly distributed tree over the Island. It is found in all kinds of soils, even in the heights, except in the very arid savannahs.

It is a quick grower and begins to produce after the fourth year.

The guasima is used in Cuba mostly to feed the pigs, for which purpose, when the forest lands are cleared up, the only tree that is left is the guasima and sometimes the ceiba.

The *portreros* (pasture land) in Cuba are characterized by the abundance of the guasima tree, under which the cattle find shade and some food. (Letter, República de Cuba, Secretaría de Agricultura, Comercio y Trabajo, Estación Experimental Agronómica. Departamento de Botanica, August 25, 1916, Juan T. Roig.)

H. J. Webber, Professor of Sub-tropic Horticulture and Director of the Experiment Station at Riverside, California, gave me the following facts. They are illuminating as to the possibilities of native trees and suggestive as to the possibilities of introducing them from other places.

I found the carob in a few places in Rhodesia and in the Transvaal, and there are some references to the production of carobs in the Department of Agriculture reports from the Transvaal, and also from the Department of Agriculture of Southern Rhodesia. Mr. Walters of Southern Rhodesia issued a bulletin on the carob which

was published by the Department of Agriculture. I had not been in
Africa long until I was impressed by the very large area in the central
part of the country in the high plateau region where it appeared
to me the carob would be an ideal crop. Few trees exist, however,
in the country at the present time. I found a group of twenty or
thirty trees on a plantation south of Untalli in Eastern Rhodesia
where the trees were about fourteen or fifteen years of age and of
fairly good size. I was told that they produced pods in abundance
and that the cattle came to the trees regularly to get the pods, and
this was evidenced by the fact that the ground was trodden down
all around under the trees. The owner of the plantation was convinced that the planting of carobs in that section would be of very
great value.

There are a number of legumes native to South Africa that have
pods similar to the carob, some of which are actually thicker and
might be even more valuable if cultivated or if planted on good
lands. I have not yet had opportunity to determine the identity of
these trees. I saw one in the forest near Victoria Falls that produced
a square solid pod about the length of the pod of the carob and an
inch square practically each way.

This passage from Dr. Webber needs pondering—especially
when one remembers that there are several million square
miles of semiarid Tropic lands on which such trees might grow,
and where famine often lurks upon the horizon.

THE OPPORTUNITY FOR TROPICAL TREE RESEARCH

The Tropic lands have great crop possibilities in their trees.

The Tropic lands have great need for a more dependable
crop than the grain crops of today.

The Tropic lands exceed the Temperate Zone in the possibilities of tree-crop development through government agencies.
Appropriations do not depend so much upon elected legislatures as they do in the Temperate Zone. Suppose the directors
of the experiment stations of Maine, Minnesota, Arizona, or
Alberta should want to do big work as outlined in this book. I
know that some of them have tried and have almost eaten their
hearts out trying to surmount the obstacles.

If the director of the station in Rhodesia or in West Africa has a big idea, he may need but to convince a small council of intelligent men who are not politicians and who do not have to get reelected by a popular majority. Really great things were done with cinchona, rubber, oil, palm, and other crop trees by the Dutch in Java before World War II. However, when we consider our own torpid slumbers, there is little reason to expect much from any independent government in the Tropics in any foreseeable time.

The Bureau of Plant Industry, U. S. Department of Agriculture, has a Division of Rubber Plants and a Division of Tropical Plants.

PART THREE

Economics, Farm Applications, and National Applications

CHAPTER XXIII

Tree Crops and Farm Management

A TREE FARMER'S FARM

R. O. Lombard, gun in hand, crept softly through the thick forest in a Georgia swamp. He was hunting for wild turkeys. He heard a cracking sound. Peering around a clump of bushes he spied some hogs crunching acorns beneath a water oak. They were miles from any house. They were fat, ready for the shambles, and it was all of their own doing. The hogs had fattened themselves on swamp produce.

As Mr. Lombard quietly watched the hogs, a thought struck him. "If they can feed themselves out here on the swamp, why can't they do it on my farm? Here they pick up a living and fatten themselves in the fall when acorns are ripe. If I were to raise other tree crops on my farm, why could not a lot of them pick up their living the rest of the year too?"

For the rest of his life Mr. Lombard had fun working out the idea of tree-crop farming, where pigs harvest the crops. When I saw him, he had two hundred everbearing mulberries, two hundred hog plums, two hundred wild cherries, three varieties of red haws, and mock oranges.

This plant [mock orange] is found growing in South Carolina to Florida and Texas. It begins to bloom in February and lasts until April. The small fruit which ripens in late summer is retained

315

throughout the winter. (Letter, P. J. Berckman's Company, Augusta, Georgia, August 21, 1915.)

Mr. Lombard thought them very fine winter hog feed. He also had a few trees of *Cudrania tribola*, introduced by Professor Wilson of the United States Department of Agriculture.

We do not think this tree has been used to any extent by the people here in the South, and we know of only two trees near our city and these are across the river in South Carolina. One of these trees bears a full crop every year, while on the other one the fruit drops before maturity. (Letter, Fruitland Nurseries, successors to Berckman's, Augusta, Georgia, July 5, 1923.)

The native persimmons, as they sprang up in his fields, were grafted to a productive variety of native American persimmon, and in September, when I saw them, the trees were bending down with unripe fruit. Mr. Lombard had a large number of water oaks scattered about the place, and had been planting them systematically for years.

As an exhibit at the county fair he had printed slips numbering twenty-six crops which were growing either wild or cultivated on his place. Some, he said, were of small value, but they were there. Here is his list:

1. Mulberries (Downing), April 20 to July 20 (dates of use).
2. Mulberries (Hix), April 26 to July 20.
3. Mulberries (White), May 4 to June 26.
4. Huckleberries (Frog Eye).
5. Huckleberries (large high bush), June 1 to July 1.
6. Huckleberries or Blueberries, June 5 to July 5.
7. Huckleberries (hog or ground), June 10 to July 20.
8. Wild Cherries, June 10 to July 20.
9. Blackberries (highland), May 26 to June 30.
10. Blackberries (swamp), June 5 to July 10.
11. Hog plum, June 20 to August 20.
12. Gooseberries, July 1 to August 15.
13. Haws, August 15 to September 30.

14. Haws, August 25 to September 30.
15. Haws, August 15 to September 30.
16. Muscadines, September 1 to October 15.
17. Dogwood berries, September 15 to December 30.
18. Black gum berries, September 10 to December 30.
19. Acorn (water oak), September 15 to April 1.
20. Acorn (post oak), October 1 to April 1.
21. Persimmons, September 15 to December 30.
22. Hickory nuts, October 1 to April.
23. Pecan nuts.
24. Chestnuts.
25. Chinkapin nuts.
26. Hazelnuts.

He had three hundred acres of fenced pasture. One-quarter of it was swamp. Some was hopeless-looking sand which Mr. Lombard said was "hardly worth the hole it filled up in the earth." Some of the farm was in Bermuda grass. When I was there in early September, Mr. Lombard had a small field of cowpeas in some of the sand. The pigs harvested these as they did all the crops which grew on the trees. He reported keeping forty hogs all the time. Acorns, he said, kept his hogs fat for five months in winter, and mulberries did it for three months in summer.

It is one of the ironies of fate that upon the day of Mr. Lombard's funeral I published a magazine article describing his work.

THE VISION OF THE HILL FARMS

I venture to enlarge Mr. Lombard's vision. I see a million hills green with crop-yielding trees and a million neat farm homes snuggled in the hills. These beautiful tree farms hold the hills from Boston to Austin, from Atlanta to Des Moines, Spokane and Edmonton. The hills of my vision have farming that fits them and replaces the poor pasture, the gullies, and the abandoned lands that characterize today such a large and increasing part of these hills.

These ideal farms have their level and gently sloping land protected by the new terraces (Fig. 136) and are intensively cultivated—rich in yields of alfalfa, corn, clover, other legumes, wheat, and garden produce. This plow land is the valley bottoms, level hilltops, and gentle slopes, where low, broad terraces hold the soil. The unplowed lands are in permanent pasture, partly shaded by cropping trees—honey locust, mulberry, persimmon, Chinese chestnut, grafted black walnut, grafted heartnut, grafted hickory, grafted oak, and other harvest-yielding trees. There is better grass beneath these trees than any which covers the hills today. The crops are worked out into series of crops to make good farm economy.

As to the terraces: upon recommendation of Dr. Hugh Bennett, I visited not long ago some farms near Muskogee, Oklahoma. Here a grassed waterway carried off the overflow of a terrace system that extended across several farms. Terraces followed contours, and corn rows followed terraces. I was told that sometimes an inch of rain fell before runoff began, and this runoff did not erode the grassed outlet. That looked like permanent agriculture, on gentle slopes with corn as a component. That place may have a future. Much of what was once corn land has no future.

It will take time to bring to pass this miracle of soil-saving tillage and tree-crop slopes. It will take time to work it out. First of all a *new point of view* is needed, namely, that *farming should fit the land*. The presence on the land of the landowner is also needed. This is not a job for tenants. Let the tenant go down to the level land which carelessness cannot ruin so quickly. Not his the beautiful home in the beautiful hills.

This tree crop farm is the place for the man who has the insurance point of view. Fortunately, insurance is now becoming one of the characteristics of this age. One of the best kinds of farmer's insurance is for him to build his hill farm over gradually to the tree-crop basis. Inflation will not ruin *this* insurance policy.

For anyone wishing really to understand this tree-crops farm idea, I refer to a new technique, and if possible I want to give a new realization.

1. The new technique is grass silage. I know a farmer who lives eighteen miles from Washington, has a farm of seventy-two rolling acres, a herd of cows, and a professional job. He hasn't milked a cow for several years. He hasn't raised an acre of corn for several years. His land is in grass for pasture, for hay, for grass silage, reseeded when necessary but never in corn. He buys grain feed from the Middle West as do nearly all eastern dairymen.

He hires two men and is away from home every day on his professional job, but is making more from his farm than the average college professor's salary.

2. The new realization is—*push really intensive use of till-able land.*

This tree-crops farm that I am advocating can have tree crops, and permanent pasture on its roughest land. Lands of considerable slope and fair surface may have grass for hay and grass silage. The levelest spots should be in really intensive cultivation, probably to corn year after year for the live-stock man. Erosion can be completely prevented by terraces. A cover crop every winter and some commercial fertilizer and stable manure from time to time will maintain fertility. This means stock farming—cows, cattle, sheep, or hogs. Fifteen acres perennially in corn at 15-20 tons of silage per acre ought to carry a lot of cows, or the land might produce 70-100 bushels of corn per acre. I wish modestly to report that my son J. Stewart Smith made 112 bushels of corn per acre in 1947 and again in 1948, and 115 bushels in 1949, on a Virginia farm he rents from me—alfalfa sod 8 years old plus cow manure and chemicals. The measurements were made by the DeKalb Seed Company in process of awarding a prize for the largest acre yield in the state.

A REAL TREE-CROPS FARM

I have pleasure in showing the field plan and crop plan of a tree-crops farm that is not a hypothesis but a fact. Mr. John W.

Hershey, R. F. D. 1, Downingtown, Pennsylvania, is a man of creative mind, enthusiasm for country living, soil conservation, and trees—crop trees. For years he has made a living operating a small nursery.

He recently bought a seventy-acre farm, with rolling surface and with soil much run down, and he has started in to make it into what he calls the No. 1 Tree-Crops Farm of America.

The accompanying map and legend practically tell the story, except that I wish to add that his intensive crop land goes first to a small nursery and secondly to grow feed. At present he is buying calves and selling heifers, cows, and young bulls. He can start tree crops all at once rather than gradually because of his nursery business.

As you examine his plan you will see that this is an owner's family farm and like almost all successful businesses the owner must love it.

Mr. Hershey says that a tree-crops farm should be a balanced program. He is trying many things to get the mistakes worked out of his system. The great secret of nature is—your security lies in a balanced land use—balanced between animal and plant production: crops for animals, and animal manure for the crops, with a margin of each for the profit book. Says Mr. Hershey:

"Now remember, we dangle no bait of a glorious income before your eyes. We're farmers—not Agriculturists. The difference is—the Agriculturist lives off the soil and off the people of the soil. The farmer lives on and with soil as a part of it. While the Agriculturist figures on what he can make off the soil and people of the soil, the farmer figures on how much he can do through 'soil love' resulting in a greater security forever.

"While the first is a parasite, the latter is a nourisher.

"While the first is a miner, the latter is a depositor.

"But I'll say this—such a program checked against any other type of land use will bring you greater profit with less work—less cost—than any other program. Oh, yes, it's a tough life getting started."

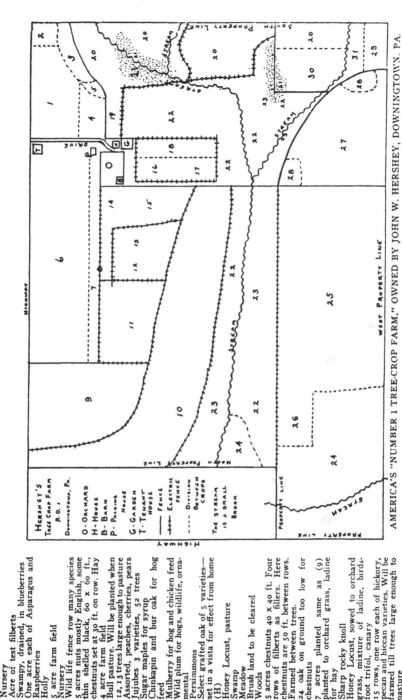

AMERICA'S "NUMBER 1 TREE-CROP FARM," OWNED BY JOHN W. HERSHEY, DOWNINGTOWN, PA.

1. Nursery
2. Acre of test filberts
3. Swampy, drained, in blueberries
4. One acre each of Asparagus and Raspberries
5. Holly
6. 5 acre farm field
7. Nursery
8. Wild life fence row many species
9. 5 acres nuts mostly English, some thin shelled black, 60 x 60 ft, chestnuts set at 30 ft. on row. Hay
10. 4 acre farm field
11. Bull pasture. Will be planted when 12, 13 trees large enough to pasture
12. Assorted, peaches, cherries, pears
13. Jujubes 3 varieties, 52 trees
14. Sugar maples for syrup
15. Chinkapin and bur oak for hog feed
16. Mulberry for hog and chicken feed
17. Wild plum for hogs, wildlife, ornamental
18. Persimmons
19. Select grafted oak of 5 varieties—set in a vista for effect from home (H)
20. Honey Locust, pasture
21. Swamp
22. Meadow
23. Brush land to be cleared
24. Woods
25. 5 acres chestnuts 40 x 40 ft. Four rows of filberts as fillers. Here chestnuts are 50 ft. between rows. Farmed between trees.
26. 24 oak on ground too low for chestnuts
27. 7 acres planted same as (9) planted to orchard grass, ladine for hay
28. Sharp rocky knoll
29. Honey Locust, sowed to orchard grass, mixture of ladine, birdsfoot trifol, canary
30. 15 rows, one row each of hickory, pecan and hiccan varieties. Will be farmed till trees large enough to pasture
31. Sprout land of nearly solid stone

The U. S. Soil Conservation Service started in with the idea
of developing hill culture, of which, of course, tree crops must
be an important part. They took a farm at Hillsboro, Georgia,
and drew up a very careful plan. It was to be a one-family farm.
They were to have 10 cows, 10 hogs, 50 chickens, and a mule
and the following acreages: Vegetable garden, fruit trees, and
home site—3; hog lot L.5; permanent hay—20; grass pasture
—25; grazed woodland—25; grain—24; farm wood lot—118;
wild life—5; total—221.5.

Planting of yellow poplar, black locust, catalpa, loblolly pine,
Arizona cypress, red oak, cherry oak, holly, and white ash;
would afford steady employment a part of each year marketing
wood, holly, and Arizona cypress. Oaks were to make acorns for
pig feed.

There was also a planting of chestnuts and test plantings of
nearly everything mentioned in Mr. Lombard's plan (see page
315) and elsewhere, and in this book, to see how they fitted in,
to make food for humans and animals and crops to sell.

A heavy slash in Congressional appropriation for soil conservation made it necessary to abandon this long-time project.
Most of the men whose brains were back of it—and back of a
host of other constructive things—had to leave the service. The
government is a bad employer for a person with imagination.
It is fine if you are just a congenital clerk, have no ideas, no
imagination, and no ambition, and do as you are told. But if you
have ideas—my name for Washington is Frustrationburg. Once
in a while someone can get through with an idea—if it's quick
and easy.

TECHNIQUES FOR THE TREE-CROPS FARM

How shall the hills be turned into enduring tree farms, since
otherwise they will be ruined sooner or later by plowing? Before this question is answered, emphasis should be laid upon
three new or little-used pieces of agricultural technique:

1. Nitrate instead of cultivation.
2. Irrigation from rain-fed water pockets or terraces.
3. Hogging down the crops.

A NEW TECHNIQUE—CULTIVATION BY NITRATE

At the mere suggestion of an uncultivated tree, some ortho-dox devotees of the plow rise up and say that trees need to be cultivated. I will agree with them and quote back for their comfort a statement given me in France that the Persian walnut tree in the pasture bears only half as much as the tree that is cultivated.

The orthodox defenders of the plow and of cultivation were telling us not many years ago that an apple orchard *had* to be cultivated. For most of the clay lands of the eastern United States, this statement is now known to be merely a kind of fetishism, as some millions of Virginia and other eastern American apple trees yearly attest. The apple trees are *fertilized*. The theory of cultivation is that it increases the available food supply. I accept the theory. Facts seem to prove it. The same theory underlies the use of the commercial fertilizer—or compost, if by chance you have it. Experiments at numerous farms and at numerous experiment stations such as those in Ohio, Pennsylvania, and Virginia show apple trees that will bear two or three bushels in uncultivated, unfertilized sod and will bear perhaps ten, fifteen, or twenty bushels if given a few pounds of fertilizer, heavy with quickly soluble nitrates. We are just beginning to realize the potency of plant hunger in field and orchard.

With an established tree in many soils and locations you may cultivate or you may fertilize, but in most naturally forested sections of the United States it is not necessary to do both, certainly not more than the Pennsylvania State system of sod shredding. I claim that tree-crop experimenters can use fence rows and corners of land to grow trees without cultivating them, but please do not quote me as saying they can do this without fertilizing the trees. Furthermore, *young* trees, if not cultivated, must be protected in the first years by mulch. I have proved in many cases, to my sorrow, that grass can rob a little apple tree and kill it, whereas three feet of mulch plus nitrate of soda will cause it to grow like the green bay tree of Scripture and bear more apples, alas, than I can sell in seasons of big crops, such as 1926 and 1937, when perfect apples rotted in hundreds of eastern orchards. Experimenters growing nut trees and other crop trees can plant them in almost any kind of place if they will give the trees a smothering coat of grass mulch three or four feet from the trunk and, in addition, give fertilizer and other attention that a cultivated tree should have. Guaranteed this treatment, most species will react as the apple does. When the top is large enough to shade the ground around the trunk, the trees can usually fight it out and make steady growth if abundantly fertilized. This statement is made for the naturally forested areas of the eastern United States. I do not know how far west it holds true. It may not be true for the Great Plains.

What I say about fertilizer may perhaps be replaced by compost by the Albert Howard school of organic fertility, *if compost can be had.* I repeat, *if* compost can be had.

AN ALMOST NEW TECHNIQUE—FERTILIZING WITH LEGUMES

Legumes of grass size, bush size, or tree size can sometimes be made to grow beside the crop-yielding nonleguminous tree which can be fertilized with nitrogen produced by the adjacent legume (p. 318). These possibilities have been as yet but little explored in the Temperate Zones, but the knowledge is collecting rapidly.

A NEW TECHNIQUE—IRRIGATION BY RAIN-FED WATER POCKETS AND WATER-HOLDING TERRACES

Our U. S. Soil Conservation Service terrace (Figs. 23, 130, 131, 132) causes water to remain upon the square rod where it falls. The effect of this is like that of cultivation or fertilizer, because it increases available moisture and increases the available food supply of the plant. Irrigation by water pockets should be one of the devices of the agriculture of the future, particularly the tree agriculture. Once the pit or water-holding terrace is made, the rains will automatically fill it, and the tree will irrigate itself for years with very little attention save an occasional cleaning out of the pit or ditch or terrace. Otherwise the forces of nature will gradually tend to level the pockets and flatten the terraces. Gravity never sleeps.

This method of irrigation seems to have been invented independently in several parts of the world. Unfortunately it has been used but little in any of them until this century.

1. It was invented at an unknown time by the Arabs of North Africa. These people still build banks around their olive trees so that no rainfall will escape. Sometimes they let a rill from a nearby hill run into the catchment basin.

2. Something of the same sort was invented by the late Colonel Freeman Thorpe of Hubert, Minnesota, who reported that it made Minnesota's black oaks grow twice as fast as their nearby neighbors that had missed the benefit of such watering.

3. Near the city of Taiyuan, the capital of the Province of Shansi, China, the natives have a practice much like that of the Arabs above mentioned, and they apply to it crops of corn and sorghum.

4. A hole catching water and irrigating trees was devised by the late Dr. Meyer of Lancaster County, Pennsylvania. He had a gully in his orchard. He put a dam of trash across it. Water collected behind this dam. The doctor observed that the trees near it grew better than the rest. This caused him to keep men at work in odd times digging holes near his apple trees. The trees prospered and Dr. Meyer thought that it was profitable.

5. A horizontal terrace holding rainfalls to irrigate trees was devised by Mr. Lawrence Lee, a graduate engineer and orchardist of Leesburg, Virginia. On his steep Piedmont clay hills, he says, "Nine-tenths of the water of a summer thunder shower runs away." Aiming to reduce this, he put rows of apple trees across the hill at equal distances apart. He laid off one row on the absolute level. The others were thirty feet apart up or down

the hill from this base row. As a result, every furrow along the row planted on the absolute contour held water, whereas it drained away from the others because they sloped a little. In a few years Mr. Lee observed that the trees of the contour row which had the water lying above the trees at every rain were distinctly the largest in the orchard.

He took the hint. In 1917 he planted another orchard where every row was on the contour, the exact level. He used a Martin grader (made in Owensboro, Kentucky) with tractor, and made horizontal terraces that would hold water above every row of trees. The trees were planted on top of the banked earth. Mr. Lee reports that there was no runoff from this orchard for several years. His orchard was called to my attention by an expert apple grower who had seen it from a distance and asked what had made the trees grow so much more rapidly than those in another orchard just across the road, which had been planted at about the same time. Mr. Lee's plan is essentially the same as that used now by the U. S. Soil Conservation Service. Near Spartanburg, South Carolina, you may see hundreds of thousands of magnificent peach trees growing thus on sloping land with virtually no erosion. (See Fig. 131.)

6. I have seen pits dug as big as a barrel to catch trash and water in coffee plantations on Porto Rican hills. When the pit is full of leaves and silt, they plant a young tree in this beautifully prepared seed bed. I am told that this practice is common also in Central America.

7. These catchment pits have reached their most systematic development, and they are most extensively used under the name of silt pits on the tea plantations of Ceylon and on the rubber plantations of Malaya (Fig. 110), where they serve the double purpose of water saver and soil saver and have been dug by cheap coolie labor in thousands of acres of plantations. They are probably used in other parts of the Far East, but I have not seen them.

8. Water catchment pits have been effectively used by the Indian Forest Service.

9. The effectiveness of these water pockets in the afforestation of denuded and gullied lands in India (along the Jumma, Chambal, and other rivers, especially in the Agra, Etawah, and Jalaun districts of the United Provinces) has been little short of miraculous. Centuries of overgrazing by cattle, goats, camels, and ponies had destroyed protective vegetation, with the result that half a million acres of alluvium had become a network of ravines. The normal rainfall of fourteen inches rushed away so rapidly that it wet the soil but six inches deep. After the digging of water pockets, water penetrated to a depth of more than four feet, and reforestation was surprisingly successful. Babul trees reached a height of twenty feet in four years. (Information from E. A. Smythies, of the Indian Forest Service).

This idea has had a most suggestive development in North Carolina, where Mr. Lee, an engineer of Charlotte, has developed a fairly satisfactory *engineering technique* to do this thing in the mechanized American way that we like to call modern. He makes the pits on a large scale on the cut-over pine land with no other object than that of saving water for his company's (Southern Power Company) power plants and increasing the growth of the timber. He calls them "terrace with black ditch." This is essentially the same device as that of Mr. Lawrence Lee (No. 4 above), although the men are unknown to each other.

The water-holding terrace is merely a slight modification of the old drainage terrace so commonly used in the cotton fields of the South. (See Fig. 97.)

The old terrace type of land management has the great drawback of dividing a tract of land into small fields of irregular boundaries because the plows and cultivators cannot cross the terraced banks. However, tens of thousands of American cotton fields have been worked that way for decades. Any other way meant speedy ruin.

Note the excellence of this terraced field for one who would convert his farm to cropping trees. His customary farming method can go on undisturbed while young trees grow on the terrace banks. It was in such a place that Governor Hardman had planted some of the honey locusts in his almost unique honey-locust orchard. (See p. 69.)

The new broad bank terrace now being pushed by the U. S. Soil Conservation Service is a great improvement, but limited to gentle slopes. This terrace lets tillage machinery cross it. (See Figs. 131, 132.)

THE SMALL TERRACED FIELD AND THE TRACTOR

The man whose mind happens to be molded in the level-land farming conception is usually much shocked by the idea of modern agriculture which employs a narrow strip of land of varying width between two terraces in the now-recommended strip-cropping system. Unquestionably it is awkward, and the cost per acre for working it is more than the cost for working a wide flat area. Perhaps good management can make this pay. If we can only get out of the land-robbing philosophy and into the land-building philosophy, the little terrace and strip cropping may come into its own. The sloping field with natural drainage almost always declines in fertility. The little hillside terrace does not lose its good soil, but it has a chance to get better and better. With care it can be made to yield fifty or seventy-five bushels of corn instead of the fifteen or twenty of the slopes. You can turn many corners for that difference. The increased yield made possible on terraced land can make it compete in cost with the wide, rolling, unterraced lands. The U. S. Soil Conservation Service has many records that reduce this advantage to crop measure and dollar measure.

If we practice the concept of making the farming fit the land and then examine this little terraced strip, we discover the following facts. Its chief disadvantage appears when crops like corn, cotton, and tobacco are cultivated in rows. On these terraces there must be much winding back and forth of man and team. This disadvantage almost disappears if the land is sown to a broadcast crop such as alfalfa, oats, and other small grain, millet, cowpeas, etc., or soybean or other legumes. Most of these can be harvested with the mowing machine and the hay rake, which require only a small amount of turning. Sometimes the combine can do it. Best of all, these crops can be harvested by the animals, which brings us to the third new technique.

A NEW TECHNIQUE—THE DISC REPLACES THE TURNING PLOW

A new soil-saving revelation is coming to the hills.

Ford-Ferguson equipment, and other makes, are now getting stubble fields and pastures ready for reseeding without plowing. The plow buries all vegetation and turns up bare ground which, not having the protection of roots and trash, is therefore ready to wash away.

The new system chops all the complete vegetation mat into the topsoil so that every straw and every blade of grass and every root furnishes a little tube to take in moisture and a thread to stop erosion. I have seen steep pastures, thus masticated in March or April, reseeded and covered with new vegetation and held by billions of roots by the first of June; and there was no erosion, although there had been heavy rains.

This system will be a great aid in putting better pasture and hay crops on hills that cannot be plowed to corn without ruin.

With grass silage and barley in place of wheat, we are on the way to a new grass economy with or without trees.

AN ALMOST NEW TECHNIQUE—HOGGING DOWN CROP

A chief (perhaps *the* chief) invention in agricultural economics (other than machines) in the past quarter of a century has been the harvesting of crops by the pigs. It has now gone far beyond the experimental stage. Two counties in southwestern Wisconsin hogged down twenty-seven percent of the corn crop as long ago as 1926. Each year in this country millions of acres of corn and hundreds of thousands of acres of clover, soybean, cowpeas, peanuts, wheat, and other crops are not touched by human hand or by machine. The pig walks in and harvests the crop without cost and with much joy to himself. Georgia and other Southern States have it worked out in many systems.

I would lay especial emphasis on the suitability of the hog for harvesting crops on the little winding strip of tilled land between two terraces. With hogging down of nontilled crops there is only soil preparation and seeding to be done on the winding little terraced fields (Fig. 136)—another reduction of the handicap.

Hogging down also permits a variety of foods to be eaten by the pig, which has a distinct advantage. See bulletins of the U. S. Department of Agriculture, Georgia, and other Southern States.

If crop trees are on the terrace edge, hogging down permits trees to have the benefit of cultivation if the owner so desires, and permits part of the hillsides to be cultivated. All parts can be profitable, because that which is not cultivated can be covered by the trees and grass.

These new techniques—nitrate instead of cultivation, fertilization by legumes, irrigation from the rain-fed reservoirs on the spot, and hogging down, taken in combination with ordinary pasture—suggest an entirely new era of heavier productivity of the unplowed and unplowable lands of the American farm. This is fortunate, because over a large section of our country the greater part of the land as now arranged (p. 11) can be regularly plowed only to its ruin. And do not forget the great new idea—grass silage!

In nearly every county in the Southern States there is from 25 percent to 75 percent of the land not in cultivation. Much of this land is of a character suitable only for the growth of grass and trees that would yield some revenue through stock feeding. (Archd. Smith, Professor, Mississippi Agricultural and Mechanical College, Agricultural College, Mississippi, letter, May 30, 1913.)

Much more than half of the land east of the Mississippi and south and east of central Ohio can be planted to row crops only to its steady deterioration and eventual ruin in no great number of generations, unless terraced. Much of it was ruined long ago.

A POSSIBLE NEW TECHNIQUE

The year after the Portuguese cork farmer strips his oak trees, the trees yield an enormous crop of acorns. Stripping off the bark injures the tree enough to scare it into yielding a big crop but gives no permanent bad result. This matter of crop stimulus through controlled injury is like whipping a lazy horse. Its possibilities are for most crops unknown, but certainly worth experimental study. It may permit substantial increase of fruiting on many kinds of trees. (See Fig. 95.)

UNPLOWED LAND AND ITS YIELD IN PASTURE

At the present time, orthodox agriculture in America recognizes but two uses for unplowed land—forestry and pasture of grass and leaves.

This book finds its chief reason for being, in the fact that pasture is a very low-grade use for land—low in return. Because of the low return of pasture, man appeals to the plow and causes ruin. Much semipoetic stuff has been written about the beautiful bluegrass of Kentucky. It is beautiful, and it is good *poetry* stuff. If well fertilized it makes good race horses, and good pasture for about six weeks in the year. Mr. C. T. Rice, Oakton, Virginia, the best pasture expert known to me, calls bluegrass a weed and plows it up when it begins to crowd out his alfalfa and ladino clover; however, it is open to question whether Kentucky's bluegrass fields are any more productive than good bluegrass fields in adjoining States, where careful test shows such pasturage produces the paltry harvest of but one hundred fifty to two hundred pounds of live meat per year. You can verify these figures by getting bulletins from Blacksburg, Virginia; Columbia, Missouri; Ames, Iowa, and elsewhere.

Figures comparable to the above one hundred fifty pounds per acre were found to be the mutton yield as a result of careful weighing and measurement of English grass yields by William

Somerville, Professor of Rural Economy at Oxford University, England.

There are millions of acres of rough pasture in the United States east of Kansas and north of the Cotton Belt which will not make over fifty pounds of beef or mutton in a year.

These figures are little short of appalling when we remember that this meat is half waste. It is much less nutritious pound for pound than most of the wild nuts that we store in the attic or permit to lie in the woods.

The unplowable lands can be classified as:

1. Steep lands.
2. Rough lands.
3. Odd corners of lands, including farm windbreaks.
4. Overflow and wet lands.

UNPLOWABLE STEEP LANDS

These lands belong naturally in grass and trees and water pockets. If not too rough or too steep, much of this kind of land can be cultivated in strips with water pockets and trees as above mentioned. In places that are too steep to be cultivated, the water pockets can still be used to save the land and nourish the trees. On land too steep to have water pockets large enough to hold all of the rainfall, small ones may be used for getting trees established and for partial irrigation. A soil conservationist wrote me that, upon reading the first edition of this book, he invented a machine with shovels like plow shovels, which rose and fell and left rows of depressions to hold rainwater.

Rough pasture land is one of the greatest wastes in land utilization in America. The low yield of these hilly pastures has just been mentioned. Nearly all such areas have undergone cultivation by plow until by the processes of erosion most of the plowable top soil has been removed, often to the depth of many inches.

See the very illuminating and discouraging article in *American Forests*, June 1927, in which Mr. Hugh Hammond Bennett, United States Bureau of Soils, states:

On the watershed of the Potomac River the writer recently checked the amount of soil wastage over some of the mountainous country from whence comes the water supply of the national capital. It was found even on the smoother plateaus that from five to eight inches of top soil had been removed from most of the cleared land. This condition obtained in many places where crops had been grown only fifteen or twenty years, and the exposed subsoil of clay and rock was so infertile that poverty grass was the principal plant seen in many pastures and abandoned fields. Remaining patches of original forest, with their undisturbed virgin soil, served as an index to what has happened.

My own personal observations corroborate this. I was born in that area and I own land there. The Soil Conservation Service has made some progress in that region during the last twenty years, but not much has yet been accomplished. The farmer is a hard man to teach conservation to.

It is so difficult for the expositor to pass from philosophy to measured fact. Experiment, the changing of *one condition only,* is so difficult for the average farmer dealing with nature, who shuffles a complex deck of cards (conditions).

Under these distressing conditions of hill tillage, the yield diminishes until finally the place goes back to pasture or thickets, to become forests if let alone long enough. The amount of this soil loss in one or two decades will astonish one who actually measures it.

USING THE LEGUMINOUS TREE AS SOURCE OF NITROGEN

I am sure that many of these pastures would have their productivity as pastures increased if they could be thinly covered with a planting of some leguminous tree whose roots would gather nitrogen from the air and leave it in the earth where the grass roots could share it. The Tennessee Station has proved it for some situations even with nonleguminous trees. More data from many locations would be very valuable and reasonably easy to obtain.

My opinion is derived from observing that good grass grows

under black locust, *Robinia pseudoacacia,* in pastures, sometimes in places where nearby lands are desolately bare. I have
seen conspicuous examples of this on some of the almost soilless
shale hills of Bedford County, Pennsylvania.

I believe the owners of such pastures in natural bluegrass
country would get more grass if their fields had compact clumps
of about twenty-five yellow locust trees set six feet apart in the
clump and the clumps one hundred feet apart. The roots would
run under all the grass in the field and nitrate it. Black walnut
trees standing on the hundred-foot spaces would eat locust
nitrogen and grow and bear grandly. The compact clump of
twenty-five locust trees would push each other up and make an
abundant supply of posts. This practice of using a leguminous
tree for nitrogen supply is much used in coffee, tea, and rubber plantations, but they have not yet used a harvest-yielding
legume.

If such pastures had an open planting of black locust interplanted with grafted walnut trees or grafted hickory trees, the
locust could furnish nitrogen for the nut trees. If the runoff
went into water pockets, so that trees and grass could share all
the water that fell, it would certainly result in a substantial increase of grass, wood, and tree crops. The water pocket would
also catch and hold the fine earth particles which the rain would
ordinarily carry away. Thus fertility would be increased, and
the land would be built up. It should be remembered that both
of the locusts and the black walnut have open tops, letting
much light through. This vegetative source of nitrogen can be
had for nut trees by keeping the pasture in clover.

THE ROUGH AND STONY LANDS

These differ but little from the steep land except that they
are usually in better condition because they have escaped the
scalping by a plow-mad race.

Under the heading of rough lands, mention should be made
of the hundreds of thousands of fruitless trees now standing in
the rough limestone fields of the great Appalachian Valley

which extends from southern New York to northern Alabama. In certain sections, such as the Cumberland Valley, Shenandoah Valley, and the valley of east Tennessee, which have long been famous for a rich agriculture, the traveler is amazed by the great amount of outcropping limestone, which makes patches where farm machinery cannot go. Here the good soil makes good grass, and often good trees have sprung up without the aid of man.

This is prime walnut land; and the walnut is thoroughly at home. Here the present system of agriculture may remain absolutely intact, but supplemented by many million bushels of Persian walnuts and American black walnuts, of hickory nuts, of honey-locust beans, or acorns produced by trees standing in spots of land that the plow cannot touch.

FENCE ROWS AND ODD CORNERS OF LAND

Every farm, even in the flattest, blackest prairie of Illinois, has some corners of land that are not cultivated, where the tree lover can make a few trials without interfering with the main business of the farm. There is the lawn. The beautiful (but worthless) elms, the fruitless sycamores and box elders and the maples—oh, Lord! how many are the maples, not golden-leafed sugar maples, that throw down worthless leaves upon our pathway and kill grass with their smothering shade! To ornament the home, few trees are more beautiful than the hickories and the pecans, and the various walnuts certainly rank high in aesthetic value. They all have the additional charms of intellectual interest and a probability of nuts. Do the American people really love maple leaves?

Then, also, there is the roadside. In France and Germany tens of thousands of roadside walnut trees and roadside plum trees belong to the local government. The annual crop is sold on the tree, and a substantial saving of taxes results. Since in America the roadside land belongs to the owner, this is a possible source of income or a place of experimentation for those with whom land is scarce.

The fence rows within the farm are even better than the

roadside for experimental planting. On thousands of American farms the fence rows are almost lined with trees. In sections of northern Virginia, where I lived as a boy, it is not at all uncommon for farms of two hundred acres to have from fifty to two hundred trees scattered along their division fences and roadsides or even standing in the fields. Many of the trees are almost worthless, although some are the post-giving, nitrogen-gathering black locust and usually enough black walnuts to supply the family with nuts.

A square farm of one hundred and sixty acres has two miles of boundary fence. If divided into four fields, it has at least one mile of inside fences. Crop-yielding trees along these fences will not interfere with the machine agriculture in any way. Of course, the trees should not be there if they cannot pay with their own crops for the reduction of crops they make in the field. In a journey from New York to St. Louis, and thence to San Antonio, Texas, I was struck by the great number of fruitless trees standing in the midst of fields that were prime for machinery.

Tens of thousands of farms have an uncultivated corner across the gully, beside a stream bank, or in a chopped-out bit of woods. In all these places, trees will grow. If the land is not pastured too hard, the seeds of walnuts and hickories and many others will grow up ready to be grafted in a few years by him who would experiment.

THE WINDBREAK

R. C. Forbes, formerly Director of the Arizona Station, said:

Oasis agriculture in the Sahara, Arabia, and other tropical desert regions is made possible, principally, by date palms, which act as windbreaks and as shade for tender plants beneath, while drawing the wherewithal for a valuable fruit crop from the basement stories of the soil.

Several hundred thousand square miles of the middle western United States are so level that the wind fairly combs the grass because there are no hills or forests to prevent or disturb its close approach to the earth.

In this region the farmers have planted windbreaks about their houses and farm buildings, and now the agricultural scientist has found the windbreak is a needed protection to the grain field.

U. S. Department of Agriculture Bulletin No. 788, *The Windbreak as a Farm Asset,* says:

Measurements made in fields of small grain indicate that the crop gain in the protected zone is sufficient to offset fully the effects of shading and sapping. In a wheat field protected by a dense windbreak the gain amounted to about ten bushels per acre where the protection was most complete, and gradually grew less as the distance from the windbreak increased. The total gain was about equal to the amount of grain which could have been grown on the shaded ground near the trees. The season in which the measurements were taken was not of high winds nor did it lack moisture. It would appear, therefore, that in a windy year when evaporation was high the total gain for the field would much more than balance the loss.

This statement ignored the value of wood or crop produced by the trees of the windbreak.

See also U. S. Department of Agriculture Farmers Bulletin 1405, where Albert Dickens, Horticulturist, Manhattan, Kansas, put it thus: "I have long had in mind that we might some day protect one-quarter section by cutting it into ten- and twenty-acre fields and protecting each field with red cedar or other windbreak."

That means that Illinois, Iowa, Kansas, Nebraska, Minnesota, the Dakotas, and other States and other similar regions in other continents need to have thousands and tens of thousands of miles of long rows of trees. Perhaps they might include fruit-, nut-, or bean-yielding trees, while stopping the wind and making wood. "Alfalfa grows almost to the base of honey locust trees," says the Kansas Bulletin 1405 above mentioned.

The element of overhead expense should be emphasized here

as a profit aspect. It seems to be established that the farmer should have a windbreak for its own sake. Therefore, every dollar of profit from fruit, nut, bean, or wood is a dollar of clear profit. Overhead expense as a business factor has not been sufficiently appreciated in American agriculture.

President Franklin D. Roosevelt boomed the Shelter Belt idea almost as presented in the above paragraphs, and the recent studies of more than a decade of experience have shown surprisingly valuable results. (See p. 337 and bulletins of the U. S. Forest Service.)

THE OVERFLOW LANDS

From Maine to Kansas and from Minnesota to Texas and Alabama there is much land which is uncultivated because it is threatened with overflow from some flooding stream at some time during the growing season. Therefore, the plow crops are unsafe and cannot be depended upon. Therefore, this, the best of land, usually remains in pasture. From Ohio and Iowa southward and southwestward this is the homeland of the pecan and the big-nut shellbark hickory. This is also the home of several other hickories and of the black walnut.

These very fertile meadows, these little Nile Valleys, are the natural home of a two-story agriculture, pasture beneath, tree crops above. Owing to the high water level from the nearby streams, the moisture supply is usually abundant. Owing to the overflowing of muddy water, fertility is so abundant that the high-headed pecans, hickories, walnuts, honey locusts, oaks, or other trees can bear their maximum crop without causing much diminution of the grass at their feet. I venture to suggest that enough pecans to supply the world's market for the next fifty or perhaps hundred years can be grown on such unplowable overflow lands in the proved homelands of the pecan. (See Chap. XIX.) This land has the lowland disadvantage of frost, but this is mitigated by the low (or absent) cost of trees that care for themselves.

STARTING THE TREE FARM

Suppose someone wishes to start a tree-crops farm. How shall he begin? The answer is plain. Few are in a position to do it as Mr. Hershey did (p. 319). For most farmers my strongest advice is—*begin gradually*. One thing this book most emphatically is *not:* it is not a recommendation to the business interests of the United States to plant out a large tract of any crop-yielding tree. Such one-crop-gambling enterprises have overhead charges which usually eat them up. A farm, on the contrary, already carries its overhead charges and an intelligent farmer can start experimenting with trees with no element of additional overhead—no purchase price of land, special tools, or anything but the trees themselves and the few things that must be especially used with the trees, and they are few.

To one who is now farming and wishes to try tree crops, the thing to do is to go on with his farming as before. Start some trees in a small way. Try a few of several species or varieties. Experiment with them. Let the business grow in the light of experience. In due time the farm can be made over as things prove themselves. It can be done gradually in the same psychology and the same way that annual payments build up a life-insurance investment. Millions of men regularly pay life insurance so that their widows and orphans may have an income in later years (in very *much later* years, we all hope as we do it). We regularly lay aside money to be used after we are dead. With this point of view in mind, I wish to point out that the grafted hickories, for example, offer tens of thousands of farmers an opportunity that probably beats five-percent bonds as a means of adding to the estate, not to mention their relation to inflation. What will my estate receive in 1955, 1960, 1965, for a big dollar I paid the life-insurance company in 1932, and for the other dollar that topworked a shagbark tree in 1932?

The man with the hickory life-insurance concept should not

go at it blindly, however. Some years ago I received cions of a number of excellent shagbarks from colleagues in the Northern Nut Growers' Association. I put them out in my Blue Ridge mountainside of northern Virginia, grafted them to nice-looking stocks which happened to be red hickory, *Carya ovalis;* pignut, *C. glabra;* mockernut, or white hickory, *C. tomentosa,* syn. *alba;* and bitternut, *C. cordiformis.* But the cions were shagbark, *C. ovata,* hostile to almost all of these other species as stocks. The result—failure, almost complete. I grafted all of these species with cions of the Fairbanks hybrid. Result—success. I regrafted the Fairbanks to shagbark, leaving about a foot or two feet, preferably the latter, of Fairbanks in the trunk and usually a small Fairbanks limb. Result—success. Even double topworking is a lot cheaper than buying a nursery tree, and much faster in its result.

This is a strange alchemy. The foot of the Fairbanks trunk receives the nourishment from pignut root and transforms it as it passes through to the shagbark above. The shagbark finds it acceptable and proceeds with life. I have many kinds of shag-barks producing thus in a very hilly bluegrass pasture.

Wherever there is a piece of rough land naturally set to hick-ory and suitable for pasture, the hickory timber can be cut down. The stumps will often throw up suckers, which grow with much speed for a few years. Cattle will not eat them, and the suckers can soon be grafted. At the height of six or eight or ten feet, the new-set graft is safely out of the reach of pasturing animals which, strange to say, will often eat a graft and let the wildlings alone. After it is grafted, a little care for a few sea-sons, such as tying in the grafts and painting the wounds, will see them safely healed over. Two or three prunings in the winter time suffice to establish completely the graft and give it a monopoly of the root. After this you can forget (unless you love trees) your investment of twenty-five or fifty cents. Then, after five or ten years, more or less, depending on plant food, your graft will probably begin to pay a big percent on the cost every other year and will increase the percentage and keep it up for several generations. Meanwhile, the field has always been a good pasture and you have been going on with your farming, while nature, slightly aided, has been turning your pasture or perhaps your fence row into a valuable property.

If the tree grower wishes to begin with nursery trees, he can buy grafted trees of the following species to be used as human foods:

1. Chinese chestnuts—grafted trees of varieties that have proved themselves over considerable areas.

2. Northern pecans and southern pecans—many varieties.

3. Shagbark—several varieties.

4. Hybrid hickories—Fairbanks, Stratford, etc.

5. Hickory—pecan hybrids. (But be sure the variety is known to produce nuts.)

6. Black walnut—several varieties.

7. Persian walnuts—several varieties.

8. Japanese walnuts (heartnut)—several varieties.

9. European filbert—many varieties.

10. American hazelnut—several varieties.

11. Hazel-filbert hybrids—several varieties.

For forage crops he can secure grafted mulberries; grafted persimmons, both native and Oriental; honey locusts, grafted varieties that have proved themselves from Pennsylvania to Alabama and probably in Iowa (see Floris, Iowa, planting of Soil Conservation Service) ; and other varieties are coming.

The best producing oak trees within ten miles of the place of residence of most people of the United States will make interesting grafting experiments.

INTERPLANTING DIFFERENT SPECIES

Interplanting of different species will be an important device in tree-crop farming. To provide early returns, quick-maturing species can be alternated with slow-maturing.

The mulberry tree is a promising filler crop. It grows rapidly, bears young, and is unusually resistant to shade.

This is shown not only by the fact that the trees bear fruit throughout the crown and even in quite dense shade, but also in the fact that the young seedlings are able to grow for a long time under the shade of other trees. (Letter, George B. Sudworth, Dendrologist, U. S. Department of Agriculture, January 23, 1923.)

Therefore, every third tree in every third row in a mulberry orchard might be a pecan ninety to one hundred feet from its nearest pecan neighbor. The pigs, as they pasture the mulberry, would hardly miss the ninth tree, which was a pecan. Gradually the towering pecan would overspread the low-topped mulberries, paying for their scalps with nuts. Similarly every fourth tree in a mulberry orchard might be a grafted black walnut, grafted English walnut, or grafted hickory. As these crowded out the mulberries, the farmer might put in another tract of nut trees with mulberry fillers to nurse along another orchard, which would be paid for almost from the beginning by the automatic harvest of the neighboring filler trees. The precocious and bushlike hazels, filberts, and hybrids thereof have interesting filler possibilities, nor should the chinkapin be forgotten.

To secure fertility, leguminous and nonleguminous trees may be interplanted so that the nonlegumes may derive nitrogen from the legumes. It is probable that clovers are a better source of nitrogen, but here is some startling evidence from Professor L. J. Young, November 29, 1948, School of Forestry and Conservation, University of Michigan, Ann Arbor, Michigan:

In 1906, black walnut was started from seed on an old field on one of our properties. By 1918, when the black locust was interplanted, most of the walnuts had a distinctly stunted appearance. The annual height gained by most of them was about three to four inches, and there was no definite terminal shoot, resulting in a broad, bushy crown.

When the locust had developed far enough to have a real effect on the soil, the annual height growth increased suddenly to between two and three feet, and the terminal shoots had a much larger diameter.

Fig. 66. *Left.* The light-colored background at lower right side of picture is a stream bed. It helps to show the extreme steep-ness of this hillside orchard of oaks (ilex) on the slopes of the Sierra Nevada Mountains (Spain). The farmer shown in picture said he pastured his small pigs here but not the large ones. He said he was afraid they would fall down the hill, which is so steep as to make climbing difficult. (Photo J. Russell Smith.) Fig. 67. *Top Right.* One of the many grafted oak trees on the stony hills of an estate at Esporlas, Majorca. The bark of stock is light, that of graft is dark. (Photo J. Russell Smith.) Fig. 68. *Bottom.* One of nature's miracles. Large, thin-shelled, moderately good, beaver hybrid from small hard-shell shagbark, left, and bitternut, right; 55 per cent of life size. (Photo Willard Bixby.)

FIG. 69. The top of a grafted bur oak only 8 feet high and bearing 69 acorns.
John W. Hershey's Tree Farm.

FIG. 70. *Left.* Is the oak slow? A Portuguese ilex that litters the ground with acorns. FIG. 71. *Center.* Sprouts of Turkey oak from Virginia Blue Ridge set up in front of five-inch weather boarding. Four seasons after forest fire. Note acorns. FIG. 72. *Right.* Seven-year-old sprouts of Virginia Blue Ridge chestnut oak bearing acorns. (Photos J. Russell Smith.)

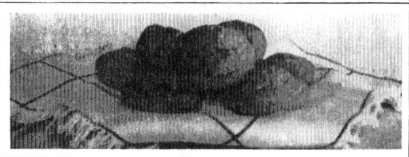

FIG. 73. *Top.* Life-size acorns (Quercus Insignis) from Orizaba, Mexico, the largest known, 1¾″ x 1³⁄₁₆″. Trees growing at Punta Gorda, Florida. (Courtesy *Journal of Heredity*.) FIG. 74. *Bottom.* Muffins made of Missouri acorns. (Courtesy Missouri Botanical Gardens.)

FIG. 75. *Top.* Caches for acorns built by Indians in the Nevada foothills of the Sierra Nevada Mountains. Acorns brought over the mountains from the California side. The winter bread for the family—the most nutritious bread in the world—possibly the oldest, perhaps a bread of the future either through machines or a return to the primitive. (Courtesy George B. Sudworth, U. S. Forest Service.) FIG. 76. *Bottom.* Acorns from which the muffins shown in Fig. 74 (opposite) were made. They are ⅓ to ¼ natural size. The largest are those of Quercus Macrocarpa Mich.

FIG. 77. *Top* Persian walnut shade trees down a monastery lane, Grenoble, France. Annual income about $150 gold, average 1910-14 dollar. FIG. 78. *Center.* A hayfield showing system of planting tall Persian walnuts with a pole in South Central France. Background, walnuts. FIG. 79. *Bottom.* Black walnut. Top-worked with English walnut in J. Russell Smith's fence row. (Photos J. Russell Smith.)

Fig. 80. *Top Row*. Japanese walnut (center) and two of its butternut hybrids, life-size. Remainder, natural variants of American black walnut. (Photo E. R. Deats.)

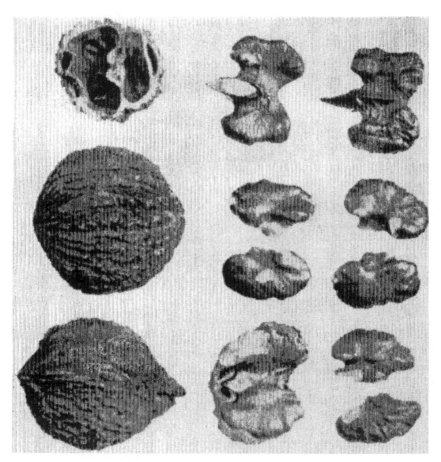

Fig. 81. Top, Stabler walnut shell and kernels; next, Thomas walnut and kernels; next, Ohio walnut and kernels. All life-size.

FIG. 82. *Bottom Right.* Cross section of Thomas Black Walnut, showing the shape of shell that permits easy extraction of kernel. FIG. 83. *Top Left.* Ohio black walnut tree, four years planted, bearing 105 nuts. The variety is a regular bearer. (Courtesy S. H. Graham, Ithaca, N. Y.) FIG. 84. *Top Right.* Black walnut trees in author's fence corner, grafted to selected variety, have borne several crops. (Photo J. Russell Smith.)

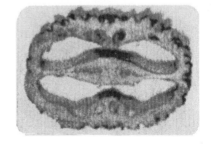

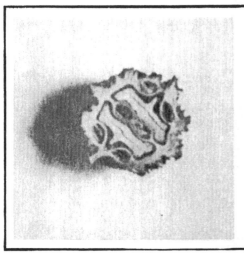

FIG. 85. *Left.* Fruit cluster of a Japanese walnut. Perhaps promising breeding stock. (Courtesy J. F. Jones.) FIG. 86. *Center.* Cross section, life size, of buart (butternut x Japanese walnut) hybrid. Yields kernel easily. FIG. 87. *Right.* Cluster of cork oak trees occupying rocky knoll in Portuguese wheat field. Typical of a method of land utilization that might produce nearly all the nuts needed in America, leaving other crop area unaffected.

FIG. 88. Some life-size hickories, suggestive of breeding possibilities. (1) Wild shagbark, Pa. (2) Shagbark x bitternut, Ia. (3), (4) Pecan x hickory hybrid, Mo. (6) McCallister hybrid, Ind. Western shellbark (12) x (8 or 9) Indiana pecans. (7) Southern pecan (Schley $1.00 a pound, 1928.) (10) Wild Illinois pecan. (11) Nebraska pecan. (12) Laciniosa, Ill. One parent of No. 6.

Fig. 89. *Top Left.* A Busseron pecan tree, one of the proved varieties, nine years transplanted, bearing 18 pounds of nuts, its third crop. Southern Indiana. (Photo J. Ford Wilkinson.) Fig. 90. *Top Right.* Pecan tree growing along the curb in Raleigh, N. C., bearing a fine crop. Why not have working shade trees? (Photo J. Russell Smith.) Fig. 91. *Bottom Left.* Life-size Butterick pecan nuts, an Illinois variety, grafted 1914, gathered 1918. Climate of Philadelphia. Typical of several varieties. And good! Fig. 92. *Bottom Right.* Pecan tree in park at Hartford, about seventy years old; southern seed; 10 feet in circumference; about 80 feet in height. (Photo William C. Deming.)

FIG. 93. *Top*. The tree near this two-story house is a pecan. Thirty years ago it paid the taxes on the farm which is located near Commerce, Georgia. (Photo J. Russell Smith.) FIG. 94. *Bottom*. Some of the offspring grown from seeds from one tree. This picture proves the worthlessness of planting seedlings of unknown ancestry, and suggests valuable results to be obtained by plant breeding. (Courtesy *Journal of Heredity*.)

FIG. 95. *Top.* Tree crop land. Rich hills of southern Ohio. Excellent for oaks, hickories, pecans, mulberries, persimmons, honey locusts, cherries, and many other crop trees. Land now almost unproductive. We have millions of hills like this. Literally millions. (Photo F. H. Ballou.) FIG. 96. *Center.* A forest of oak (ilex) on a mountain in Majorca. One product is charcoal, made by the continuous thinning and trimming necessary to keep it open, as it now is, for large acorn production—an acorn orchard. FIG. 97. *Bottom.* A harmless-looking wash in a cotton field in the red clay hills of Georgia. Note that no cotton grows in or near it. All the top soil is gone. The terrace on which the boy stands did not hold the water. In a few years . . .

These three pictures suggest a combination of timber and grafted nut trees. The crop of timber first will make tall fruiting trees which will give maximum leaf (fruiting) surface. FIG. 98. *Top*. Grafted chestnut tree seventeen years old assuming pole form when competing successfully with others in coppice. Man's hand is at one graft, another is just above the handkerchief on tree in the center. FIG. 99. *Bottom Right*. Chestnut tree which had practically the pole form in 1897 but developed heavy growth of lower branches below the fork in trunk in eighteen seasons after light-robbing neighbors were removed. FIG. 100. *Bottom Left*. Oak trees growing along a French roadside. Every few years the branches are cut off for firewood. Trees shaped like this and like FIG. 99 produce maximum bearing surface per unit of land. (Photos J. Russell Smith.)

FIG. 101. This pecan tree standing in the field near Evansville, Indiana, shows the gigantic size. This tree is about 100 feet high, although there are some that are much larger. If you want your place to become a landmark, plant pecan trees. They are good for 100-300 years. Note man. (Photo J. Russell Smith.)

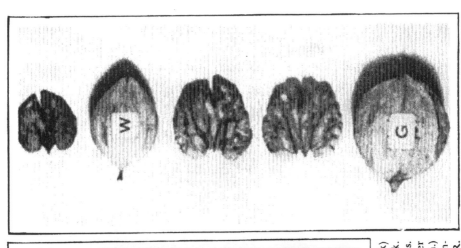

FIG. 102. *Left.* This Fairbanks hickory, a hybrid (bitternut x shagbark) from Iowa, grafted to pignut, C. Ovans. Large leaves are a part of the stock. Grafted 1918; photographed 1924; bearing good crop of nuts. Note the rocks. Blue Ridge Mountains. (Photo B. W. Gahn.) FIG. 103. *Center.* This Beaver hickory grafted on pecan, 94 inches first season. (Courtesy S. H. Graham.) FIG. 104. *Right.* Shagbarks, life size. G—Grainger, in North Carolina Mountains. W—Weschke, N. E. Iowa. Two of my best producers. (Photo E. R. Deats.)

Fig. 105. *Left.* Typical wild shagbark tree; typical rocks; typical Appalachian pasture—thousands of such pastures. There might be millions more of the trees. Good for many bushels of valuable nuts in your lifetime. Fig. 105A. *Center.* Young bearing trees. Steepness of bank shown by slope from base of tree to white paper in lower right-hand corner. Fig. 105B. *Right.* An olive tree in the garden of Gethsemane, and the claim might be true that under it Jesus spent hours of agony. (Photos J. Russell Smith.)

Fig. 106. *Top.* Pistache tree on the Lord Nelson estate on the slope of Mt. Etna, Sicily, growing on a tumble of volcanic bowlders where dirt is rarely visible. The occasional crops were worth $20 gold at prices for 1900-1913. The extremest, rockiest agriculture thus far seen by the author. (Photo J. Russell Smith.) Fig. 107. *Bottom.* These gigantic pistache trees growing in South China, taken in combination with the growth of the tree in Turkey and other Mediterranean countries and Kansas, indicate wide range and great adaptability of the species. (Courtesy U. S. D. A.)

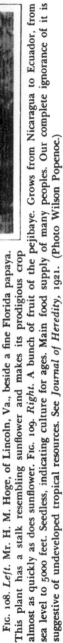

Fig. 108. *Left.* Mr. H. M. Hoge, of Lincoln, Va., beside a fine Florida papaya. This plant has a stalk resembling sunflower and makes its prodigious crop almost as quickly as does sunflower. Fig. 109. *Right.* A bunch of fruit of the pejibaye. Grows from Nicaragua to Ecuador, from sea level to 5000 feet. Seedless, indicating culture for ages. Main food supply of many peoples. Our complete ignorance of it is suggestive of undeveloped tropical resources. See *Journal of Heredity*, 1921. (Photo Wilson Popenoe.)

FIG. 110. Silt pits dug by hand labor in rubber plantation, Ceylon. Relative cost decides whether it is to be done by hand or machine (terrace). Some machines are now being used. (Photo J. Russell Smith.)

FIG. 111. *Top.* Majorca. Heavy trees in the foreground, grafted oaks; feathery trees scarcely visible in left center are almonds. The heavy trees on the steep hill at the back are seedling oak (ilex), thinned out to make acorns for hog-fattening. FIG. 112. *Center.* An Appalachian valley in Central Pennsylvania. The steepness of the hillside field is submitted as evidence of widespread agricultural insanity that is reducing our resources at a rate that should be appalling to the sane. FIG. 113. *Bottom.* A grove of locusts (Robinia pseuda-cacia) whose nitrogenous roots support a good stand of bluegrass on a shaly Allegheny Ridge in Central Pennsylvania, where grass is rare and a plow could not make a decent furrow. (Photos J. Russell Smith.)

FIG. 114. Bare eroded lands on the upper Jumna were quickly reforested by digging these pits to catch rain water and irrigate the trees. Pits increased water penetration severalfold, from 6 inches to 40 inches. Pits dug by famine relief funds. In a few more decades a hungry world will dig these pits by the million. (Courtesy E. A. Smythies, Indian Forest Service.)

Fig. 115. *Top.* This typical Old World terraced hill in the Apennines of Italy prevents erosion at the price of much hand labor. These terraces have grass banks built up gradually by strip plowing. I have a garden so terraced. Machinery can cross the device shown in Figs. 130 and 131. Fig. 116. *Bottom.* Top of Appalachian shale hill. Soil too shallow to plow, yet supports grass and trees, because roots penetrate the deeply fissured and partly decayed shale. An invitation to tree culture. (Photos J. Russell Smith.)

FIG. 117. *Top.* Black Belt of Alabama. Shallow soil, one of best in the world, washed away to bedrock of white chalky limestone in a few decades. (Courtesy U. S. D. A.) FIG. 118. *Center.* Sorrento peninsula, Italy. Background, olives; foreground, walnuts. Every tree grafted. Most of the land utterly unplowable. FIG. 119. *Bottom.* Central Tunis. Rainfall less than 10 inches. Olives, spaced far apart, as far as the eye can see. Tree agriculture is the best of desert agricultures.

FIG. 120. The fissured limestone on this mountain top on the Island of Majorca leaves almost no earth visible. But wild olives grow in the crannies, send their roots far down into the fissures, look healthy, are grafted to good varieties, and yield well. There are also some small pickings for sheep in the same enclosure, which was so fenced with thorn that I could not enter. (Photo J. Russell Smith.)

FIG. 121. *Top.* These three magnificent native American chestnut trees, now dead with blight, suggest the effective use that can be made of millions of odd corners of land on our American farms. Indeed, millions of such places are now occupied by brambles, bushes, and other worthless or almost worthless vegetation and might just as well be occupied by some noble tree yielding nuts or other useful crop. (Photo J. Russell Smith.) FIG. 122. *Bottom.* The windbreaks around the prairie farmstead invite the tree breeder to discover trees that produce crops as well as wind resistance. (Courtesy U. S. D. A.)

FIG. 123. *Top*. Tall-headed pecan trees planted by owner in cow pasture of rented dairy farm. The two stakes support the tree. The barbed wires keep the cows from rubbing the stakes. The pieces of old rubber hose by the man's finger protect the tree from the stakes. This invention is freely given to the public. The trees were mulched and manured. They are thriving in the *pasture* of a *rented farm*. No overhead cost. The latest improvement in this technique in land too rough to do as in Fig. 22, is to dig a two-bushel hole above tree, plow furrows leading water to it for shower irrigation. Try it. P.S. I'm now eating pecans from this tree. FIG. 123A. *Bottom*. The rough olive orchards of Judea, north of Jerusalem. Excellent land use. (Photos J. Russell Smith.)

FIG. 124. *Top*. Northern Algeria. Rainfall, 20 inches. Pasture land scattered with wild olives. FIG. 125. *Center*. Same locality as Fig. 124. Hill which was like Fig. 124 has had all its trees grafted to become like the larger trees in lower center. FIG. 126. *Bottom*. Abandoned farm buildings. Abandoned house. Abandoned Pennsylvania hills. The level land agriculture would not pay. Typical of thousands and thousands of American tragedies where the farming did not fit the land. (Photos J. Russell Smith.)

FIG. 127. The maple trees down the lane, with their sap buckets, produce a toothsome and wholesome sugar. They are suggestive in their income, and suggestive in the fact that they have not yet been subjected to scientific improvement. (Courtesy H. R. Francis, N. Y. State College of Forestry.)

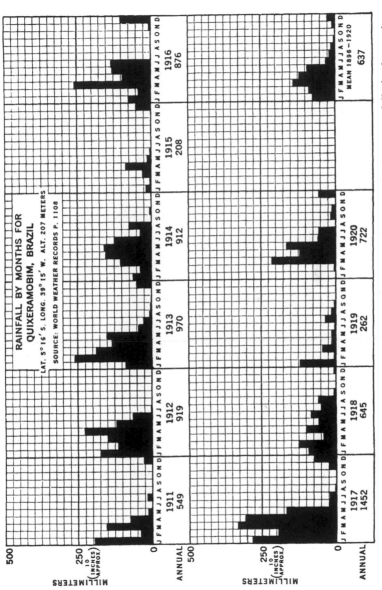

FIG. 128. This graph shows a very undependable rainfall. There are many such places, especially in the tropics. Can grain agriculture survive such irregularities? The answer is probably negative. Can tree-crop agriculture do better? The answer is probably affirmative. This book is written in the hope that tests will be made to increase our knowledge. This town is in the Brazilian state of Ceará, where populations have been fed by the government in dry years.

RAINFALL BY MONTHS FOR
QUIXERAMOBIM, BRAZIL

LAT. 5° 16' S. LONG. 39° 15' W. ALT. 207 METERS

SOURCE: WORLD WEATHER RECORDS P. 1108

Fig. 129. This wide channel was all cut by the Iowa rains April 15—July 17, 1948. A good field gone to nothing. Corn as we grow it is one of the great enemies of man's future.

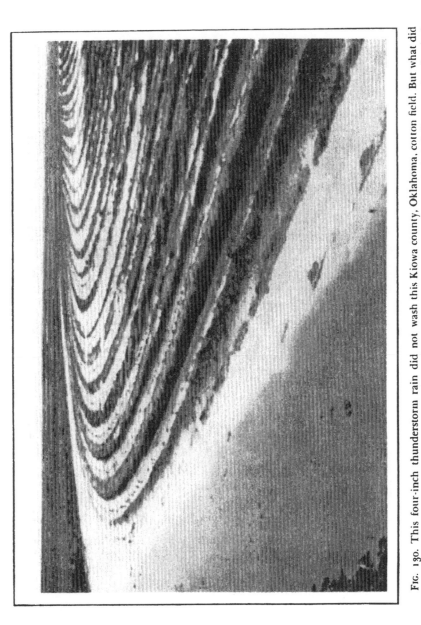

FIG. 130. This four-inch thunderstorm rain did not wash this Kiowa county, Oklahoma, cotton field. But what did it do to the ones with rows up and down the slope? Oh my! See Fig. 134. (Photo U. S. Soil Conservation Service.)

FIG. 131. Florida's first contoured orange orchard. Slopes 2-8% cover crop of indigo sown spring, 1944. Volunteer seed produced present cover. humus! humus!! humus!!! (Photo Bordon Webb, U. S. Soil Conservation Service.)

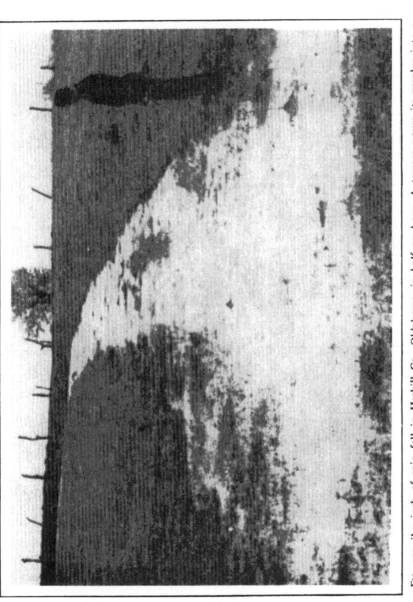

FIG. 132. An inch of rain fell in Haskill Co., Oklahoma, in half an hour. A terrace pours its surplus into a grassed channel that will not erode, and spreads the water over a pasture. (Photo Hufnagle, U. S. Soil Conservation Service.)

FIG. 183. Mesquite beside a wash near Safford, Arizona. Note dam and brush water spreader to let water soak in rather than run away. See Yearbook, U. S. D. A., 1948. This is an invitation to put crop trees along the gulch for shower irrigation. (Photo U. S. Soil Conservation Service.)

During the first two years of this stimulated growth, many of the trees doubled their height and developed a well-defined single trunk. The walnuts on an adjacent area that were not interplanted with locust are still growing slowly and show very poor form. Quite a number have died and others are dying.

In all cases where plantations of other species have adjoined those of locust, there has been a marked stimulation of growth of the other species. . . . The most marked effect upon the height of the other trees occurred at a distance of about 15 feet. At 30 feet the growth of other trees was distinctly better than on trees still farther away. Age of locust, 19 years.

TREE CROPS FOR THE DRY FARMER

Tree crops have unusual merits for agriculture in some lands too dry for plow farming. If a competition were opened for the driest farmers in the world, I should enter as promising contestants the Berbers who live in the Matmata section of central Tunis. Their average rainfall is about eight inches a year. It is of course often less than this, yet they are the owners of the finest olive trees I have seen in my journeys, from Gibraltar at one end of the Mediterranean to Gethsemane at the other. These trees are of record-breaking excellence, though growing in a climate of almost record-breaking aridity. Why? The Berbers build dams of dry stone wall across gullies in a limestone plateau (Fig. 119). At every sudden shower, water rushes down the gullies, sweeping a certain amount of loose soil. This catches behind the dams. Olive trees are planted in this soft earth. Every shower that produces a runoff in the gullies soaks this evergrowing mass of collected top soil, so that one-half inch of rain may give these trees in the rich gully pockets the equivalent of six, eight, or ten inches of rainfall because of the thorough soaking of the collected soil mass.

This practice of gully-shower irrigation could be used in the arid parts of America and every other continent. In a certain modified sense, it has already been copied in America. A Montana bulletin describes the building of barrages and the impounding of runoff gully water in grain and alfalfa fields, to the great improvement of the crop.

The U. S. Soil Conservation Service has done suggestive work at *water spreading*, by which an hour of gully water from a local shower is spread over a meadow by means of furrows and rock-diversion dams.

This device, however, has much wider possibilities with trees than with grain and hay, for the reason that grain and hay require wide areas, while a tree can stand at the bottom of a gully in a ravine.

Here is an interesting possibility for the million or more square miles of arid lands between Kansas and California, and between northern Alberta and southern Mexico. At present, millions of arroyos (gullies) waste their rushing waters at the time of occasional rains. These gullies might become rows of useful trees fed and watered by the gully itself.

Perhaps the water of gullies could be led into long horizontal trenches or field reservoirs reaching long distances across the face of the slope and lined with the fruit- or bean-yielding trees which would be watered every time the gully flowed. This practice would be closely akin to the one common in parts of Shansi, a Chinese province southwest of Peking with scanty rain. There, in the agricultural villages, one of the common sounds of summer nights is the booming of the temple gong as the priests walk through the streets awakening the people to the fact there has been a shower on the village or on the hills, and the gullies are running. The population scurries out, armed with shovels, and diverts water from gully to field.

I have seen these fields months in advance all prepared, the field banked around, and gullies with dams, so that when the first rain came it would irrigate field number one. Shovel out the dam, and the water would flow down and irrigate field number two, and so on. This seemed to be a widely established practice. How long has it been in use? And how widespread? Ask the man who has traveled through China with agricultural and soil-conserving eyes—if you can find one.

May not the use of cement, accurate leveling devices, road-scraping machines, tractors, and dynamite in America put rain-catching devices on a basis that is not only mechanized but almost automatic, so that the gully water might fill our field reservoirs while we sleep?

Plant breeders might produce hardy and productive strains of almond, olive, apricot, oak, mesquite, honey locust, walnut, or other fruiting trees which could be planted out in the arroyos of the arid Southwest, so that the sheep ranches of our own country might be dotted with productive trees in a manner identical to that of the Berbers, who have dotted their sheep and goat ranges with olive trees and date palms. Enough of this has been done by our Soil Conservation Service and others to merit much more attention than it has received.

In 1927 Albert Dickens, Horticulturist, Kansas Agricultural College, Manhattan, Kansas, wrote as follows:

Taking advantage of runoff water in order to give trees an increase in moisture over the natural fall is, as you doubtless conjecture, an old and common method. The Experiment Stations at Colby and Tribune have taken advantage of this, and it is quite the common thing in the parks. The park at Colby was planted for this purpose, and the trees have made very satisfactory growth. . . .

The past twenty years I have visited every county in the state, and the forty western counties would have from five to twenty units and the average perhaps fifteen. The area would be harder to estimate. Probably a couple of acres for each location.

The value of trees has been more from ornamental and esthetic points of view than from production of food. The mulberry and apricot are quite generally grown in the dry territory and produce some food. The black walnut is generally planted more for the production of nuts than for timber. Mulberry, of course, is not a high-class food, but it attracts the birds and is quite commonly planted.

Examination of the map (Fig. 138) will show that every continent has large areas of grassland or scrub land on which this is about the only possible form of agriculture.

The world needs immediately eighteen or more, especially more, experiment stations whose staffs are experts in the breeding of drought-resisting trees—in Alberta, Texas, Arizona, Mexico, Argentina, Brazil, Russia, Turkestan, northwest China, India, eastern Palestine, Turkey, Sudan, Rhodesia, Cape Colony, Queensland, New South Wales, and other areas. (See map, Fig. 138.)

Each of these stations would have, on the average, a half million square miles of land to serve. Each station could possibly increase the productivity of perhaps fifty thousand square miles or thirty-two million acres. Suppose five dollars per acre per year were added to this thirty-two million acres—a rather large result to follow from an endowment of a half million or perhaps a million. This value could only arise from trees that would have to be scattered over wide areas. This would enhance their value exactly as the value of the irrigated land of western North America is enhanced by the fact that it is scattered. As the irrigated land is chiefly in alfalfa, the haystacks scattered over a million square miles combine with the adjacent range land to support animals and the family in a way that could not result if all the irrigated land were in one block.

Similarly, millions of scattered fruit, nut, and bean trees in the ranch country would meet their greatest need in helping to feed both the family and the flocks, with occasional small surpluses or specialties for export. The chief cash value would reach the world market in the form of meat and wool and hides, in exchange for an infinitude of manufactures.

The variation of plants of the same species has received frequent mention. One more point needs to be presented in connection with the idea of breeding desert trees. They sometimes vary fifty percent or more in the amount of water required to produce a given result. This variation *within the species* suggests an interesting line of experimentation with different species and strains, to find the most efficient for particular places.

See "Water Requirements of Plants as Influenced by Environment," L. J. Briggs, and H. L. Shantz, *Proceedings,* Second Pan-American Scientific Congress, Washington, 1915.

The variety of wheat having highest water requirement was eighteen percent above the lowest, corn thirty-one percent, vetch thirty-five percent, alfalfa forty-eight percent, sorghum sixty percent, and millet seventy percent. I regret that Messrs. Briggs and Shantz did not test trees. I should expect similar results among them.

TWO-STORY AGRICULTURE FOR LEVEL LAND

Lastly, I wish to submit the thesis that the trees now unused or little used as crop mediums may be the best kind of crop for some of our levelest and most arable lands. I have in mind a two-story agriculture with tree crops above and tilled crops beneath. By analogy I would recall the French practice (Fig. 78) of scattering walnut trees all over the farm while going on with the farming. This does not sound alluring to the machine-using American, but let us consider it. Suppose you had a farm on the sandy plain somewhere between New York and Galveston. You are growing hogs and letting them harvest a series of crops with, perhaps, cotton in the series as a cash crop. You plant out most of your farm with pecan trees in rows two hundred feet apart, fifty or one hundred feet apart in the row. This is not much of an interference with plowing, harrowing, planting, or tillage. Save for the cotton and probably some corn, you have no harvesting operations; the pigs do that, and little trees do not interfere with them. Little trees may not interfere very much with harvesting cotton or corn by machines.

After you have planted your pecans, walnut, hickory, honey locust, grafted oak, or other large-growing productive trees as just described, you go on with the hogging-down crop rotation. Gradually the trees grow to gigantic size and maximum productivity. Meanwhile, the hog farming goes on beneath the trees to the benefit of the trees, but the crops from the trees more than make up for the reduction in the forage and cotton-crop series.

Concerning this two-story agriculture, it is a little-used fact that some plants do not require full sunshine for maximum growth. Mr. H. L. Shantz of the U. S. Department of Agriculture states that experiments with artificial shading showed that when the light was so decreased as to range from one-half to one-seventh of normal illumination, a general increase in growth resulted in potato, cotton, lettuce, and radish. Corn made its best growth in full light. (Bulletin, No. 279, Bureau of Plant Industry, U. S. Department of Agriculture, *Effects of Artificial Shading on Plant Growth in Louisiana,* H. L. Shantz.)

When we know more about this subject, we may be able to work out a crop rotation that will actually do better when taken from full sunshine to the partial shade made by some kinds of crop-yielding trees, especially in our southeastern area of abundant rain—Cotton Belt and Corn Belt.

CHAPTER XXIV

Plan or Perish—Tree Crops—
The Nation and the Race—
A New Patriotism Is Needed

Considered from the standpoint of the future, and no long-distant future, either, a large part of the United States is on the road to economic Hades, going rapidly, by way of gullies, and few there are who seem to realize the significance of the catastrophe or the speed of its approach.

For example, the hillside shown in Figure 3 is in Virginia, within one hundred miles of Washington. It is fairly typical of thousands in the whole Piedmont area that reached from New York to Alabama, Kansas City, and the southern Ozarks. George Washington and Thomas Jefferson were mightily concerned about soil erosion, such as shown in Figure 3, but the State Director of Extension Work in Virginia told me in 1920 that he did not know there was need for such a thing as mangum terrace in that area, which is typical of thousands of square miles of rolling hills impoverished and gullied by erosion. It would interest you to check upon the Pennsylvania official agricultural attitude toward gullies and soil conservation.

If there were danger that a foreign country might get possession of some little island on the coast of Maine, Florida, or Texas, thousands of Americans would jump to their feet, willing to fight and perhaps to die that this speck of land should not pass to the possession of another nation. If it did pass to some other national ownership, it would still be the same piece of land. It would still have the same good for humanity that it had before the fight. It would, in west European and American eyes, even continue to be the private property of its previous owners, yet these same men who would fight to prevent change in national government of a piece of land have little compunction about destroying land in their own country. By neglect, they will destroy an acre or two in a season. Thousands of them *are* doing it *yearly, now.* In a single generation, each of tens of thousands of Americans destroys enough land to support a European farm family for unknown generations of time.

These land-wasters think they are patriotic citizens. We need a new definition of patriotism and a new definition of *treason!* You are still free to destroy all of the land that you can buy or (in most cases) rent.

THE DEAD NEIGHBORHOOD

Take as an example the hills of New England and the Appalachian hills and ridges. This area has one of the most wholesome climates in the world, a climate that helps man to be healthy and vigorous both in mind and body.

It has one of the most agriculturally dependable climates in the world. The land is not visited by the droughts and famines that are so often and so feelingly referred to in the Old Testament and which desolate so vast an area of South America, Africa, Australia, and Asia. These American hills have one of the best climates to feed man's body with food and his mills with raw material.

These American hills are variegated with beautiful flowers in spring, clothed with green in summer. The glory of autumn foliage, its red, brown, yellow, and gold set off by the evergreens, makes one of the most beautiful landscapes in the world, fit to inspire man's spirit and lift it above the prosy but useful bellyful of nuts that lies beneath the falling leaves. Yet this wholesome, dependable, and beautiful land is in agricultural decline. Much of it is desolated and abandoned. The old agriculture of the level land has been tried upon the hills—tried and found wanting.

The hills are gullied. The fields are barren. Tenantless houses, dilapidated cabins, tumbledown barns, poor roads, poor schools, and churches without a pastor—all are to be found in too many places. No wonder whole townships are for sale and at a cheap rate. But these neighborhoods might be transformed through tree-crop agriculture and become like the chestnut communities of Europe described in Chapter II and in the chapter on the chestnut.

Professor Stephen S. Visher, Indiana University, puts it this way:

One phase of the rough land situation which has interested me greatly here in southern Indiana is the present hopelessness of the economic condition of the people who try to make a living there by the use of "flat land methods." They can't be good American citizens—they are too poor. Your program affords hope for such people and regions.

In 1946 a professor in Ohio University, Athens, on the same Ohio Valley Hills wrote me as follows:

At present I am making an effort to find some method of saving the wasting counties of southern Ohio. I am taking an actual inventory of the financial income of the so-called farmers of this depleted area. I find that almost all of them have some industrial income not only to supplement their farm income but to finance their farms. It is only a matter of time until everything "goes down the river."

I feel that somewhere there must be an answer to the plight of these men. None of them are able to—or they cannot see the tangible return of plain lumbered trees.

And I add, they are right about that in most cases. But why must they kill their trees to use them? Why not crop-yielding trees?

TREE CROPS AND TENANCY

The tree-crop agriculturist must almost certainly be a home owner, not a shifting tenant. A half million square miles possessed by landowning small farmers is a greater basis of national strength and endurance than the landless and roving tenant who drifts from farm to farm, skinning them as he goes, or the still more landless and rootless crowds of humans, who, in the cities, shift from apartment to apartment and surge back and forth on the trolleys, subways, elevated railways. Can a nation survive on the basis of rented farms and rented apartments? There is small evidence that it can. For one thing look at the vital statistics. Our city dwellers have seventy babies per hundred funerals.

THE GAME PRESERVE

What a game preserve a collection of good crop trees would make! The trees would both shelter and feed the animals. This job is worth doing for that purpose alone by a naturalist and big-game hunter of great means or by a sportsman's club owning game preserves.

WORLD APPLICATIONS

An examination of the world regions map (Fig. 138) will show that every type of climate that is found in North America recurs in other continents. Some of them recur in every continent save Antarctica. Therefore, this book is one of world-wide application. Since I am an American and have spent only two and one-fourth years in foreign lands, my philosophy is naturally illustrated chiefly with American facts. But the *philosophy* is of world-wide application. The regional map shows in what parts of the world a given American tree has some chance of thriving. Conversely it shows the parts of the world in which we have a chance of finding trees that are likely to thrive in a given area within the United States.

In most cases a crop that is a success in one continent has several more continents in which to spread itself. For example, experiments by government agriculturists in India seem to indicate that mesquite seeds from California and Hawaii have been planted in several localities in India with apparent success. Since most of India suffers from drought, this is a fact of vast significance. The chapter on the Tropics gave some inkling (p. 298) of the valuable but unused tree crops that nature has already developed in arid lands.

Suppose we should work out a tree-crop agriculture along lines indicated and suggested by this book. What might it mean for the United States?

TREE CROPS CAN INCREASE CROP AREA

This table shows that in 1925 less than one-fifth of Massachusetts was improved land in farms and less than one-eighth of the land of the State was in crop; that West Virginia had even less of her land in crop; while Iowa, a state blessed with much

	Total crop land and pasture land other than woodland, 1925		Total harvested crop land, 1925		Total harvested crop land, 1944	
	Acres (thousand)	Percent of area	Acres (thousand)	Percent of area	Acres (thousand)	Percent of area
Massachusetts...	1,072	20.8	625	12.1	581	11.2
New York......	12,467	40.8	8,290	27.1	6,922	22.6
West Virginia...	5,304	34.4	1,677	10.8	1,490	9.6
Ohio..........	17,979	68.9	10,703	41.05	10,837	41.6
Iowa..........	29,505	82.9	21,466	60.3	21,562	60.6
Illinois........	26,700	74.4	19,755	55.07	20,302	56.6
Tennessee......	10,922	40.9	6,209	23.2	5,844	21.9
Oregon........	10,385	16.9	2,592	4.2	3,276	5.3
California......	21,045	21.1	5,723	5.7	7,536	7.5

All figures from *Yearbook*, U. S. Department of Agriculture, except columns 5 and 6, from *Agricultural Statistics*, 1946.

level land, has four-fifths of her land improved and three-fifths actually in crops, and the War Boom and tractors of 1944 showed that there was little room for expansion.

The decline in harvested acres in Massachusetts, New York, West Virginia, and Tennessee is a plain result of the gully and of mechanization. Farm machines seek good topography, and the era of tree crops has not yet come.

Now, the soil and climate of Massachusetts and West Virginia are such that certainly ninety percent of their land area would grow crop-yielding trees of some productive variety, after the manner of the chestnut orchards of Corsica, described in this book. Therefore it seems fair to assume that tree crops may easily increase fivefold or sixfold the crop-yielding area of New England and of the Appalachian region of which West Virginia is a type.

When one adds to this the large amount of rolling land, too steep for permanent agriculture of the plow type, to be found in the nonmountainous parts of eastern States and the rolling sections of Ohio, Indiana, Illinois (Figs. 51, 52, 95, 116, 117, 129), Iowa, Wisconsin, Missouri, Kansas, Oklahoma, and other States, it seems a conservative statement to say that tree crops, by utilizing steep, rough, and overflow lands, could double the crop-yielding area of that part of the United States lying east of the Mississippi and perhaps all east of the one hundredth meridian.

By utilizing the same sort of land in the foothills of the Rockies, the Sierras, the Cascades, and the Coast Ranges, it would seem probable that tree crops could double, perhaps more than double, the present crop-yielding area of the Rocky Mountain and Pacific States. Note the California figure, less than one-thirteenth of her area in harvested crops. This use of crop trees on the slopes of mountains in semiarid lands (Fig. 125) indicates that there are many tens of thousands of square miles waiting in the arid parts of the United States, Mexico, Central America, South America, Asia, Africa, and Australia. However, it should be pointed out that, in the irrigated West, it is not fair to compare unirrigated tree-crop acres with irrigated acres on a basis of equality.

USE OF TREE CROPS IN ARID LANDS

If the available gully water that now runs away in arid regions should be utilized for isolated crop trees in the manner indicated in the last chapter (Fig. 133), at least one million square miles of semiarid land west of the one hundredth meridian in the United States might possibly have its productivity doubled.

THE WOOD SUPPLY

The continuance of geological surveys and technical invention seems to reveal unexpected supplies of some of the mineral resources, especially oil (from shales), but no credible estimator finds any optimism in, or any quick alleviation for, the declining timber supply. In other words, our Western civilization, with its vast use of raw material, seems to be inevitably moving into a shortage of wood and timber. The United States has the best there is. We are cutting timber faster than we grow it, and we are still exporting, for a time, at least, to wood-hungry neighbors east, west, and south, but the timber famine looms.

In our present civilization the refuse of the grain agriculture is straw, almost worthless and quite generally wasted. In contrast to this, the tree crops, once established, will leave a substantial annual by-product of wood. Only a little of it will be saw timber, but this fact is of declining importance today, when the use of wood in the form of pulp, paper, carton, and even paper board, is increasing rapidly.

TREE CROPS, THE WATER SUPPLY, AND NAVIGATION

Suppose that three-fifths of the hill lands east of the one hundredth meridian were in crop trees whose productivity made it profitable for the farmers to cover their lands with water pockets or terraces with back ditch (Fig. 130). This would mean a greatly increased amount of water held in the ground from rainy season to dry season. This would make decreased flow in the spring of the year, with the result that many streams would be carrying in the minimum period several times their present flow. This would mean improved water supply for cities and for river navigation and quite likely for lowland irrigation.

TREE CROPS AND WATER POWER

We are entering the age of almost universal distribution of electric power. To make this power we are rapidly building reservoirs to store mountain water for power purposes. And these reservoirs are being filled up by silt at a rate which promises an untimely end, and in some cases much sooner than the builders expected. It is now estimated that Lake Meade, back of the gigantic Hoover (Boulder) Dam in the Grand Canyon, will be full in a century. And what will that do to Southern California?! The people of California seem quite unconcerned over the fact that nearly half of the people live in a land that has prospect of greatly reduced water supply a century hence because of overpasturing and wild erosion in the Colorado River Basin. The future? What has the future done for us?? And what are we doing to the future?? Reservoirs built in Algeria by the French have been completely filled. The same thing has happened in our Cotton Belt. Most of this silt in eastern America is produced by wastage of fields through preventable erosion. The tree-crop agriculture, with field water pockets or Soil Conservation Service terraces, would keep most of the silt on the land where it is needed. This would greatly prolong the life of reservoirs. The storage of water in little field reservoirs would increase the minimum stream flow and therefore the minimum water supply and would therefore increase power output at minimum seasons. Since this low peak in power development is a very damaging factor, these water pockets would have a great influence on the capital value of power installations. It should be remembered that it has been done already for its power and timber value alone (p. 327) and for its agricultural value alone (p. 325).

TREE CROPS AND FLOOD CONTROL

If three-fifths of the hill lands east of the one hundredth meridian had water pockets large enough to store all ordinary rains, the flood problems on our rivers, including the Ohio and the mighty Mississippi, would possibly be so much reduced

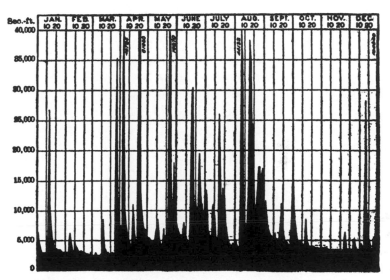

DISCHARGE OF YADKIN RIVER AT SALISBURY, N. C., 1901

This graph shows (in part only) the actual amount of water day by day for a year in a small North Carolina river. In this particular year the maximum discharge per second was 104,640 cubic feet; the minimum was 2420; the average was 8636. Suppose masonry reservoirs had raised the minimum to 6000, while water terraces raised it to 7500 and protected the masonry reservoirs from filling—at the same time that they doubled the agricultural output and quadrupled the agricultural valuations. It is high time that we quit skinning this Continent.

in size as to cease to be a serious economic menace to property situated in their flood plains.

Every continent can use these advantages. America has no monopoly on the possibilities of increasing the proportion of crop land, the usable resources of wood, of water, of navigation. Other continents also may mitigate the extremes of high and low water in their rivers by detaining the rain upon their uplands in water pockets and terraces with back ditch.

After the great flood of 1907, Pittsburgh created a Flood Commission. The Commission investigated and recommended a series of reservoirs in the mountain defiles upstream from Pittsburgh to hold flood waters until the flood danger had passed. These expensive reservoirs, if built, are destined to rapid filling, if the short-lived mountain farming of the present type continues with its gullies. But if the agricultural land were in water pockets or horizontal terraces, these would catch the silt which otherwise fills the reservoirs, would hold back much more water than the reservoirs themselves hold back. It would thereby increase the available resources of flood control, of water power, of navigation, and would benefit every town from Pittsburgh to New Orleans. Too bad some Pittsburgh millionaire does not make a demonstration of this on a few thousand acres.

The Ohio River floods are charged with destroying a thousand lives and a billion dollars' worth of property (1940 dollars), and they are getting worse rather than better.

TREE CROPS AND THE WORLD'S FOOD

The ability of tree crops to increase the world's food is suggestively shown in a table invented by W. J. Spillman (Appendix). The high rank of nut trees compares most favorably with the animal products, because the animals eat our crops before we eat the animals.

PART FOUR

Who Will Do the Basic Scientific Work Which This Book Calls For?

CHAPTER XXV

Who Is Working Now?

In Part I of this book I laid out a plan for an institute or institutes to investigate trees, breed them and create a permanent agriculture to prevent soil erosion which is destroying the United States and many other countries with a speed which, if continued, promises almost to wipe them out in a few generations or, at most, in a few centuries.

Part II presents a great variety of fact and suggestion leading to the conclusion that the job might produce results of incalculable value to the United States and other countries.

In Part III I have pointed out how present crop trees, and others that might be found or made, might be used in the economy of the individual farm and how such farms might influence the national wealth and especially the stability and permanence of our nation and our civilization.

I have long believed this to be an important work. I have tried to promote it by letting the idea be known. After writing in some smaller publications, I put the idea with what seemed convincing examples in the *Saturday Evening Post* in 1909. This was followed by articles in *Harpers* (several issues), *Atlantic Monthly*, the old *Century Magazine*, the *Country Gentleman* (a series), the *Farm Journal, Science,* and a number of other scientific journals. These all sum up millions of copies. Then, in 1929, I brought it all together in book form in the first edition of this book, containing 315 pages. The book was published by

Harcourt, Brace and Company, one of the leading publishing houses. It was highly praised in reviews. I am reliably informed that the book is known in many experiment stations and U. S. Department of Agriculture libraries.

I did all this in the sweet faith that if it just became known, an idea so good as this would be taken up by the U. S. Department of Agriculture, the State experiment stations, the arboreta, as well as by private individuals, and that something would be done about it. I think I complimented humanity, or exposed my own ignorance by assuming too close a relation between knowledge, motivation, and action.

To see what had been done, I sent the following letter, each copy a signed original, to the directors of the agricultural experiment stations in each of the forty-eight States of the United States.

COLUMBIA UNIVERSITY

NEW YORK

October 20,
1947

To the Director,
Agricultural Experiment Station,
Auburn,
Alabama [and every other state]

Dear Sir:

For more than thirty years I have been a member of a group of enthusiasts known as the Northern Nut Growers' Association. The members of this organization have been gathering information concerning the desirability and possibility of nut trees as a crop for the north central and northeastern sections of the United States.

I am writing to you in the hope of learning what work has been done, is being done, or is on the schedule to be done in this field at your Station or the substations under your control, or by experimenters in your State who are known to you.

I am inquiring, therefore, as to what, if anything, has been done with regard to:

| Any breeding experiments | { | Black Walnuts
Butternuts
English Walnuts
Shagbarks
Shellbarks
Hazelnuts
Chinese Chestnuts
Japanese Chestnuts | Breeding | { | Japanese Walnuts (Heart-nut)
Filberts
Pecans
Acorn-yielding oaks for pig food or human food
Honey-locust bean pods for stock food
Persimmons for human food or stock food
Paw Paws for human food or stock food
Any other new or little-used crop trees |

Have you experimental plantings of any of these? If so, I would appreciate knowing what species are included.

Perhaps you may have seen my book *Tree Crops, a Permanent Agriculture* covering this field, but now unfortunately out of print.

I know this is a slightly bothersome letter, but I hope you will find time to answer it or to hand it to someone who can answer it. It will not be consigned to the waste basket, because I am a rather active disseminator of information, and am now revising the book *Tree Crops*.

Very truly yours,

[SIGNED]   J. Russell Smith

J. Russell Smith,
Professor, Economic Geography,
Author of *Tree Crops, a Permanent Agriculture*
Men and Resources
North America
Industrial and Commercial Geography

A somewhat similar letter was sent to all of the six hundred ninety-three members of the Northern Nut Growers' Association. This organization is composed almost entirely of amateurs with nut trees as an avocational interest. The results from these two letters are tabulated and, along with salient excerpts from the replies, are being published in the 1949 annual report of the Northern Nut Growers' Association, J. C. McDaniel, Secretary, State Office Building, Nashville, Tenn.

The tabulation of results showed that one hundred two members of the Association were testing ten or more varieties of nut trees, persimmons or honey locusts.

Fifteen experiment stations in ten states were doing the same.

Among the experiment stations, New York led with one hundred three varieties and fourteen species. The ten highest stations average thirty-nine varieties and ten species.

Among the members of the Northern Nut Growers' Association, Benton & Smith, of New York, had one hundred seventy-nine varieties, nine species; Shessler, dirt farmer of Ohio, one hundred seventy-two varieties, seventeen species; Etter, of Pennsylvania, one hundred seventy varieties, sixteen species.

But I think the spiritual prize goes to Dunstan, of Greensboro, North Carolina, a professor of Spanish, with one hundred twenty-eight varieties, seventeen species; and he has so little land that he has to put half a dozen shagbark varieties on one tree.

Among the members of the Association the ten highest averaged one hundred forty varieties and fourteen and eight-tenths species, compared to State stations with thirty-nine varieties and ten species.

The most conspicuous things about the survey were the following:

1. Very small amount of breeding that is being done. When one considers the nature, organization, supposed objectives and financial backing of the experiment stations, it seems to me fair to say that it is mere piffle if a station contents itself with mere variety testing and does no breeding work.

2. Second conspicuous result—the complete ignoring of the oak.

My search brought up no oak data except the work done under the Tennessee Valley Authority by John W. Hershey, who has not had the disadvantage of an agricultural college course, or the disadvantage of being a botanist, or a forester, or an experiment-station staff member, so that without these limitations he has been able to use his mind in a practical way. It is especially interesting to note that, whereas he started this important work in the T.V.A., he left in a few years on account of ill health, and the work with acorns seems to have slumbered or perished after his departure, although the T.V.A. has done excellent work with black walnuts and is doing some other useful nut tree testing, and something with honey locust.

Lest someone should think I am belittling the whole field of organized agriculture, so far as nut trees are concerned, I wish to mention the roll of honor of men in this field: L. H. Mac-Daniels, Cornell University; George L. Slate, Geneva Experiment Station; George F. Gravatt, U. S. Department of Agriculture; C. C. Lounsberry, Iowa Experiment Station, at Ames, and Ray Watts, Penn State College, who reports three butternut trees. I mention these men because, as I understand it, they have shown they are interested in nut trees by having some trees *on their own grounds*. Especially do I want to emphasize Messrs. Slate and MacDaniels, because without any specific appropriations for this field, they have considerable collections on station grounds and on their own grounds—one of nut trees and the other of persimmon trees, in their own gardens. J. C. McDaniel, of the Tennessee Staff, is doing excellent work as secretary of the Northern Nut Growers' Association.

The U. S. Department of Agriculture has done a magnificent job with the Chinese chestnut, having got it to the point where an industry is ready to start with good varieties, but they have not any area yield tests, and they have neglected the promising Japanese chestnut.

Upon the whole, we have really an interesting mystery. Why have the experiment stations neglected anything so important as breeding oak trees? This becomes all the more mysterious when it is known that in 1903 a quaint old gentleman named Ness in the College Station, Texas, tested out the then recently announced Mendel's Law by hybridizing some oaks. He published his results in the *Journal of Heredity* and they are truly remarkable, but neither Texas nor any other station has followed up this most suggestive lead.

It is true that they can rarely get definite appropriation for such work as breeding oak trees, but the excuse has been much overworked. Let us examine the situation.

THE POLITICIAN AND HIS NECESSITIES

The United States is a democracy and therefore we are governed by politicians. A politician is first and last a vote getter. He does not need to be a man of vision. Look around you and see if this is not true. Perhaps you have never fully appreciated the politician's position. We fool citizens make him fight for a chance to serve us. His first necessity is to get elected, often a back-breaking labor. His second necessity is to get reelected, and that is a constant care. If, between these two grinding necessities, he has time, inclination, and capacity to be a statesman, we are lucky indeed. We, the citizens, devil him so much by asking favors that he has not half a chance to be a statesman even if he had the necessary knowledge and understanding.

Why has so little been done in the matter of crop-tree breeding? I laid the matter before Secretary of Agriculture Jardine, and he told me very simply and directly why no important step had been taken in his department. He said his Department was busy with urgent matters, curing the troubles of crops that are already established. In marches a congressman carrying a wilted tobacco plant. He is followed by a delegation of tobacco growers. They want that new disease stopped and that congressman will raise heaven, earth, and hell to get it investigated *and keep this Tobacco Belt vote* upon which his reelection may depend.

And thus Mr. Secretary Jardine explained why they had no money to expend on breeding crop trees for the future or for saving the soils by which future generations must live. We should not criticize the Secretary or the Department too harshly. Sometimes there are men there who yearn to do such work, but we must remember that the Department of Agriculture gets its money by way of the lord of the budget and by appropriations by Congress. Budget makers are not primarily men of creative instinct. They must be money savers, and that encourages the all-too-common instinct to discourage and block new enterprises. Budget makers, like members of Congress, are not rewarded by any distant future. If the man who made the budget should happen to be convinced of the efficacy of tree crops, something might be started; but the succeeding budget chief would probably have a mind quite impervious to trees, and the experiment might be blocked.

And we should not forget the potency of the man with a notion, or a prejudice, or a grudge, who also happens to be chairman of a committee or subcommittee or is influential in a committee or a subcommittee in Congress or in a State legislature.

Do not forget that great mutual benefit service of the legislators, the pork barrel, and the wonders it performs as it appropriates something for every district, even if it is to dredge channels where no boats will go, or to chuck millions into the Missouri River to "stop floods." Many can howl and also be easily fooled by those who appropriate our money.

Votes do not grow on trees. Congressmen must get elected, and if they don't get reelected they do not get the experience needed to be really useful.

State experiment stations, here and there, have men who would like to do this kind of work; but the stations are dependent on State legislatures for the money they get, and it should not be forgotten that the first necessity of the legislator is to keep his eye on the *next* election, else he may wake up on a certain November morning and find that he is no longer a legislator. In view of these urgent and even pressing necessities, we can see that it is a rare accident when public money is to be had for the prolonged task of developing new crop trees. Neither by theory nor by practice are we justified in expecting democratic legislation to look far forward in its appropriations save for education (too little) and for military defense (usually too late).

As an evidence of the low intellectual and civic level of legislation in America, I will not refer to the psychology and facts of municipal elections in Chicago, for example. Instead, I cite the widespread American practice of taxing young forest. Tax the forest, and you put a premium on cutting the forest and discourage planting. This continues despite the well-known fact of coming lumber scarcity and much preaching. If legislation should tax the *lumber* and exempt the forest, there would be a premium on its preservation. Most American State legislatures cannot be made to see so simple a proposition as this. No, it is foolishness to expect State legislatures to do the big thing in crop-tree creation. Here and there, of course, one will do something, but it is the exception.

SOME BLOWS FROM THE LEGISLATORS

Our Soil Conservation Service, under the leadership of Hugh H. Bennett, its devoted and most diligent creator, had one of the most intelligent, devoted, enthusiastic, well-selected, well-managed staffs any government ever had, but it, too, must live on appropriations. In 1936 it made a brave start at developing tree agriculture for the hills.

From 1936 to 1940, there were about thirty research people studying new economic plants and how to use them. I have seen some of their studies. Most suggestive! Then came war. In June 1947, Congress cut Soil Conservation Service funds twenty-seven percent, just as prices were climbing. Research funds went down to a third of the 1940 figure. By the processes well known to readers of *Alice in Wonderland*, Congress at this time ordered higher salaries but gave no money. Departments were supposed to reduce numbers to get money to raise pay of the lucky ones who stayed.

There is precedent. A Central American country stated in its constitution that education was free and compulsory, although it happened that there were no schools.

When Congress made this remarkable economy and pay-raise move, nearly a hundred valuable, trained research personnel had to leave Soil Conservation Service for pastures new, and the research in new crop trees stopped. I have been told by persons not in government employ, but well informed on what goes on in Washington, that this 1947 cut in Soil Conservation funds appears to have been the venting of a personal grudge. You know you really can't do important things without stepping on some small toe.

About the time of this Soil Conservation slash, the Library of Congress had a similar cut, and semichaos of books and papers had to follow.

Here we see two legislative blows at the very base of civilization—muddying the waters in the head fountain of knowledge and neglecting to save soil—the one thing which the country *is*. And that same Congress appropriated millions and pledged billions to be spent in and on the Missouri Basin. There were bad floods there and sore voters. They had to be appeased. Congress didn't know what to do, so they appropriated our money for flood control to be spent by the Army Engineers. And the Army Engineers have proved over and over again that they don't know what to do in the civilian world. They are fine gentlemen and honorable, but their lack of horse sense is one of the phenomena of the 20th century.

Tree crops are a new thing for America, a new idea. No elected legislature can possibly be expected to appropriate regularly for such creative work during the decades necessary to do the big things. Was there a State appropriation or a Congressional appropriation back of Morse when he created the telegraph, or of Edison working wonders with electricity, or of Orville Wright, the first to make successful flight, or of Lindbergh when he flew from continental mainland to continental mainland? No! These things were done with private money urged by an *idea*. By this means also, I fear, must come most of the creative work in the beginnings of a tree-crop agriculture.

Readers of this page should not confuse the breeding of crop *trees* and the breeding of annuals, such as wheat, corn, cabbage, or tomato. The State can see as much as a year ahead and support such breeding. It has done wonders. See the *Yearbook*, 1936, U. S. Department of Agriculture, for a thousand pages of it.

If a farm crop is established and gets into trouble, the State may help it or give it a boost. The high-bush blueberry is a perfect example of this and also the work with pecans. Government work with both started with an industry that was established. When an industry is big enough to have votes back of it, it has a chance at government aid.

Persons of prophetic vision are needed to finance institutes of mountain agriculture to breed crop trees. There, in such places as the Arnold Arboretum, protected from the variations due to elections and change of officials, new things could be developed to the point where experiment stations and governmental bureaus could take them up, test them, and pass them along. Please note the distinction between *creating* a new crop tree by the slow process of breeding, and the much quicker process of testing the new tree in experimental grounds. After the tests of experimental grounds, progressive farmers would take hold and start production. Then, if it became big enough to have a block of voters interested, it might get experiment station or federal help with a new disease.

The short-time view is shown by the following: A State agricultural extension service representative was chided by a citizen for not starting anything new. He said his job was to teach proved and well-established practices.

I asked a State experiment station man what they were doing about honey locusts. He replied that Hill Culture work should be limited to work which will be rather immediately beneficial to the hill farmer, and therefore no great outlay should be spent on honey locusts.

The difficulties of long-time creative work with plants do not end with the discussion of appropriations. One of the sad facts of civilization is that there is an institution called bureaucracy, and the mixture of bureaucracy and human nature on its sadder side is often sad indeed as to its influence upon progress.

One of the aspects of bureaucracy is that almost every young man has somebody over him, and his boss may have somebody over him, and this second boss up may have somebody over him, and when these superiors publish they often ignore mention of the name of the underling who really did it.

Now, suppose the young man happens to be more able than his boss. How many bosses are pleased to help their underlings to pass on up above them?

Suppose the young man has a new idea. Will his chief always help him to develop the new idea and thereby of course possibly make the young man more famous than his boss? You could fill a book with tragic stories that have that plot. Perhaps one of the best known is the record of General deGaulle of the French Army.

He learned something in World War I, wrote a book describing in detail the thing which about 1940 got the name of "Blitz Warfare"—lightning dispatch of ground forces aided and guided from the air. But deGaulle, who conceived it and wrote the book, was only a major in the French Army. Lots of fat, crusty, rusty, beribboned colonels, brigadier generals, major generals, lieutenant generals, and Parisian bureaucrats were above him in rank. And here this puppy, only a major, wrote a book! The book showed that all of the prevailing ideas about war were antiquated and therefore the holders of these ideas must be antiquated, too. The impertinence of it! It wouldn't do. So the Brass Hats delegated one of the higher-ups to blast the book to pieces with a savage review. However, words can't destroy an idea, and the German army did not happen to have such thoughts about the major's idea of blitz warfare. They seized upon his plan, and with it crushed France to earth in 1940 and rubbed the French face in the mud. They might have been unable to do this if deGaulle's superiors hadn't wanted to keep him down, probably because he was young, and under them in rank, and by his originality had really showed them up.

A few generations ago there was a group of six men working in the British Civil Service. One of them had an idea. He devised a plan whereby five men could do the work of the six. He gave it to his superior. The superior accepted it and fired the young man, Henry Bessemer by name, later Sir Henry Bessemer, millionaire steel magnate. He invented the process that bears his name and made a fortune out of it. Lucky thing he got fired.

I knew a young man who was employed fifteen years ago in a great railroad office—a railroad which speaks very well of itself indeed. There were three men working together. In the course of a year, the young man devised a scheme for greatly simplifying the work. He showed it to his boss. The boss read it, asked, "How many copies of this have you, Sam?"

"Two more."

"Give them to me."

The boss got them and in Sam's presence took the three, put them together, and with fierce gestures tore the three to little shreds and hurled them into the waste basket. "You damn fool," said he to Sam, "don't you see that would make me lose my job?"

A member of the agricultural experiment staff in a western university discovered that he could grow plants in solutions of plant food in water without soil. It became known a little bit. A newspaper man got hold of it. He rushed to the station and interviewed the young man—took a picture or two and wrote an article. Promptly it went to the ends of the world. Journalists flocked there; distinguished foreigners from many lands came to see. They didn't want to see the chief of horticulture—they didn't want to see the director of the station—they wanted to see the young man and his new technique, which a journalist had called "Soilless Agriculture." The discoverer wrote it up in a careful and accurate manner, but the journalistic account of it spread around the world. "Undesirable publicity," said the elders, most of whom had never had much publicity. Result? Those immediately over the young man finally managed to get the president of the university to consent to his being dismissed, a broken man. His offense? He had done something really new, which the world acclaimed, and which the U. S. Government used to good advantage during World War II. Hydroponics is the proper name.

How often have things like this happened in the American and other agricultural experiment stations? Only God knows. It would keep the FBI busy for some years to get them all traced out, and I fear there would be a long sad bookful if they were described in some detail.

And the new ideas that the young men never dared express?

NEW IDEAS AND DARING

Fortunately, some new ideas are not fatal.

Here's a story that pleases me. Some twenty-five or thirty years ago, E. V. McCollum, a young man in a midwestern agricultural experiment station, said to his chief, "This theory of food values that we are acting upon is based on laboratory experiments. Let's try a feeding experiment running through several generations of animals and see how it really works." He asked for an appropriation to start a herd of cows and got turned down.

Then he had another thought. Rats don't eat much and they breed rapidly. He bought some cages, got some white rats and kept them in his office, and fed them out of his personal funds. He got results in a fraction of the time that the cows would have given it, he established a world reputation and was called up higher, but not with the help of his boss.

In spite of all these difficulties of appropriations, bureaucracy and the boss, I have one devastating criticism to level at the bureaucrats both young and old. Too few of them, far too few of them, keep rats in their offices. Too many of them really lack the creative urge or they would reach out and do something on the side over and above their jobs. Success in this world comes to those who are bigger than their jobs and show it by doing something over and above the job. And they may get called up higher—if jealousy does not strangle them. But mediocrity is the safest thing in the world. If you want to be perfectly safe, get a government clerkship, do as you are told, and never have an idea of your own, or if you do, keep it quiet, they are a bit dangerous.

"WE HAVEN'T ANY APPROPRIATION"

That is a, perhaps *the*, stock excuse of the government man, but Cornell University has a fine and valuable collection of nut trees under test. Professor L. H. MacDaniels tells me there never was any appropriation for nut trees, but he happened to be interested, as was John Craig, who was there before him.

I don't believe Helge Ness had any special appropriation when he did that hybridization of oaks in Texas over forty years ago, with its remarkable results. It's no great mystery to hybridize oaks. Why hasn't it been done by a dozen or two of station men *on the side?* I think the answer must be found in the character of the institutions or the staffs, essentially torpid, lacking in the vital quality of curiosity. Curiosity is one of the parents of scientific progress, perhaps it is *the* parent.

No appropriation? Consider the one hundred two private experimenters in the Northern Nut Growers' membership. In Greensboro College, North Carolina, an institution that is not famous for high salaries, R. T. Dunstan—Professor of Spanish, mind you—is testing one hundred twenty-eight varieties of nut trees in his yard. The yard is so small that he has to put several varieties on one tree, but that tests them so far as that climate is concerned. Appropriation?

That professor of Spanish has curiosity and enthusiasm, and his example puts to shame the experiment-station staffs. If, for the last forty years since Ness hybridized the Texas oaks, two dozen of the station men had as much interest in economic trees as does the Greensboro professor of Spanish, they would have had trees in their yards or they would have stuck them out in

Fig. 134. A neglected orchard in Santa Clara County, California, shows how quickly ruin may follow folly and neglect. You are still free to destroy the United States piece by piece. (Photo Radford. U. S. Soil Conservation Service.)

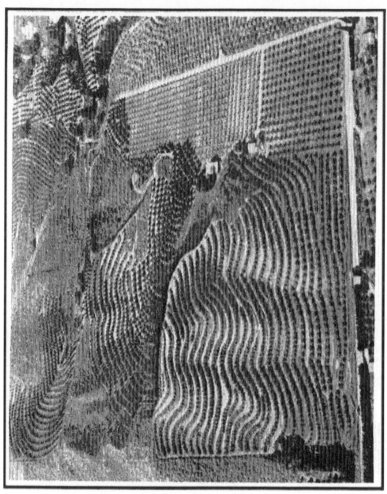

FIG. 135. Mr. Harry E. Reddick's citrus orchard four miles northwest of Santa Paula, California. Slopes vary from 20% to 40%. Some rows are short, adjusting to the slope. The rows themselves slope from 1% to 5%, so that irrigation water will flow along the rows after being pumped up. Cover crop is grown. 11½ acres of lemons have lost only a few hundred pounds of top soil in four years. When we get acorns, carobs, honey locust, and mesquite growing forage crops in this way, we will have entered a new epoch. (Photo J. T. Allison, U. S. Soil Conservation Service.)

Fig. 136. This New Jersey peach orchard is planted on contour with plan for surplus water to escape at left. (Photo U. S. Conservation Service.)

FIG. 137. Highbush blueberry accepts cultivation, moves from woods to field, and yields a crop worth millions. Variety Concord, largest berries 19 millimeters. This is a monument to F. V. Coville, one of the few constructive botanists who ever saw a tree as anything but two Latin names. (Photo U. S. D. A.)

some corner of the station grounds and they would have slipped out at lunch time and done a bit of pollination and might have been called to high places long ago. The young fellows don't appreciate the reputational value of something really new!

Fayette Etter of Pennsylvania supports himself by being a line-maintenance man for a power company, but his private experiment station, supported by curiosity and enthusiasm, makes the Pennsylvania State Experiment Station a laughing stock and rivals the collection of any State station. And, thank God, Etter is still young!

I strongly suspect bureaucracy of doing one or both of two things:

1. Causes some men of originality to shun the bureaucracy.

2. Suppresses the originality of some of those who happen to get into the trap.

I do not mean to say that this is always the case, but I am suggesting that there may be an ever-present danger which probably cannot be entirely removed even by magnanimity of spirit —one of the great needs of all human organizations.

THOSE IN A POSITION TO KNOW

There is another danger that besets the young man or the old one with a new idea—*"those in a position to know."*

When Edison first demonstrated his incandescent electric lamp, the President of Stevens Institute of Technology, Henry Morton, sneered, "Everyone acquainted with the subject will recognize it as a conspicuous failure."

You could fill a book with examples like that. If every new idea had to wait for its approval by *those in a position to know,* no new idea would ever have a chance. Young man! Stand up, follow your imagination if you have one.

OH! THE BOTANISTS! WON'T THEY DO IT?

The botanists are another group whose inactivity amounts to mystery. Why have not a dozen or two of them pointed out and built up the crop values of a score of unused trees?

As an example, consider the honey locust tree, *Gleditsia tria-canthos*. For more than two hundred years American school-children have been picking up honey-locust beans and eating the sugary pulp that is contained in the pods. This has happened on a large area of the eastern United States, but so far as I know, no botanist has paid any attention to this fact of sugar content. Yet we have one or more professors of botany in several hundred ordinary colleges, also one or more in every one of forty-eight agricultural colleges and a greater number of experiment stations.

Nothing about sugar, but Detwiler's report on the honey locust reported on page 74 reports a veritable botanists' field day. He refers to paper after paper by good men, wrestling with this question: does the honey locust, which is a legume, have nodules or lumps on its roots, or branches, or some other means of gathering nitrogen from the air, as do most legumes? Paper after paper, and experiments, but no mention of sugar or protein, in both of which these pods and beans are rich. *No mention of economic value.*

Apparently the botanist is almost immune to economic values. I upbraided a good botanist about this, and he confessed to having the pure-science disease. It was, he said, widely held among botanists that it was not their function to point out that things were useful—they should leave such things as that "to others"—these were the exact words, and that botanists should deal in pure science. Pure science says that oak or hickory trees will hybridize, but a proper botanist should not mention the possible use of a hybrid. Taxonomy, scientific classification, is the proper thing for botanists. Names—what's its name? Its name is "this"! Oh no, it's "that"! That starts a botanists' holiday—greatness is achieved if at long last he has a plant named for him.

Botanists do not wish to go to heaven. All the trees there are properly named, properly labeled, and the botanist can't tell God He is wrong—no fun for the botanists. Indeed that would be hell for them.

I wish to pay a tribute of respect to one off-color botanist, the late F. V. Coville. He gave us a new example of the old method that needs to be used on an indefinite number of species. He hunted up the best wild highbush blueberries, crossed them and lifted to a high level an old industry of picking wild berries.

The nonproductivity of the botanists is an even greater mystery than that of the experiment-station men. How could they *all* be so blind for so long? If 1% of them had happened to have imagination!

A CASE HISTORY: THE HONEY LOCUST, THE BUREAUCRATS, AND THE SCIENTISTS

The record of what happened to the honey locust tree at the hands of the botanists and the bureaucrats is illuminative, and I fear prophetic of what these pleasant gentlemen can and will do.

Search of the *Botanical Index* in the U. S. Department of Agriculture Library at Washington shows the earliest mention of the honey locust (*Gleditsia triacanthos*) to be in Bulletin 1194, 1923, A Contribution of the U. S. Bureau of Chemistry from its Cattle, Feed and Grain Investigation Laboratory, "A Chemical Study of Mesquite, Carob, and Honey Locust Beans" by George Pelham Walton, Assistant Chemist. On page 3 he said:

It grows in western New York and over a large area south from that region. The pods and young trees are relished by livestock. The ripe pods, with their syrupy pulp, are popular with human beings, especially children. Because of the richness of the fruit in sugars and protein, the plant may prove to be of economic value.

As a measure of his enthusiasm for the honey locust, he took beans from a tree that was growing in the Agricultural Department grounds, and apparently looked for nothing better, even after he discovered the above-mentioned merit.

If I remember correctly, this bulletin was called to my attention by the late W. J. Spellman, one of the most creative intellects that have thus far graced the Department of Agriculture. He and Assistant Secretary of Agriculture Willet M. Hays early encouraged me with the tree-crops idea.

In 1926 I offered prizes in the *Journal of Heredity* for sugar-producing beans (p. 68) and got a honey locust which analyzed twenty-nine percent sugar. The results were published in the *Journal of Heredity,* in a number of magazine articles, and, in 1929, in the first edition of this book, which went into a good many experiment-station libraries. A good many experiment-station men tell me they have read it. However, *no one seems to have done anything about the honey locust up to 1934.*

Shortly after the Tennessee Valley Authority was created, I managed to get a copy of the book *Tree Crops* onto the desk of the Chairman of the Board, Arthur E. Morgan. He was sufficiently interested to ask whom they could get to do that work for them. I named John W. Hershey and said, "He hasn't much formal education, but he is interested in this thing. He is a born salesman and knows how to connect with the common man, especially the countryman, and he will make things hum." Morgan took him and made him chief of a tree-crops section, which was unfortunately attached to the Division of Forestry. There was no good philosophic reason for putting it under forestry.

It should have gone to horticulture, not forestry. By professional occupation, the forester bears the same relation to a horticulturist that a butcher does to a dairyman. The dairyman pets a cow, gives her what she needs for years, that she may give him a continuous harvest. In an hour, the butcher cuts her throat, skins her, and sells her carcass.

The horticulturist pets his trees all the days of their lives, that they may yield him many annual harvests. He forgets them when they die.

The forester looks after them that he may at some distant day cut them down, burn the slash, saw them up and sell the pieces.

By every act of his professional life, the forester is foreign to the tree-crops idea. No man can get away from his past. Give a tree-crops experiment to a forester, and his past gives him no reason to be sympathetic to beans, berries, acorns, or fruiting trees.

I do not want anyone to think I'm agin foresters—*as foresters*. We need them terribly. The first magazine article I ever wrote was booming forestry as a new career about fifty years ago, when the Division (I think it was Division) of Forestry at Washington had a seven-man staff, if I remember correctly.

Hershey's tree-crops section in the T.V.A. offered prizes for the best acorns, the best honey-locust pods, the best persimmons, the best blueberries, and other wild fruits. These berries grow wild in the Appalachian Highlands on the upper waters of the Tennessee. Dr. Morgan told me some time later that if they would let Hershey alone he would soon "have the whole valley in trees." Hershey understood the rural mind and had a flair for publicity. His message had so much more for the farmers than had that of the foresters, that the foresters were almost mute beside him—and he not a forester, not a college man! That was hard on the human spirit. Perhaps you scent lack of forester enthusiasm.

One of Hershey's finds was some sweet acorns, reported by a farmer in southwestern Pennsylvania, whose interest is measured by the fact that he took his ten-year-old grandson Jerry with him on his tree hunts and showed the boy how he was marking the special trees with a loose piece of wire around them, the wire passing through a hole in a block of wood on which saw-marks XI to XV ranked the quality of the acorns produced by the trees. Then he said to Hershey, "I have explained the whole thing to Jerry and the importance of the work, and any time after I am gone he can supply you with seed." Did the old man "have an appropriation"?

Hershey's most important find was the Milwood Honey Locust and the Calhoun—both of which produce above thirty percent sugar (p. 76).

In a short time Hershey's health failed. He retired to his

farm. The foresters took over, and Hershey's plans seem to have had a considerable check, according to T.V.A. replies to my requests for information. Nothing appears to have been done about the oak, probably the most promising, eventually, of all tree crops. Despite its perfection as a crop plant, they so nearly neglected the honey locust that they have no results yet ready to report, and this, too, despite the fact that honey locust trees from the T.V.A. nursery went to Alabama, performed their miracles, and were slaughtered.

Meanwhile, the Soil Conservation Service which, under Hugh Bennett, had a genius for getting men with a flame in them, put its research department into the hands of Walter Lowdermilk and the Hill Culture (tree crops) work under Samuel Detwiler, both of them old friends of mine through our community of interest. Both of them had read the book *Tree Crops* and thought its idea worth trying.

The Soil Conservation Service entered into cooperative relations with the Alabama Station, got honey locust trees from the T.V.A. nursery and produced the astounding results recorded on page 76. Then Congress slashed the appropriation for the Soil Conservation Service, and the Alabama Station unbelievably cut down the honey locust trees.

Similar cooperative ventures in Iowa and Ohio left considerable plantings of these trees to the fate of rival brush and fires.

However, the T.V.A. reports two belated cooperative pasture plantings of honey locusts, at Virginia and Tennessee agricultural experiment stations, which should give enough results in a few years to keep the interest from dying out. Also, Hershey and I have each sold several hundred trees in small lots to many scattered buyers in the eastern United States.

Now, to see the deadly implication of this sad little story of the beans, it should be noted that this honey-locust business started with a hardy tree native to a million square miles of the United States, which in its wild state produced beans capable of being an important crop, *at once*, by the same simple process by which we get an orchard of a new variety of apple trees. But after fourteen years the mighty T.V.A. cannot quote an acre yield.

This complete honey-locust story belongs high up in the history of human stupidity. I cite it in final support of my low estimate of government as an agency for *long-time creative experiment in making new crop trees* for a tree-crop agriculture.

The point of this book is that we need long years of research and experimentation to make many new crop trees. That is the way most of our present crops have come—by the chance experimentation of primitive agriculture. But the honey locust tree was prepared by nature, all ready to propagate, and we Americans have trampled on it and neglected it for three hundred years—and that, too, with all our equipment and public-supported organization for the promotion of agriculture.

THE GOVERNMENT SLEEPS

Two blades are necessary before scissors will cut. Ideas and money are both necessary before a government can do much. Some thirty or more years ago the imagination of David Fairchild, backed up a bit by some private money, resulted in a start at plant exploration in foreign lands.

The Division of Plant Exploration and Introduction (U. S. Department of Agriculture) still lives, but in near coma. If it were well financed and adequately manned, something might happen, but it seems that most of our travelers these days are soldiers. The present status of this potentially important division is suggested by the fact that, according to a statement from a Department member, we have had no plant explorer in China since 1929; and southeastern Asia, of all places, with the lands of China and India reaching upward from sea level to snow line in Tibet, and with a multitude of valleys, is perhaps the greatest natural botanical garden in the world. The chances for species differentiation to fit almost any climate are not elsewhere matched by any such a multitude of little climate areas. A well-nigh infinite number of varied habitats and plant associations await us, and here we sit!

CHAPTER XXVI

The Great Hope and
the Many Little Hopes

The great hope is an endowed foundation. It should not be many years until the increasing realization of our childish resource waste comes like a revelation in the mind of John Doe, multimillionaire, and he sees a double vision. He sees the beautiful green acres of the John Doe Foundation, which has started the Doe Institute of Mountain Agriculture. Also he sees the name of the John Doe Foundation appearing month after month, year after year, with increasing frequency in farm papers, magazines, Sunday supplements, scientific journals, and finally in treatises on a wide variety of subjects.

In the early years these references will include two kinds of information. The first will be the reports of the erosion-survey expeditions in foreign lands. This will shock the reading world with carefully measured results of erosion, which has destroyed communities and nations in Europe, Asia, and Africa, and is still doing it, and promises to wipe out half the farm land of the United States unless we do something about it, and quickly.

Along with the erosion survey will be reports from half a dozen plant explorers who are scouring the ends of the world and sending back useful new varieties of trees for crop production, but especially for breeding better varieties.

In a few years after this gets started, geneticists and economic botanists will flock to the grounds of the Doe Foundation and walk down the long rows of new hybrids, to see what nature's dice roll out when shaken by the tree breeders of the Doe Institute. Wide-awake professors of botany and students of genetics will bring groups of graduate students to see the relation of these things to scientific theory. As these new hybrids begin to produce crops, the scouts from the big farm papers will come around to get material for articles showing how these new things at the Doe Foundation Grounds are going to make new crops in various corners of this and other countries. Foreign scientists and government experts will come to see. Intelligent farmers, fruit growers, conservationists, and progressive teachers will arrive by the bus loads on vacation trips to see the works of the Doe Foundation.

By this time the Doe Foundation has given its founder a world reputation, but it has only started. The work of such an institution builds up like a rolling snowball. There might be hundreds of cooperating farmers testing out this new icthionomite, a hybrid from a tree that a Doe explorer sent in from Shangri La.

Here and there one of these cooperating experimenters will succeed so well that he starts commercial growing of a new tree crop. His neighbors copy. A small industry arises. It gets a problem, calls on the State experiment station, and now the station hears their cry, for here is a block of votes. The station experts come to investigate, and a new industry has started down the accredited road. John Doe has passed over the River, but his idea and his endowment live after him and are benefiting his fellowmen and his native land.

Meanwhile, the men in the laboratory at Doe Foundation Headquarters, undisturbed by the whims and changes of political control, are working on problems of pure science so called, the seemingly useless things that sometimes turn out to be useful to the point of revolution—such as the work of Mendel, when he found what we call Mendel's Law. This is the key that unlocks the permutations and combinations of qualities of plants and animals, and make breeding a science.

Some years after his death, John Doe's expanding reputation begins to approach that of Thomas Jefferson and the other real founders and builders of our American civilization. Magazine articles tell about his boyhood and how he came to start a thing so profoundly creative and so important as the Doe Foundation has grown to be. A new biography of Mr. Doe appears.

The Doe Foundation has the Rockefeller and Carnegie foundations clearly beaten for publicity. The Rockefeller and Carnegie foundations do good work, but a bit stodgy from the news standpoint. These new plants, new crops, new farming methods, new conservation methods from the Doe Foundation have news value, popular news value, and will produce vast and continuing publicity, crops of publicity.

The above-mentioned biography of John Doe impresses a magnate of that day so favorably that he sets aside millions to start the Richard Roe Foundation in *his* native State, near the town where he was born. He has a corking good time planning and starting it, as Boyce Thompson did, and he wills it his factory and business as an endowment. Mr. Roe has as much fun as MacManus had with cork (pp. 172-176).

We need about twelve such institutions in the United States, located so that each can work out tree crops and techniques for two or three hundred thousand square miles with a given climate and its problems. This is for tree crops only. The States and private business can see far enough to do the annuals.

I suggest the following locations for these crop-tree centers of re-creation. One each within a hundred miles of: Albany; Washington, D. C.; Augusta; Asheville; Minneapolis; Kansas City; Dallas; Portland, Oregon; Los Angeles; Santa Fe or Tucson; Salt Lake City, Boise or Spokane.

The map of world climate regions (see Fig. 138) will suggest areas from which useful plants may come for use in the different sections of the United States, and in return it will show what regions may profit by the findings within the various sections of the United States.

THE MANY LITTLE HOPES

And now, John Q, my fellow citizens of small means, such as the farmers, lawyers, manufacturers, doctors, teachers, spinsters, villagers, who are the members of the Northern Nut Growers' Association, and others like you: don't wait for the Government to do it. It may help, but probably can't do the long-distance stuff. Don't wait for the millionaire. There is room for him when he comes and also for you, too, and right now. Have you an intellect? Or just a kind of high-animal brain? Do you know the thrill of creative work? Of learning or creating something really new?

Use your imagination constructively. An *idea*, a *small yard* and one tree may be made to produce astounding results. You might start with one burr oak tree grafted by your own hand to the most promising variety you can find. You might then import pollen and hybridize, or better, you might graft five branches with each of five promising varieties and have the material for big hybridizing work right in your yard on your own tree.

Sprouting hybrid-oak acorns in pots in the house might easily be just as interesting as growing blooming plants; I even suspect you could sneak two acorns into any five-inch flower pot that had a blooming plant, and they could sprout and come through the first winter in the house and go out of doors after one or two years in pots. Put them a foot apart in a corner of your backyard and you could raise them to be three to five feet high. You might use honey locusts instead of oaks.

Now comes the real problem of getting results. If some of these little trees were very promising, you could preserve fifteen or twenty of them grafted onto branches of your big mother tree and carefully labeled. Also some cooperative organization like the Northern Nut Growers' Association would perhaps have persons who would give them space to grow up and prove themselves. You would have to expect some loss here, but then that is the nature of experimental work. You could have a very interesting time, and great results might come out of your backyard with its one oak tree or honey locust tree.

Oaks are of two families, white and black. Bur oak, *Q. macrocarpa*, is a white oak, a fast grower, and not difficult to graft with other white oaks. There is great need of good northern honey locust.

THE NEIGHBORHOOD EXPLORER

I have mentioned the private experimenter, but there is another kind of private worker, the explorer. George L. Slate, Geneva Experiment Station, Geneva, New York, gives three interesting examples:

Mr. W. W. Adams, Union Springs, New York, was a private citizen who, I am told, used to ride about the countryside with his horse and buggy looking for large-fruited elderberries. He also raised seedlings in an attempt to improve the fruit. [Adams elderberry plants are now being sold.]

Mr. Loup is a fruit grower near North East, Pennsylvania. He apparently has his eyes open and notices all sorts of things about plants. In addition to finding elderberries, he finds bud sports in his orchards, and is somewhat of a variety tester. Mr. S. H. Graham, Bostwick Road, Ithaca, New York is another man of this same type. He has hunted up such things as the Tasterite Walnut, and some other walnuts and a hickory nut or two.

These men are types. Scores like them have made a beginning by creating and joining the Northern Nut Growers' Association and many other scientific societies.

There is no reason to think that we have found the best wild trees of any species in the United States. We especially need good honey locusts from north of latitude 40°.

Small cash prizes will usually be announced as news by local newspapers; such prizes may start observation by Boy Scouts, rabbit hunters, and other fence-corner naturalists.

My investigation of the work of the Governmental agencies and that of the members of the Northern Nut Growers' Association causes me to have a new vision of the possibilities that may result from amateur effort. Its great need is, of course, extension, but a greater need is more careful recording of its results. How regularly do your trees bear, and how much? Varieties differ greatly from place to place. We need measured facts.

The experiment stations, as previously stated, have their limitations, but they have permanence, and experiment-station staffs might, with great profit, visit each year, or every other year, the leading private experimenters in their States, gather their results, put them in permanent form, and furthermore get results that the private experimenters had not noticed or had not had the opportunity to observe.

For example, the McCallister Hybrid Hickory (see Fig. 18) is the largest nut known of the hickory genus, unless some of the newly bred pecans can rival it. The parent tree seems to have a record of producing nuts that were well filled—good crops of them. I propagated this tree and sold it for several years in my little nursery, but they never bore nuts, and I never saw well-filled nuts on mine, so I quit selling these trees. Now here comes Mr. Fayette Etter, of Lemasters, Pennsylvania, with a McCallister that bears and fills its nuts. This tree is in the presence of a great variety of other hickories—one or more of which evidently pollinates the McCallister. Which is it? No one knows.

Five years after it happened, Bill Wiley, a very careful observer who works with my trees, told me that one year a small McCallister tree in my nursery had four well-filled nuts. At the time this happened I had forty or fifty varieties of hickory within fifty yards of the McCallister. The question is: What tree pollinates the McCallister? Someone should camp in Fayette Etter's planting every spring for a week or ten days and note what trees were shedding pollen while the McCallister blooms were receptive. If there were several blooming, as may easily be the case, a few hand pollinations should settle the matter, and a marvelous yard tree and a possible commercial crop might quickly result. The McCallister is a beautiful, vigorous, fast-growing Pecan x Laciniosa (hickory) hybrid.

The great danger with private experimenting is the scattering of the trees, on the death of the experimenter, just as they have reached the age to produce educational results.

MAKE YOUR WILL

Wherever possible, I urge private experimenters to make some provision, while still alive, for the preservation of your trees for as long as possible after your passing. As an example of this loss to the world through death of tree owners, I cite the fact that twenty years ago Dr. Robert Morris had about fifty varieties of nut trees under test. J. F. Jones, near Lancaster, Pennsylvania, had eighty-eight, and Willard Bixby, Baldwin, Long Island, had one hundred ninety-nine varieties—indeed he had a marvelous layout. He spent a fortune starting and operating a private, specialized, economic botanical garden, but he suddenly died without the ability to leave it endowed. Within a few years these three large collections of great possible value were scattered, neglected, and mostly lost—and their lands were owned by persons to whom a tree was just a tree.

State experiment stations might with great profit, from their point of view, lease such collections for a few years to harvest the scientific results—provided they had staff members with time and mental competence. That means a tree-crop specialist, perhaps more than one. You can't just go out and pick up competence.

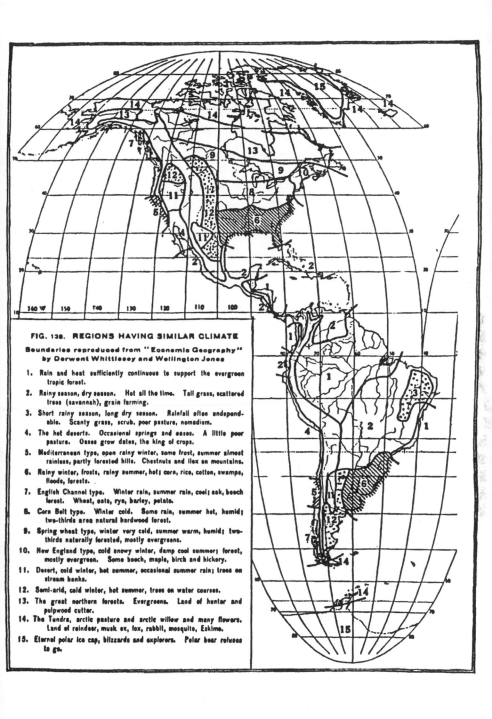

FIG. 138. REGIONS HAVING SIMILAR CLIMATE

Boundaries reproduced from "Economic Geography"
by Derwent Whittlesey and Wellington Jones

1. Rain and heat sufficiently continuous to support the evergreen tropic forest.

2. Rainy season, dry season. Hot all the time. Tall grass, scattered trees (savannah), grain farming.

3. Short rainy season, long dry season. Rainfall often undependable. Scanty grass, scrub, poor pasture, nomadism.

4. The hot deserts. Occasional springs and oases. A little poor pasture. Oases grow dates, the king of crops.

5. Mediterranean type, open rainy winter, some frost, summer almost rainless, partly forested hills. Chestnuts and ilex on mountains.

6. Rainy winter, frosts, rainy summer, hot; corn, rice, cotton, swamps, floods, forests.

7. English Channel type. Winter rain, summer rain, cool; oak, beech forest. Wheat, oats, rye, barley, potato.

8. Corn Belt type. Winter cold. Some rain, summer hot, humid; two-thirds area natural hardwood forest.

9. Spring wheat type, winter very cold, summer warm, humid; two-thirds naturally forested, mostly evergreens.

10. New England type, cold snowy winter, damp cool summer; forest, mostly evergreen. Some beech, maple, birch and hickory.

11. Desert, cold winter, hot summer, occasional summer rain; trees on stream banks.

12. Semi-arid, cold winter, hot summer, trees on water courses.

13. The great northern forests. Evergreens. Land of hunter and pulpwood cutter.

14. The Tundra, arctic pasture and arctic willow and many flowers. Land of reindeer, musk ox, fox, rabbit, mosquito, Eskimo.

15. Eternal polar ice cap, blizzards and explorers. Polar bear refuses to go.

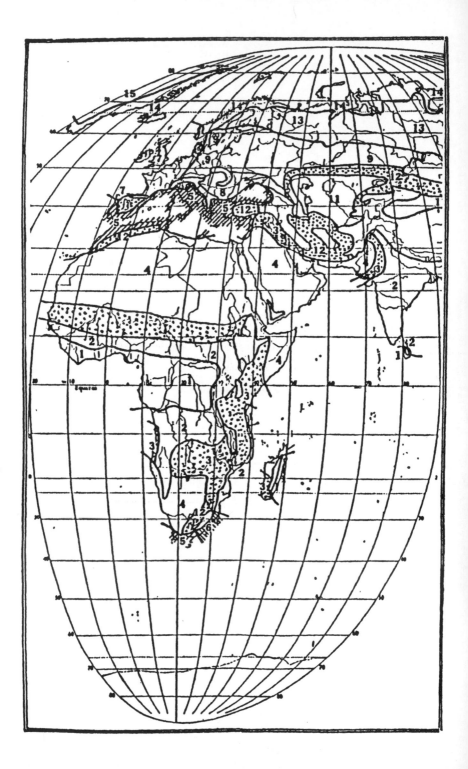

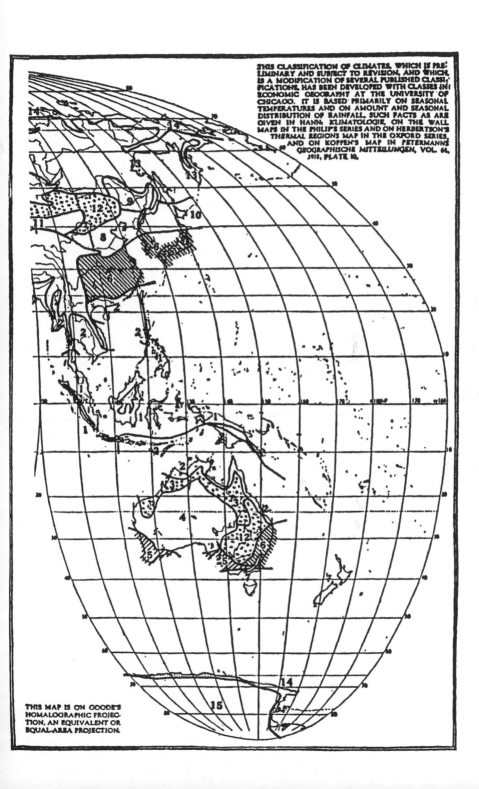

THIS CLASSIFICATION OF CLIMATES, WHICH IS PRE-
LIMINARY AND SUBJECT TO REVISION, AND WHICH
IS A MODIFICATION OF SEVERAL PUBLISHED CLASSI-
FICATIONS, HAS BEEN DEVELOPED WITH CLASSES IN
ECONOMIC GEOGRAPHY AT THE UNIVERSITY OF
CHICAGO. IT IS BASED PRIMARILY ON SEASONAL
TEMPERATURES AND ON AMOUNT AND SEASONAL
DISTRIBUTION OF RAINFALL, SUCH FACTS AS ARE
GIVEN IN HANN, KLIMATOLOGIE, ON THE WALL
MAPS IN THE PHILIP'S SERIES AND ON HERBERTSON'S
THERMAL REGIONS MAP IN THE OXFORD SERIES,
AND ON KOPPEN'S MAP IN PETERMANNS
GEOGRAPHISCHE MITTEILUNGEN, VOL. 64,
1918, PLATE 16.

THIS MAP IS ON GOODE'S
HOMALOGRAPHIC PROJEC-
TION, AN EQUIVALENT OR
EQUAL-AREA PROJECTION.

APPENDIX A

ANALYSIS OF FEEDS FOR FARM ANIMALS

Feed	Moisture	Ash	Crude Protein	Crude Fiber	Nitrogen Free Extract	Fat or Ether Extract	Digestible Protein	Digestible Carbohydrate Equivalent
Barley [1]	9.6	2.9	12.8	5.5	66.9	2.3	10.4	63.8
Corn [1]	12.9	1.3	9.3	1.9	70.3	4.3	7.1	74.8
Wheat [1]	10.6	1.8	12.3	2.4	71.1	1.8	9.8	63.3
Wheat bran [1]	9.6	5.9	16.2	8.5	55.6	4.2	12.5	48.7
Cottonseed meal (good) [1]	7.3	5.8	36.8	13.5	30.0	6.6	30.9	42.1
Alfalfa hay [1]	8.3	8.9	16.0	27.1	37.1	2.6	11.5	42.0
Alfalfa [1]	72.9	2.6	4.7	8.0	11.0	0.8	3.6	12.8
Potatoes [1]	78.9	1.0	2.1	0.6	16.3	0.1	1.3	16.3
Turnips [1]	90.6	0.8	1.3	1.2	5.9	0.2	1.2	7.4
Honey Locust [2]								
U.S.D.A. Grounds	4.1	3.7	13.4	16.3	61.3	1.2		
New Mex. Agr. Col.[3]	5.20	3.58	4.50	14.56	69.94	2.22		
Carob: Entire Bean [6]								
Italian	11.3	2.9	5.1	6.0	74.4	.3		
Portuguese	8.3	3.1	4.3	7.9	76.1	.3		
Algaroba or Keawe [4][7]								
Sample No. 1	....	2.14	10.84	26.48	56.40	.77		
Sample No. 2	....	....	9.88	31.29	53.13	.62		
Mesquite Beans [5]								
Hawaii; 5 samples	12.3	3.3	9.0	23.4	51.4	.6		
Arizona: 4 samples	6.3	4.5	12.7	24.5	49.5	2.5		
Calif.: 2 samples	11.4	4.0	9.7	22.6	51.3	1.0		
New Mex.: 1 sample	4.8	3.4	12.2	32.0	45.1	2.5		
Texas: 7 samples	6.9	4.4	12.4	25.7	47.9	2.7		
N. M. Tornillo beans	5.1	3.0	9.8	19.3	61.8	1.0		
Mesquite Tree [5]								
No. 1343 [8]	7.25	4.31	12.48	25.67	55.51	2.03		
No. 1345 [1][10]	6.21	5.24	14.12	22.17	54.80	3.69		
No. 1313 [10]							*Sugars*	
Pods, 70%	5.48	5.71	5.70	30.70	55.46	2.40	*Reducing*	*Sucrose*
Seeds, 30%	7.69	3.38	37.54	5.75	46.89	6.45		
Carob Bean [6][7]								
Pods and Seeds								
1704	11.91	1.67	7.96	5.60	44.96	1.00	12.94	13.96
Minimum	9.12	1.67	3.26	4.98	26.99	1.00	3.25	6.39
Maximum	19.81	3.46	15.22	17.42	43.57	3.82	18.69	41.56
Average	13.28	2.57	6.75	9.29	39.80	2.17	11.08	19.44
Pods without seeds								
2200	12.27	2.50	3.77	9.96	40.28	2.64	6.88	21.70
2201	18.08	2.39	3.33	8.24	37.54	2.86	20.54	7.02
2371	5.70	3.87	3.40	13.62	18.36	3.08	13.04	8.93
2493	8.21	2.71	7.18	4.73	24.48	.71	8.36	43.62
Minimum	3.70	1.76	2.02	3.14	24.48	.22	3.00	7.02
Maximum	24.70	3.87	7.18	15.31	48.36	4.02	20.54	43.62
Average	11.50	2.72	4.50	8.78	36.30	2.37	11.24	23.17

[1] United States Department of Agriculture, Bureau of Animal Husbandry. The carbohydrate equivalent shown in the last column of the table is the sum of the digestible crude fiber and nitrogen-free extract, plus 2.25 times the digestible fat.
[2] United States Department of Agriculture, Bulletin No. 1194.
[3] Analysis No. 12053 Misc. Div. United States Department of Agriculture, Bureau of Plant Industry, Washington, D. C., of an unusually broad-podded variety of honey locust pods obtained near grounds of New Mexico Agricultural College, Mesilla, New Mexico.
[4] Hawaii Agricultural Experiment Station, Bulletin No. 13, Edmund C. Shorey, chemist.
[5] Composition of Entire Mesquite Beans. Analyses from "The Mesquite Tree" by Robert H. Forbes, Bulletin No. 13, Arizona Experiment Station.
[6] Bulletin No. 309, University of California, "The Nutritive Value of the Carob Bean."
[7] Prosopis juliflora: Entire beans.
[8] Gathered August 1 on Rillito River.
[9] Sample furnished by N. R. Powell of Pettus Bee Company, sent by W. J. Spillman.
[10] Gathered October 7, on Santa Cruz River.
[11] It should be noticed that these maximum and minimum figures refer to a particular element in a number of different samples. They are not complete analyses of one sample like Nos. 1704, 2200, 2201, 2371, and 2493.

APPENDIX B

FOOD VALUE OF THE PERSIMMON AND OTHER FRUITS

From Farmers' Bulletin No. 685 U. S. Department of Agriculture

COMPARATIVE ANALYSES OF FRESH FRUITS, SHOWING THEIR FOOD VALUES IN PERCENTAGES OF THE WEIGHT OF THE FRUIT.[1]

Fruit	Total Solids	Ash	Protein	Sugars	Crude Fiber
	%	%	%	%	%
Apples.	13.65	0.28	0.69	10.26	0.96
Blackberries.	13.59	.48	.51	4.44	5.21
Cherries.	22.30	.65	.81	11.72	.62
Currants.	15.23	.72	.51	6.38	4.57
Dates [2][3].	66.86	1.20[4]	1.48[5]	56.59[6]	3.80[7]
Figs.	20.13	.57	1.34	15.51	....
Grapes [8][2].	21.83	.53	.59	17.11[9]	3.60
Oranges (Navel).	13.87	.43	.48	15.91	....
Peaches [7][2].	10.60	.40	.70	5.90[6]	3.60
Pears.	16.97	.31	.36	8.26	4.30
Persimmons [10][2].	35.17	.78	.88	31.74[9]	1.43
Plums.	15.14	.61	.40	3.56	4.34
Raspberries.	13.79	.49	.53	3.95	5.90
Strawberries.	9.48	.60	.97	5.36	1.51

[1] Data, with exceptions as noted, from Bureau of Chemistry Bulletin, No. 66, pp. 41–42.
[2] Dry matter.
[3] Average of eleven analyses. See "Chemistry and Ripening of the Date," Arizona Agricultural Experiment Station Bulletin No. 66, p. 408.
[4] See "Principles of Nutrition and Nutritive Value of Food," Farmers' Bulletin No. 142, p. 18.
[5] Adapted from the two publications mentioned in footnotes 3 and 4.
[6] Fats and carbohydrates.
[7] See "Use of Fruit as Food," Farmers' Bulletin No. 293, p. 14.
[8] See "The American Persimmon," Indiana Experiment Station Bulletin No. 60 (1896), p. 52.
[9] Nitrogen-free extract.
[10] Average of six analyses in "The American Persimmon," Indiana Experiment Station Bulletin No. 60 (1896).

APPENDIX C

AVERAGE COMPOSITION OF NUTS AND OTHER FOODS[1]

Kind of Food	Refuse	Water	Protein	Fat	Carbohydrates Sugar, Starch, Etc.	Crude Fiber	Ash	Fuel Value per Pound
	%	%	%	%	%	%	%	Calories
Nuts and Nut Products:								
Acorn, fresh [2]	17.80	34.7	4.4	4.7	50.4	4.2	1.6	1,265
Almond	47.00	4.9	21.4	54.4	13.8	3.0	2.5	2,895
Beechnut	36.90	6.6	21.8	49.9	18.0		3.7	2,740
Brazil Nut	49.35	4.7	17.4	65.0	5.7	3.9	3.3	3,120
Butternut	86.40	4.5	27.9	61.2	3.4		3.0	3,370
Candle Nut	.....	5.9	21.4	61.7	4.9	2.8	3.3	3,020
Chestnut, fresh	15.70	43.4	6.4	6.0	41.3	1.5	1.4	1,140
Chestnut, dry	23.40	6.1	10.7	7.8	70.1	2.9	2.4	1,840
Horn Chestnut, or Water Chestnut	.....	10.6	10.9	.7	73.8	1.4	2.6	1,540
Cocoanut	34.66	13.0	6.6	56.2	13.7	8.9	1.6	2,805
Filbert	52.08	5.4	16.5	64.0	11.7		2.4	3,100
Ginkgo Nut (seeds)	.....	47.3	5.9	.8	43.1	.9	2.0	940
Hickory Nut	62.20	3.7	15.4	67.4	11.4		2.1	3,345
Peanut	27.04	7.4	29.8	43.5	14.7	2.4	2.2	2,610
Pecan	50.10	3.4	12.1	70.7	8.5	3.7	1.6	3,300
Pine Nut, Piñon	40.6	3.4	14.6	61.9	17.3	...	2.8	3,205
Pine Nut, Spanish, or Pignolia (shelled)	.....	6.2	33.9	48.2	6.5	1.4	3.8	2,710
P. edulis [3]	.....	3.1	14.8	60.6	18.7 [5]	1.8 [4]	2.8	....
P. pinea [3]	.....	4.2	37.0	49.1	5.5 [5]	1.0 [4]	4.2	....
P. gerardiana [3]	.....	8.7	13.6	51.3	23.4 [5]	0.9 [4]	3.0	....
Pistachio	.....	4.2	22.6	54.5	15.6		3.1	3,250
Walnut (Persian)	58.80	3.4	18.2	60.7	13.7	2.3	1.7	3,075
Walnut (Amer. Black)	74.1	2.5	27.6	56.3	11.7		1.9	3,105
Almond Butter	.....	2.2	21.7	61.5	11.6		3.0	3,340
Peanut Butter	.....	2.1	29.3	46.5	17.1		5.0	2,825
Malted Nuts	.....	2.6	23.7	27.6	43.9		2.2	2,600
Cocoanut, desiccated	.....	3.5	6.3	57.4	31.5		1.3	3,125
Chestnut Flour	.....	7.8	4.6	3.4	80.8		3.4	1,780
Cocoanut Flour	.....	14.4	20.6	2.1	45.9	10.1	6.9	1,480
Hazelnut Meal	.....	2.7	11.7	65.6	17.8		2.2	3,185
Other Foods for Comparison:								
Meat, Round Steak	.....	65.5	19.8	13.6	....	....	1.1	950
Cheese, Cheddar	.....	27.4	27.7	36.8	4.1	....	4.0	2,145
Eggs, boiled	11.20	65.0	12.4	10.7	....	....	.7	680
Wheat Flour, high grade	.....	12.0	11.4	1.0	74.8	.3	.5	1,650
White Bread	.....	35.3	9.2	1.3	52.6	.5	1.1	1,215
Beans, dried	.....	12.6	22.5	1.8	55.2	4.4	3.5	1,605
Potatoes	20.00	78.3	2.2	.1	18.0	.4	1.0	385
Apples	25.00	84.6	.4	.5	13.0	1.2	.3	290
Raisins	10.00	14.6	2.6	3.3	73.6	2.5	3.4	1,605

[1] Unless otherwise stated, from "Nuts and Their Uses as Food" by M. E. Jaffa, Professor of Nutrition, University of California, U. S. Dept. Agr. Farmers' Bulletin, No. 332. In studying the column marked "refuse" it should be remembered that many of the nuts are based on poor wild produce of the present markets rather than selected strains or improved strains which can be grown.

[2] For further acorn comparisons, see pages 189 and 180, 203.

[3] Information from Frederick V. Coville, U. S. Department of Agriculture.

[4] Fiber.

[5] Carbohydrates plus fiber.

APPENDIX D

FOOD VALUES OF CROP AND LIVESTOCK PRODUCTS PER ACRE

Food Products	Yield per Acre		Calories per Pound	Pounds Protein per Acre	Calories per Acre	Acres to Equal One Acre of Corn
	Bushels	Pounds				
Field						
Corn [6]............	35	1,960	1,594	147.0	3,124,240	1.00
Irish Potatoes [6]....	100	6,000	318	66.0	1,908,000	1.64
Wheat [6]..........	20	1,200	1,490	110.4	1,788,000	1.75
Dairy Products						
Milk [6]............	....	2,190	325	72.3	711,750	4.39
Cheese [6]..........	....	219	1,950	56.7	427,050	7.32
Meat	(live lbs.)	(dressed)				
Pork [6]............	350	273	2,465	22.7	672,945	4.64
Beef [6]............	216	125	1,040	18.5	130,000	24.00
Poultry Crop [6]						
Meat and Eggs....	66 lbs.	111 eggs	....	27.5	149,000	21.00
Nut Crops		Pounds				
Chestnuts (fresh)...	....	1,600	1,140 [7]	....	1,824,000	1.71
Persian Walnuts...	....	1,000 [1]	3,075 [7]	....	1,266,900	2.47
Black Walnuts.....	....	1,000 [2]	3,105 [7]	....	776,250	4.03
Hickory Nuts......	....	1,000 [3]	3,345 [7]	....	1,672,500	1.86
Pecans...........	....	1,000 [4]	3,300 [7]	....	1,650,000	1.89
Acorn............	....	1,400 [5]	1,265 [7]	....	1,455,762	2.12
Keawe [8]...........	....	....	....	....	....	....

In comparing these nut crops with corn it should be remembered that the figures for the nut crops are supposed to be annual averages, whereas corn, even on the best of land, is almost always put in rotation; and therefore there is rarely less than one crop in three years, often one crop in four or five years. Page 347 shows that many of these tree crops *might* have side crops also.

[1] Based on California production (page 215). Edible portion see page 393.
[2] Quantity is estimated yield. Calories are for edible portion (p. 393). Assume 25 per cent. edible. See 1919 report Northern Nut Growers' Association for tests of weights, also 1927 report of same. Some yield more than 25 per cent. kernel.
[3] Yield estimated by Dr. W. C. Deming for full stand grafted trees. Calories for edible portion. Assume 50 per cent. edible. See 1919 report Northern Nut Growers' Association where many nuts were more than 50 per cent. edible.
[4] Quantity, Georgia Experiment Station Estimate. Calories for edible portion. Edible portion estimate 50 per cent. Some yield more than 50 per cent. kernel.
[5] Quantity, author's estimate. See chapters on the oak. It is probable that the figure for yield is too low. Edible portion taken from p. 393.
[6] From U. S. Department of Agriculture, Farmers' Bulletin No. 877: *Human Food from an Acre of Staple Farm Products*, by Morton O. Cooper and W. J. Spillman, from whom the idea came.
[7] Analysis on page 393.
[8] Counted as calories and considered as stock food it outranks corn. See Chapter V.

Index

About Island Press

Island Press, a nonprofit organization, publishes, markets, and distributes the most advanced thinking on the conservation of our natural resources—books about soil, land, water, forests, wildlife, and hazardous and toxic wastes. These books are practical tools used by public officials, business and industry leaders, natural resource managers, and concerned citizens working to solve both local and global resource problems.

Founded in 1978, Island Press reorganized in 1984 to meet the increasing demand for substantive books on all resource-related issues. Island Press publishes and distributes under its own imprint and offers these services to other nonprofit organizations.

Funding to support Island Press is provided by The Mary Reynolds Babcock Foundation, The William H. Donner Foundation, Inc., The Ford Foundation, The George Gund Foundation, The William and Flora Hewlett Foundation, The Joyce Foundation, The Andrew W. Mellon Foundation, Northwest Area Foundation, The J.N. Pew, Jr. Charitable Trust, Rockefeller Brothers Fund, and The Tides Foundation.

Island Press
Board of Directors